ICELAND:

ITS SCENES AND SAGAS.

a

INDIAN BIVOUAC

PL. VI

ICELAND:

ITS SCENES AND SAGAS.

BY

SABINE BARING-GOULD, M.A.,

FELLOW OF S. NICHOLAS COLLEGE, LANCING; MEMBER OF THE NORSE
LITERARY SOCIETY.

WITH NUMEROUS ILLUSTRATIONS AND A MAP.

"Farewell heat, and welcome frost."
Merchant of Venice.

LONDON:

SMITH, ELDER AND CO., 65, CORNHILL.

M.DCCC.LXIII.

[*The right of Translation is reserved.*]

TO

MAJOR-GENERAL EDWARD SABINE, R.A.,

PRESIDENT OF THE ROYAL SOCIETY.

MY DEAR UNCLE,

It is with great pleasure that I dedicate this book to you, as a token of my gratitude for the kind interest you have taken, both in my expedition to Iceland, and in the progress of this volume.

I remain,

Ever yours sincerely,

SABINE BARING-GOULD.

Ísland far-sœldar frón,
Ok hag-sœldar hrím-hvíta móðir!
Hvar er þín fornaldar frœgð,
Frelsið ok manndáðin best?

Iceland, thou fare-blessed spot,
Thou use-blessed rime-whitened mother!
Where are thine olden fame,
And freedom, and manliness best?

CONTENTS.

	PAGE
PREFACE	xii
INTRODUCTION	xix

CHAPTER I.

COASTING.

First Sight of Iceland — Hjörleif's Head — Dyrhólar — The Westmann
Islands — Life of the Islanders — The Yankee's Story — My first
Saga : " The Red Rovers "—Cape Reykjanes— The Mealsack—
The Great Auk 1

CHAPTER II.

REYKJAVIK.

Arrival—Description of Reykjavík—" A very Antient fish-like Smell ! "—
The School — The Residence — The Cathedral — Thorwaldsen—
Icelandic Silverwork—The Merchant Stores—Mr. Briggs and the
Porters—Dinner—I visit the Rector—A would-be Suicide—The
Catholic Mission—The Origin of Iceland 26

CHAPTER III.

MÓSFELL.

Leave Reykjavík—Rich Soil—Leiruvogr—Accidents—Camp out — A
lazy Guide — Arctic Midnight — Rank Breakfast—The Story of
Sonartorrek—Alpine Road—Icelandic Dogs—A Farmhouse—Filth
—On the Heithi—Birds—Icelandic Indifference—Arrival at Thing-
vollum—A Quarrel 48

CHAPTER IV.

THINGVELLIR.

PAGE

The Plains of the Council—The Almanna-gjá—The Hill of Laws—
History of Althing—Lava Broth—Icelandic Food—Butter—Curd
—Lichen—Sunday—The Burning of the Hostel: a Saga.   .   67

CHAPTER V.

AMONG THE JÖKULLS.

Leave Thingvellir—Meyjar-skarth—A lovely View—The Big Bed breaks
loose—Grettis-tak—A mysterious Vale—Attempts to re-discover
it—Ascent of Geitland's Jökull—Brunnir—Dearth of Fuel—
Kaldidalr—The Old Woman of Bones—A Vale of Desolation—A
Tract of Ash—Kalmanstúnga—Robbers   86

CHAPTER VI.

THE EAGLE TARNS.

Eiriks Jökull—Taxation—The Sheep's Disease—A Swan's-nest—Lava
—Lose our Way—Camping out—Glorious Scene—Cooking
Arrangements—Rare Skua—Great Northern Diver—Storm on the
Heithi—Gunnarssonarvatn—The Great Eagle Tarn—Desert—
Hólma-kvísl   99

CHAPTER VII.

THE VAMPIRE'S GRAVE.

Barren Wilderness—A Cairn—The Valley of Shadows: a Saga—A
magnificent Gorge—Waterfalls—The Vatnsdalr—Arrival at Grímstúnga   115

CHAPTER VIII.

THE VALE OF WATERS.

Vatnsdalr—Forsœludalr—A dangerous Predicament—Thorolf wi' the
Dead-face—An ancient Fort—Colonization of Vatnsdalr—Hóf—
The Story of Hrolleifr: a Saga—MSS.—Old Nick—Icelandic
Endearments—Music—Haukagil—Undirfell   .   .   .   133

CHAPTER IX.

FROM HNAUSIR TO EYJA FJORD.

A Tract of Slag Cones—An Icelandic Doctor—Part from my Friends—
Giljá—Conversion of Iceland—Svínavatn—Icelandic Churches—A
miserable Lodging—Slang—Harlequin Duck—Ford the Blandá—
Vatnsskarth—Vithimyri—Purchase a Horse—Ford the Heradsvatn
—Mikliboer—Oxnadals Heithi—Steinstathr—A Caravan—Strange
Merchandise—The Princess Alexandra—A Death .   .   . 155

CHAPTER X.

AKUREYRI.

A little Town—New Church—Site of the Town injudiciously chosen
—The Post—Grimsey—An Island Cure—Danish Hospitality—The
Newspaper Office—Supper—English Vessels—Jack Tar in Iceland
—Trees—MSS.—Icelandic Poetry—The Dream-Ballad .   . 171

CHAPTER XI.

UXAHVER.

A second Guide — Icelandic Horse-calls — A theological Candidate —
Vathlaskarth—Wood in the Fnjoská-dale—Háls—Ljósavatn—The
Raven—Myth regarding it—Extent to which the Myth has spread—
Góthafoss— Flowers — Cross the Flood of Quivering Waves —
Acquaintances—Lava Cracks—Grenjatharstathr—Uxahver—Boil-
ing Springs—The Wild Huntsman—Origin of the Myth—Myvatn. 184

CHAPTER XII.

A REGION OF FIRE.

Myvatn—Lava Streams—Reykjahlith—Sand Columns—A Plain of Boiling Mud—Chaldrons—Krafla—Obsidian Mountain—A Ride over
a Desert—Eylif—A wretched Farm—Dettifoss—A magnificent Fall
—Volcanic Cones—Return to Reykjahlith   .   .   . 205

CHAPTER XIII.

A SNOW PASS.

I hold a Levee—My Plans upset—The Church of Thverá—Taking French
Leave — Swimming across a River — MSS. Sagas—Jón's Bills—
Runaway Ponies—Rocky Spires — Perverseness of Grímr — A
Mountain Sel—A fearful Pass—Snowbridge—A desperate Scramble
—Stone Bog—Midnight on the Snow—A crumbling Snowbridge—
Arrival at Hólar—An alarming Proposal 220

CHAPTER XIV.

HÓLAR.

PAGE

The Cathedral—Altar Vestments—Triptych—Portraits and Tombs—
MSS.—The Ancient Possessions of the Church—Repairs—Flowers
—The Skagafjord—Drángey—Birds—Gannet—Puffin—Skua   . 235

CHAPTER XV.

The Outlaw's Isle : a Saga   246

CHAPTER XVI.

FROM SKAGAFJORD TO MITHFJORD.

Svathastathr—Heathen Charity—Church of Vithimyri—An odd Cow—
Borg, an old Castle—Queen Victoria's Cousin—Icelandic Duck—
The long-tailed Duck—Red-necked Phalaropes—Deep Bogs—Melr
—A happy Meeting—A Wedding—Female Dress—Music—National
Anthem—Michaelmas Hymn—Mr. Briggs in Love—Church Service
—Sacred Music—Ecclesiastical Position of the Icelandic Establish-
ment   278

CHAPTER XVII.

THE MIDDLE FRITH.

Glacial Action—Bjarg : the Home of Grettir—Cairn—Spear-head—
Carved Stone—Magical Characters—The Story of the Banded-men :
A Saga   297

CHAPTER XVIII.

THE HRUTAFJORD.

Leave Melr—Icelandic Etiquette—A beautiful Frith—Mr. Briggs's Story
—Thorodds-stathr—Row across the Fjord—Merchant Vessels—
Hospitality—Mr. Briggs recovers his Heart—Trouble with the Pack-
saddle—Holtavörthu Heithi—Sclavonian Grebe—Arctic Foxes—
Icelandic Traditions concerning the Fox—Icelandic Mice—Tra-
vellers' Tales — Swans —Terns—Difficult Pass—Baula—MSS.—
Church of Hvammr   317

CHAPTER XIX.

THE VALE OF SMOKE.

Runaway Horses—Grjóthâls—Glorious View—I succeed in Mastering the
Horses—Cruelty to Birds—The Future of Animals—Ptarmigan—
Whimbrel and Plover—A Heithi—Icelandic Way of Sleeping—
Names—Tunguhver—Boiling Jets in a River—Reykholt—Snorro's
Bath—The Church—Snorro Sturlason   .   .   . 331

CHAPTER XX.

ÓK.

PAGE

Ascend the Side of Ók—Flowers—Strange Sight—Skogkottr—Meet old
Friends — The Skrimsl — Mermaids — Francesco de la Vega—
Smoking in the Tent   344

CHAPTER XXI.

GEYSIR.

Flowers — A natural Chimney — Extensive Plain — Laugarvatn — An
eccentric Bridge—Uthlith—Sleeping in a Church—Position of the
Geysir District—Description of the Springs—The Little Geysir—
Jack in the Box—Boiling Wells—Strokr—Blue Ponds—Experi-
ments—Great Geysir—Keeping Watch—Magnificent Explosion—
Mr. Briggs misses seeing it—Theory of the Geysir   .   . 354

CHAPTER XXII.

Thorgils' Nursling : a Saga about Greenland   368

CHAPTER XXIII.

CONCLUSION.

Leave Geysir—Last View of Heckla—The rumoured Eruption of Skapta
—Return to Thingvöllum—Latin Conversation—Seljadalr—The
Plague of Flies—Halt at a Farm—A fair Haymaker—The Spell
broken—Return to Reykjavík—Sale of Horses—Icelandic Ponies—
Their strong and weak Points—Leave Iceland—The Captain's Joke
—Reach England—Advice to Travellers   385

APPENDICES.

APPENDIX A.—Ornithology of Iceland, by A. Newton, Esq.   .   . 399
    „   B.—Advice to Sportsmen, by J. W. R.   .   . 422
    „   C.—List of Icelandic Plants   .   .   . 424
    „   D.—List of Icelandic published Sagas   .   . 439
    „   E.—Expenses of my Tour in Iceland   .   . 445

LIST OF ILLUSTRATIONS.

PLATE									PAGE
	Map of Iceland	.	.	.	.	.	.	to face	xix
I.	Panorama of South Coast and Thingvalla	.	.			.	.	,,	6
II.	Almannagjá	.	.	.	.	.	.	,,	67
III.	Tindarskaggi	.	.	.	.	.	.	,,	86
IV.	Geitlands Jökull	.	.	.	.	.	.	,,	88
V.	Vatnsdalr from Undirfell	.	.	.	.	.	,,	133
VI.	Öxnadals-heithi	.	.	.	.	.	.	frontispiece	
VII.	On Öxnadals-heithi	.	.	.	.	.	.	to face	166
VIII.	Akureyri	.	.	.	.	.	.	,,	171
IX.	Myvatn and Hverfjall	.	.	.	.	.	,,	205
X.	Hlitharfjall	.	.	.	.	.	.	,,	208
XI.	Hörgadalr	.	.	.	.	.	.	,,	228
XII.	Mælafellshnukr	.	.	.	.	.	.	,,	279
XIII.	Bjarg	.	.	.	.	.	.	,,	297
XIV.	Near Brunnir	.	.	.	.	.	.	,,	345
XV.	Kálfstindar	.	.	.	.	.	.	,,	355
XVI.	The Great Geysir	.	.	.	.	.	.	,,	362

ILLUSTRATIONS IN TEXT.

Sketch Map of Reykjavík	.	.	.	.	.	.	.	27
Plan and Elevation of a Farm	.	.	.	.	.	.	.	59
Nicolásagja	.	.	.	.	.	.	.	69
Great Northern Diver	.	.	.	.	.	.	.	109
Kitchen at Vithimyri	.	.	.	.	.	.	.	163
Mountains in Öxnadalr	.	.	.	.	.	.	.	167
Runic Stave at Grenjatharstathr	.	.	.	.	.	.	194
Uxahver: the Second Spring	.	.	.	.	.	.	197
Boiling Mud Chaldron, Námar-hlith	.	.	.	.	.	211
Dettifoss	.	.	.	.	.	.	.	217
Diagram of Triptych, Hólar	.	.	.	.	.	.	237
Skagafjord	.	.	.	.	.	.	.	246
Carved Stone in Mithfjord	.	.	.	.	.	.	299
Icelandic Magical Characters	.	.	.	.	.	.	299
Tunguhver	.	.	.	.	.	.	.	340
Runic Tombstone at Reykholt	.	.	.	.	.	.	343
The Skrimsl	.	.	.	.	.	.	.	346
Bird's-eye View of the Geysir District	.	.	.	.	.	358
Gyr-Falcon	.	.	.	.	.	.	.	393

PREFACE.

My object in visiting Iceland was twofold. I purposed examining scenes famous in Saga, and filling a
portfolio with water-colour sketches.

The reader must bear this in mind, otherwise he
may be disappointed at finding in these pages little
new matter of scientific interest.

The landscape painter will thank me for having
opened to him a new field for his pencil; and the
antiquarian will be glad to obtain an insight into
the habits and customs of Icelanders in the tenth
and eleventh centuries.

My illustrations faithfully represent the character
of the country, though they necessarily fail in render-

ing the wild beauty of colouring. I invariably submitted them to my guide, and found that he at once
recognized the spots, so that I am satisfied with their
fidelity. Some of the panoramic views have unavoid-
ably suffered in being contracted to the compass of the
book, but if the reader will imagine them to be pulled
out like bits of india-rubber he will obtain a correct
notion of the scenes. I refer to the panoramas on
Plates I. and XIV.

My specimens of the Sagas have been selected
with a view towards illustrating the voyages, quarrels,
litigations, and superstitions of the ancient Icelanders.

It must be remembered that the Sagas from which
I draw my extracts are not mere popular tales; they
are downright history. To quote the words of our
highest English authority on the subject when speaking of the Njala, but which apply equally to the
Gretla, Aigla, Bandamanna Saga, Vatnsdœla Saga,
&c. :—" We may be sure that as soon as each event
recorded in the Saga occurred it was told and talked
about as matter of history; and when at last the
whole story was unfolded and took shape, and centred
round Njal, that it was handed down from father to
son as truthfully and faithfully as could ever be the
case with any public or notorious matter in local
history. But it is not on Njala alone that we have

to rely for our evidence of its genuineness. There are many other Sagas relating to the same period, and handed down in like manner, in which the actors in our Saga are incidentally mentioned by name, and in which the deeds recorded of them are corroborated. They are mentioned also in songs and annals, the latter being the earliest written records which belong to the history of the island, while the former were more easily remembered, from the construction of the verse. Much passes for history in other lands on far slighter grounds, and many a story in Thucydides or Tacitus, or even in Clarendon or Hume, is believed on evidence not one-tenth part so trustworthy as that which supports the narratives of these Icelandic story-tellers of the eleventh century. That with occurrences of undoubted truth, and minute particularity as to time and place, as to dates and distance, are intermingled wild superstitions on several occasions, will startle no reader of the smallest judgment. All ages, our own not excepted, have their superstitions; and to suppose that a story told in the eleventh century, when phantoms, and ghosts, and wraiths, were implicitly believed in, and when dreams, and warnings, and tokens, were part of every man's creed, should be wanting in these marks of genuineness, is simply to require that one great proof of its truthfulness should be wanting, and that, in order to suit the spirit of our age, it should lack something which was part and parcel of popular belief in the age to which it belonged." *

I do not mean to say that *all* Sagas are of equal authority ; some are mythological, others fabulous or romantic ; but there is no difficulty whatever in distinguishing fact from fiction in these works of a bygone age.

I give these specimens of the Sagas to the world with great diffidence, as I am by no means a proficient in the Icelandic tongue.   I have worked at it for three or four years, and have arrived at the conclusion that both language and literature require the devotion of a lifetime for their proper mastery.   The language is full of obscure idioms, and to these there is no tolerable dictionary.   That of Bjorn Haldorson, which is the only lexicon, is out of print and rare, so that I had considerable difficulty in procuring a copy.

I have not hesitated to make a few very trifling alterations in the stories (they consist chiefly in names) for the advantage of the reader.

I have used a few antiquated and provincial words in my versions of the Sagas, where such words closely resembled the Icelandic.   The principal of these

---

* Burnt Njal, vol. i. 6.

are : — *Chapman,* a good English word, signifying merchant ; *gill,* a narrow glen ; *fell,* a mountain ; *byre,* a farm ; *bonder,* a farmer ; *to busk,* with its past participle, *boune,* to make ready ; *hight,* called.

With regard to my fellow-travellers, I have so altered their names and the incidents related of them, as to prevent the possibility of their identification.

Finally, my thanks are due to my friend, Mr. G. G. Fowler, for much information with regard to Icelandic birds, and especially to Mr. Alfred Newton for his invaluable paper on the ornithology of the island, inserted in my Appendix ; also to Mr. W. Boyd and Mr. W. Ardley for their harmonies to the Icelandic melodies I brought home with me.

SABINE BARING-GOULD.

b

ERRATA.

Page 7, line 4, *for* " shall " *read* " will."
" 23, " 14, *after* " and never shall " *insert* " be."
" 28, "   6, *for* " that he has " *read* " that he should have."
" 40, " 26, " " overlay " *read* " overlie."
" 63, "   5, " " farm " *read* " tarn."
" 67, " 17, " " lead " *read* " leading."
" 107, " 23, " " index " *read* " appendix."
" 112, "   4, " " water " *read* " waters."
" 191, " 20, *dele* " are."

Pour moi, parmi des fautes innombrables,
Je n'en connais que deux considérables,
Et dont je fais ma déclaration,
C'est l'entreprise et l'exécution ;
A mon avis fautes irréparables
       Dans ce volume.
                        BENSERADE.

INTRODUCTION.

ICELAND lies just south of the Arctic circle, touching it in the north. It is situated between lat. 63° 25′ and 66° 30′ north, and long. 13° 38′ and 24° 40′ west. Geographical and physical features.

The shape is peculiar. It is that of an irregular ellipse, with a considerable excrescence in the north-west, which is united to the mainland by a neck only 4½ miles across at the narrowest part.

The island is one-fifth larger than Ireland, and contains about 37,000 square miles. Its greatest length is 308 English miles, and greatest breadth 190. It is deeply indented with fjords on all sides except the south. It owes its upheaval entirely to volcanic agency, and is composed exclusively of igneous rocks.

The interior of the island consists of an elevated band of Palagonite tuff, pierced by trachyte veins; on either side of this formation is basalt. It has been generally held that the island was traversed by a broad trachytic valley, hemmed in between chains of trap mountains; but this view is erroneous. Instead of a vale, we have the great jökulls of the centre formed of tufa, and only the fells and smaller ice-mountains on the north coast composed of basalt.

The mountain system is in the south, and takes the shape of a triangle, having for base a line drawn from Ók to Eyjafjalla, and for apex, Thrándar jökull, which towers above the Beru-fjord. A glance at the map would convey the idea that extensive plains occupied the area of the lower portion of this triangle, but such is not the case. The space intervening between Blá-fell, Hekla, Torfa jökull, and the vast ice regions of Hofs and Vatna jökulls is, in fact, occupied by ground rising gradually in rolling "heithi" sweeps, till it meets the snows of Skapta. Towards the apex of the triangle, the glacier mountains form a compact mass called Vatna, or Klofa jökull, covering an area of 3,500 square miles of unexplored snow recesses. North of this triangular mountain system is a triangular elevated plain, with the base towards the east, extending from Thrándar jökull to Langanes, and with its apex at Baula. This plain is a complete desert, covered with vast lava beds, as the Odatha Hraun, with extensive tracts of black sand, as the Stori and Sprengisandur, or with rock and mud sprinkled with lichen and moss in sheltered hollows, as the Arnarvatn Heithi. This wilderness is traversed by three main routes, the Thvidœgravegr, the Kjalvegr, and the Sprengisandur way.

These two triangles form a parallelogram including 20,000 square miles of country perfectly barren and uninhabitable, and only partially explored. It has been estimated that but 4,000 square miles of Iceland are inhabited; the rest of the country is a chaos of ice, desert, and volcanoes.

Mountains.   The mountains of Iceland are divided into two classes, the fells and the jökulls. The former are, for the most part, free from snow during the hottest portion of the summer; but the latter are shrouded in eternal ice. The conformation of the mountains is very varied. The jökulls have generally rounded heads of ice resting on abrupt flanks of rock. This ice is formed by the pressure of enormous superincumbent masses of fresh snow, converting this at any depth to blue glacier ice. Sections such as may be seen in Kaldidalr and on the flanks of Eireks jökull give to these beds a thickness of about one hundred feet.

As there is a constant thrust from the highest points of the mountain exerted upon the ice, it moves slowly over the rocky ledges, and breaks off in crags of green ice, which fall to the bottom of the precipices with a roll like thunder. On the south of the Vatna and Myrdals jökulls, where the mountains shelve gently to the sea, glaciers resembling those of Switzerland may be seen, but they are entirely absent from the centre of the island. The principal jökulls are—the Orœfa, height 6,241 feet; Eastern Snœfell, 5,160 feet; Eyjafjalla, 5,432 feet; Herthubreith, 5,290 feet; and Western Snœfell, 4,577 feet. The latter has alone been surmounted. Mr. Holland, in 1861, ascended to a great height on the Orœfa, but did not gain the summit. The fells are mountains of different character; they may be rounded at top, but they support no ice-fields, and are inferior in elevation to the jökulls. In shape they vary considerably: some are saddlebacked, others conical or pyramidal. Some, as well as the jökulls, are volcanoes, and have often caused much havoc. Hekla, for instance, is a fell, whilst the terrible Skapta is a jökull.

Volcanoes.   The most violent volcanic action seems to lie in a band from Krafla to Reykjanes.

In this belt lie the following principal volcanoes—Krafla and Leirhnukr, both near Myvatn ; Trölladyngja, in the Odatha-hraun ; Skapta and Orœfa, the westernmost points of the enormous Vatna jökull, Katla or Kötlugjá, and its twin moun-tain Myrdals jökull, and Hekla.

The total number of recorded eruptions is :—

Eldborg, date uncertain.

894. Katla erupted for the first time.

984. Katla for the second time.

1000. Katla again (the date is not certain).

1004. Hekla for the first time.

1029. Hekla for the second time.

1104–5. Hekla for the third time.

1113. Hekla for the fourth time.

1150. Trölladyngja for the first time.

1157. Hekla for the fifth time.

1188. Second eruption of Trölladyngia.

1204. Hekla.   Sixth eruption.

1210. At sea, near Reykjanes.

1219. At sea, near Snœfells jökull.

1222. Hekla.   Seventh eruption.

1222–26. At sea, off Reykjanes.   Four outbreaks.

1237. Seventh eruption at sea.

1240. Eighth eruption, off Reykjanes.

1245. Myrdals jökull, south-west flank, called Sólheimar jökull.

1262. Sólheimar jökull again.

1294. Hekla.   Eighth eruption.

1300. Hekla.   Ninth eruption, and one of the most violent.

1311. Rauthukambar, in Austur Skaptarfells sysla.

1311 or 1332. Katla.   Fourth eruption.

1332. Eruption of the Orœfa jökull.

1340. Second eruption of the Orœfa; and, in same year, the tenth of Hekla ; also, the only known eruption of Móafell, in Kjósarsysla.   Herthubreith also vomited ; so also Trölladyngja.

1359. Fourth eruption of Trölladyngja.

1362. Eruption of the eastern portion of the Orœfa ; and the third of the western head.

1874. Hekla.   Eleventh eruption.

1390. Hekla.   Twelfth eruption.

1416. Katla.   Fifth eruption.

1422. Ninth eruption in the sea, off Reykjanes,

1436. Hekla.   Thirteenth eruption.

1475. Trölladyngja.   Fifth eruption.

1510. Hekla. Fourteenth eruption; at the same time the second eruption of Herthubreith, and the sixth of Trölladyngja.

1554. Hekla. Fifteenth eruption, from a side crater.

1580. Sixth eruption of Katla, which then formed the huge rift between itself and Myrdals jökull. The volcano has since been called Kötlugjá.

1583. Tenth eruption in the sea off Reykjanes; and the sixteenth of Hekla.

1587. Bursting of ground near Thingvalla lake; clouds of smoke and streams of lava gushed forth.

1597. Seventeenth eruption of Hekla.

1612. Eruption of Eyjafjalla jökull.

1619. Eighteenth eruption of Hekla.

1625. Seventh eruption of Katla. In the same year, the nineteenth of Hekla.

1636. Hekla. Twentieth eruption.

1660. Kátla. Eighth eruption.

1693. Hekla. Twenty-first eruption.

1716. Eruption of ash and smoke from Ball jökull.

1717. Eyjafjalla jökull. Second eruption.

1720. Fourth eruption of the Orœfa.

1721. Ninth eruption of Katla.

1724–30. Eruption of Krafla, near Myvatn.

1725. Eruption of lava in a grass-grown plain at Hítaholl, near Myvatn; followed by a similar outburst in the grass-land of Bjarnaflag. In the same year, an eruption of Leirhnukr; and also of Skeitharár jökull, west of the Orœfa. That of Leirhnukr lasted till 1729.

1727. Second eruption of Skeitharár jökull, and the fifth from Orœfa jökull.

1727–28. Tenth eruption of Katla.

1728. Second eruption in the grass-land at Bjarnaflag, and, in the same year, a slight outburst from Hekla. Another eruption of lava took place in Horsadalr, near Myvatn, and filled the valley; at the same time, the earth opened near Reykjahlith and threw up ash, fire, and lava. Sithu jökull, in the south, also erupted.

1748–52. The Hverfjall was thrown up.

1753. Second eruption of Sithu jökull.

1754. Hekla. Twenty-third eruption.

1755. Eleventh eruption of Katla.

1766. Four-and-twentieth eruption of Hekla.

1772. Hekla. Twenty-fifth eruption.

1788. Eleventh eruption off Cape Reykjanes. In the same year occurred the most appalling eruption on record — that of Skapta.

1821. Third eruption of Eyjafjalla jökull.

1823. Katla. Twelfth eruption.

1845–46. Hekla. Twenty-sixth eruption.

1860. Katla. Thirteenth eruption.

1861. Skapta threw up ash and sand.

1862. Trölladyngja erupted ash. There is much uncertainty about this explosion.

From the above-given list it will be seen that the number of eruptions from—

Hekla, since 1004, inclusive, have been 26.

Katla, since 894, inclusive, have been 13.

At sea, since 1210, inclusive, have been 11.

Trölladyngja, since 1150, inclusive, have been 7.

Oræfa, since 1332, inclusive, have been 5.

There have been eighty-six eruptions, including outbreaks of lava, mentioned in some of the Icelandic histories, but the date of which it is impossible with accuracy to determine.

These outbursts have taken place from twenty-seven different spots. Some of these vents have been active at several different periods, whilst others have erupted but once.

The intervals between these explosions have been most irregular. Those of Hekla have varied from six to seventy-six years, and the intervals between the eruptions of Katla from six to three hundred and eleven years.

At periods of peculiar activity more volcanoes than one have vomited simultaneously, as, for instance, in the year 1840, when the Oræfa, Hekla, Mósfell, Herthubreith, and Trölladyngja erupted together; or, in 1510, when Hekla, Herthubreith, and Trölladyngja poured forth fire and molten rock at the same time. From 1724–30 was the period of greatest activity, twelve eruptions having taken place in those six years. The interval of greatest length between outbreaks of subterraneous fire was between the eruptions of Trölladyngja, in 1510, and Hekla, in 1554, a period of forty-four years. Between those of Hekla and Trölladyngja, in 1436 and 1475 respectively, thirty-nine years elapsed.

Lava breaks forth not only from mountain sides, but from the grass-land under foot. The earth gapes and pours forth a flood of fire or casts up scoriæ, where meadows had previously existed. Such eruptions took place near Myvatn, in 1725, and at Thing-vellir, in 1587.

The traveller notices many instances of the lava having thus welled up and overflowed older strata.

Beds of lava of this nature exist in the midst of the Storisandr, at Olfus, and throughout the Gullbringu sysla.

Immediately after an eruption has taken place, the volcanoes of Iceland relapse into quiescence, and no smoke or steam rises from them, as it does from Vesuvius and Etna.

The great majority have no circular craters; some have split themselves in the fury of explosion, and the lava has flowed from their sides, whilst the fire and water have found vent at the gashed crown. The perfectly symmetrical craters are Eldborg, Hverfjall, Vilingafjall, Borg in Vithidal, and every mountain named Skál.

Lava and obsidian.

The lava stream from Eireks jökull we distinctly traced to the mountain roots. It had not flowed from the summit, but from a chasm at the base. The Odatha Hraun is the most extensive lava bed, covering a space of 1,160 square miles. It has flowed from Trölladyngja and Herthubreith, and its recesses are quite unknown. It extends farther north than has been represented by Gunnlaugson, and reaches indeed as far as Burfell. The second largest bed is more broken and intercepted by hills and lakes. It extends from Skjaldbreith and Hlöthufell to Reykjanes, a distance of seventy-three miles. The tract around Hekla covers an area twenty-five miles long by ten broad.

Obsidian has flowed from some of the volcanoes, as well as lava. The most important streams are near Krafla, Hekla, and at As, in Hvitárdalr. I found coarse obsidian on Ók.

As already stated, the great mountain system of Iceland is formed of Palagonite tuff. The principal places where this rock has been pierced by trachyte are Baula, Thorishöfthi in Kaldidalr, a portion of Ók, and Laugarfjall, above Geysir.

Surtur-brand.

One of the most singular formations in Iceland is the surturbrand, a species of lignite, which lies in beds between clinkstone and trap. The wood is brightly glossed and black, free from all admixture of sulphur, very splintery in fracture. Logs and branches are preserved with their knots and roots; the circles denoting the age of the tree are very distinct at the ends of the fragments. In several places a layer of leaves overlies the coal in beds of four to six inches in thickness. The impression of the leaves, with all their delicate fibres, is perfect and very beautiful. The leaves belong to the poplar, willow, and birch.

The alternation of basalt and surturbrand deserves peculiar attention from geologists, as the existence of leaves, and absence of marine shells in the deposit, seems to point out that there may have been a too hasty generalization, in concluding that all basalt is of submarine formation.

Surturbrand is to be found in Borgarfjordar and Myra sysla. At Hersthvatn, west of Northurá, is a bed two feet thick, black and lustrous.

At Tandarasel there crops out a considerable amount, so also in the Hitárdalr.

In the Barthastrandar, Dala, Isafjarthar and Strandar sysla, are three beds at different levels, extending through the whole peninsula at the heights respectively of 150 and 600 feet above the sea. At Lœkir on the Barthastrand, at Forsdalr, in the Arnarfjord, they are found alternating with beds of leaves. Surturbrand has also been noticed at Svínadalr, Gnúpurfell, Barmahlith, Ranthasand, Stigahlith, Steingrims-fjord, &c.

In Skagafjarthar, Eyjafjarthar, and Thingeyjar sysla it occurs; at Hofsgil in Gothdalr, Tinná in Skaga-fjord, Ulfá in Eyja-fjord, near Húsavik, the headland Tjörnes, between Skjalfunda and Axar fjords, in the Vapna-fjord, and at Thussahöfthajóta, near Eski-fjord.

Few metals have hitherto been met with in Iceland. Copper is found in some places, but in small quantities; plumbago has been discovered near Krafla. Magnetic iron is undoubtedly very widely dispersed through the volcanic rocks of the island; it occurs at Esja; and I found some near Eylifr, north-east of Myvatn.

Minerals.

A vast and inexhaustible supply of sulphur is deposited by vapour jets in four spots—Hengill, near Thingvalla Lake, Krisuvik, Hlíthar-námar, and Fremri-námar, near Myvatn. None of this is now exported.

The most remarkable boiling springs in Iceland are the Geysir at Haukadalr and at Reykir, Uxahver, the numerous fountains of Reykholtsdalr, and those of Hveravellir. These by no means constitute all in the island, for there is hardly a valley without hot springs; and the natives have learned to distinguish their varieties by appropriate names.

Boiling springs.

Hver is a general term expressing a warm or boiling spring; *geysir* is one which spouts; *reykir*, one which sends up clouds of steam; *laug* is a warm fountain which will serve as a bath; *ölkelda* is a mineral spring; and *náma* a pit of boiling mud.

The characteristics of these fountains have changed within historic times. Tunguhver, which Sir George Mackenzie mentions as sending intermittent jets twelve or fourteen feet high, no longer alternates: it boils furiously, but its jets have been spoiled by travellers, who have choked its bore with stones. Hveravellir, spoken of by Olafsen and Povelsen as the most wondrous sight in Iceland, with its roaring mountain of steam, is now reduced to a dozen caldrons of boiling water. The geysir which Henderson saw in the crater of Krafla plays no longer; and its place is occupied by a still green pool of cold water. Some further instances will be adduced in the course of the narrative.

Lakes.

Numerous lakes, either single or in groups, are scattered over the surface of the country. The largest are Thingvalla-, My-, and Hvitár-vötn; the principal groups are those of the Arnar-vatn-heithi, and those at the foot of the Skapta, the remains of a considerable lake which existed previous to the great erup-tion of 1788.

Myvatn was formerly considered the largest lake in the island, and so it may have been till it was nearly filled and dried up by the influx of lava from Krafla in 1724–1730. Thingvalla-vatn is now the most considerable sheet of water—it is ten miles long and between four and five miles wide. The contorted shape and irregularities of outline in Myvatn preclude one from giving any correct account of its dimensions. The Lagarfljót, in the east, is thirty miles long, but its width is only from half a mile to a mile and a quarter. On the Arnarvatn-heithi, Gunnlaugson marks fifty-three lakes; but, from what I saw there, I am satisfied that not a fifth of those actually exist-ing appear on the map, which, with regard to that district, is somewhat inaccurate.

Rivers.

The rivers of Iceland are both numerous and large. The Jökulsá is the longest—it rises in the Vatna jökull, receives a tributary from Herthubreith, and, after a course of 125 miles, reaches the sea in the Axa-fjord, having passed exactly ten houses in its way, and having plunged into a chasm in a water-fall, the like of which is not to be seen in Europe. Other fine rivers are the Thjorsá, Skjálfandafljót, Hvitá, and Jökulsá á brú. A very curious phenomenon is the broad short river which is found in the south of Vatna jökull. In that portion of the island violent torrents, a couple of miles wide, and only eight or ten from their source to the sea, whirl down with frightful velocity, carrying with them masses of ice dislodged from the glaciers which are their feeders, and volumes of sand from the volcanic mountains which they drain. In passing Icelandic rivers, the traveller trusts either to his horse or to a ferry. The ponies swim well; but if the current be too strong there is considerable danger of their not being able to carry their rider across. Fords are continually shifting; and it is of the utmost importance for a stranger to secure a guide from a neigh-bouring farm, before venturing into the river. The beds are, in many cases, composed of quicksands, and the pebbly bottom on which the horses can find a sure footing changes with every spring.

The fjords into which the rivers empty themselves may be *Fjords.* divided into two classes—the friths proper, and the bights or bays. The former are indentations in the line of coast, extending for a considerable distance into the land between precipitous mountains, whose tops are snow-covered, or continually veiled in mist, which the sea-breeze brings up with it. The noblest of these is the Isa-fjord in the north-western peninsula, fifty-two miles long, and winding between magnificent mountains, rising in inaccessible walls of basalt many thousand feet above the water's edge. Ten lesser friths open out of it, piercing the barrier crags, and stretching to the roots of the great barrel-headed jökulls of Dránga and Glámu.

Other magnificent fjords are Hvamms, Skaga, Eyja, and Arnar fjords, all with distinct characteristics.

The bays of Iceland are very extensive; the noble Faxafjord, sixty-five miles across, opens between capes Reykjanes and the silver sugarloaf of Snœfell. Breithi-fjord, studded with innumerable islets, the home of myriads of eider and wild duck, is forty-five miles wide; and Húnaflói, into which the Arctic sea rolls without a break, is forty-six miles long and twenty-seven wide. Other bights are the Axar, Skjálfanda, and Thistil fjords.

A peculiar feature of Iceland is the gjá, pronounced gee-ow. *Chasms.* This is a fissure in the crust of the earth, formed by earthquakes, or volcanic upheavals and sinkings of the land. These zig-zag rents run from north-east to south-west. The most remarkable are the Allmanna and Hrafna gjás, at Thingvellir, the huge chasm in Katla, the rift into which pours the Jökulsá at Dettifoss, and the Stapa, Hauksvörthu, and Hrafna gjás, in Gullbryngu sysla. The first-mentioned extends for four miles, and is, in one spot, a hundred and thirty feet deep. The Hrafna gjá, or Raven rift, is somewhat longer, but only fifty feet deep.

In 1728 there opened a chasm in the Oræfa of immeasurable depth. The Archdeacon Jón Thorlaksson, who visited it, found a large stone at one spot crossing the lip of the gulf. He and a companion dislodged it, and sent it down into the abyss, but, though they listened attentively, they could not hear it reach the bottom. The great fissure of Katla has never been properly examined. It runs south-west to north-east, then turns at a right angle from south-east to north-west. This is probably the crater of the volcano. The only person who has been near the chasm is an Icelandic priest, Jón Austman, who ascended the mountain in 1823. He describes it as quite inaccessible, all progress being stopped by enormous walls of basalt and obsidian; while other profound chasms radiate from the grand trunk or primary fissure.

Morasses. The general aspect of Iceland is one of utter desolation. The mountains are destitute of herbage, and the valleys are filled with cold morasses. Grass springs on the slight elevations above the swamps, in the dells, and around the lakes. By drainage, a large percentage of marsh might be reclaimed; but some must always remain hopeless bog. The extraordinary amount of swamp is due to the fact that the ground is frozen at the depth of six or eight feet, so that, when there is a thaw, the valleys are flooded, and the water, unable to drain through, rots the soil. In many places a stream is thus completely absorbed, and a considerable tract of land rendered impassable, where the labour of a few weeks would give it a channel, and transmute marsh into productive meadow land. Many bottoms are filled with an amazing depth of rich soil, the wear of volcanic rock, abounding in the constituents necessary for vegetable life. Yet the ignorance of agriculture prevailing in the island has deterred any from turning them to advantage, by draining off the icy water which nips and destroys the tender grass, ready enough to spring.

Besides these swamps, there are stone bogs on all high land, caused by the breaking up of the tufa rocks, through the united action of frost and snow: a bed of soft mud and stone is thus formed, which is particularly trying to the horses, who sink in it to their knees, and cut their hoofs with the rocky splinters.

Caves. There are a considerable number of caves in Iceland, formed in the lava by the generation of gases during the process of cooling. Few of them have been explored; and, indeed, they hardly repay the labour of investigation. Their bottoms are strewn with immense angular fragments of vitreous rock, making the toil of traversing them very considerable. The few caverns which have been examined, are those already known as having been resorted to by outlaws and bandits in historic times. Of these the most interesting are — Surtshellir; that of Bárdr Snœfelsás; one in the Hallmundar Hraun; and Paradísarhellar, long regarded by the superstitious as the entrance to regions like those in S. Patrick's purgatory. The openings to similar caves are visible near Myvatn, and in the lava tract above the Raven rift, near Thingvellir.

Erratic blocks and glacial grooves. Over the whole surface of the country are to be seen blocks of stone placed in singular positions, much resembling the Logan rocks of Cornwall. These go by the name of Grettistaks, and are perched on high moors or in valleys. That they have been brought by ice can hardly be doubted; the uplands bear many evidences of having been covered by water, and traversed by floating icebergs. In one spot alone did I find unmistakable glacial grooving, and that was along the hill above Bjarg, in Mith-fjord: a complete description of this will be given in Chapter XVII. One of the Grettistaks I have sketched; it will be seen in Plate III. In no case did I find them belonging to other rock formation than that already existing in the island. There are no traces of moraines, except at the skirts of modern glaciers.

Rock needles, which abound on the coasts, are named Drángir *Rock needles.* by the natives. Some of these are very noble. The entrance of the Isa-fjord is guarded by one such, standing up from a platform of basalt high above the water; it goes by the name of "the Sentinel." A curious spire of rock above the Hörgárdalr is illustrated in Plate XI. There are needles in the Skaga-fjord off Drangey, and in the Breithi-fjord.

Of roads, there is not one in the whole island; tracks are all *Roads.* that mark a *vegr* or way, and these are obliterated at every thaw. The routes are, consequently, indicated by *vörthur* and occasional *kerlingar*. The former are heaps of turf, or simply a stone or two placed on a rock, in a manner which the eye will recognise as artificial. The latter are stone pyramids, bearing a fanciful resemblance to old women. Many of these marks are out of repair, and others are too far apart to be of much practical advantage. A few feeble attempts have, in some spots, been made to clear the path of the larger stones, but with little result. The natives complain that the Danish Government does nothing for the roads; but surely each *hrepp* ought to look after its thoroughfares; and Government is like Providence, it only helps those who help themselves. It is essential for the prosperity of the island that these ways should be kept open for traffic; and Althing might well devote its session to a consideration of the means by which money might be raised for improvements of this nature, instead of frittering away its time in idle grumblings against the mild and merciful rule of Denmark.

In certain spots on the surface of Iceland are forests, *skogar,* *Forests.* as they are termed by the natives. These consist of low coppices of birch; the trees being mere shrubs, from one to twelve feet high. The Icelanders believe that in former times the growth of birch was much loftier: woods were undoubtedly more abundant, as the Sagas mention forests where no trees grow at the present day, and the underwood still existing rapidly diminishes to supply the neighbouring farmers with fuel. We read in the *Gretla* of some boors slinging a rope over the fork of a tree, for the purpose of hanging the outlaw Grettir. No tree of sufficient size exists in the island at the present day, to support the weight of a large Newfoundland dog. Every *skog* is marked on Gunnlaugson's map. The rapid destruction may be exemplified by the instance of a forest in Öxnardalr, marked on the map published in 1844, and which is now completely destroyed. By far the finest woods in Iceland are those of Fnjoskadalr, and Thverárhlid in Myra sysla. One very remarkable forest, completely surrounded by snow mountains, except at one point where a rivulet escapes, exists in the south of Vatna jökull, at Nupsstathr. I understood from a Danish merchant, that a singular forest is found near Rautharhöfn, completely encompassed by high lava walls, and that the only admission to this secluded recess is through a hole in the lava.

Tempera-ture. The temperature of Iceland is varied. The north is far colder than the south-west, which is washed by a branch of the Gulf Stream. The average temperature of Reykjavik is about the same as that of Moscow, the whole year included. At Reykjavik the average summer heat is 53° 6′ Fahrenheit; winter, 29° 8′; and that for the whole year is 39° 4′. At Akureyri, in the north, on the Eyja-fjord, the average summer heat is 45° 5′; that of the winter, 20° 7′; and the mean for the year is freezing point (*Almanak um ár*, 1868, *af Schjellerup*). According to Humboldt, the mean temperature of the year is 40°; of the winter, 29° 1′; of the spring, 36° 9′; of the summer, 53° 6′; of the autumn, 37° 9′; of the warmest month, July, 56° 3′; of the coldest month, February, 28° 22′. Horrebow, in Bessa-stathr, found the hottest day, in the years 1749—1751, to have been 70° 5′, which was the 30th July, 1751; and the coldest to have been 13° 75′, on the 25th January of the same year. The maximum heat at Akureyri in the summer is 75° 2′, and in winter the thermometer sinks as low as—29° 2′.

It will be seen that there is a difference of seventeen degrees between the average temperature of Reykjavik and Akureyri; so that whilst the mean of the former is very nearly the same as that of Moscow, the mean of Akureyri is about that of Julians-haab, in Greenland. The isotherm of 32°, which is that of Akureyri, touches the north Cape, on the continent of Europe, under latitude 71° N.; from which point it turns suddenly to the south-west, running along the Dovrefjeld. It then takes a bend towards the south-east, and returning to the Arctic Circle, touches Tornea at the head of the Baltic, passes Uleaborg in

Finland, dips towards the interior of Russia, more and more south, almost touching Statoust. It then passes the Ural, and leaving Tobolsk on its north, runs nearly parallel with the lines of latitude, reaches Irkutsk, then turns again towards the sea, cuts through the middle of Kamschatka, and reaches the Polar Circle on the north-west of America. In the interior of this continent it makes a rapid descent towards the south, following the Rocky Mountains, touching lake Winnipeg, and cuts the southernmost sweep of Hudson's Bay. From the eastern coast of Labrador it stretches northward once more, and traversing the snowy promontory of Greenland above Julianshaab, returns to the north coast of Iceland.

The coldness of the winter depends upon the formation of Greenland ice. Periodically large masses break away and float off south, producing cold summers in England. The Gulf Stream is then able to take a higher sweep, and a succession of mild winters ensue till the ice-fields have again recovered their southern position. This advance and shrinking of the ice is an infallible index to the changes in the temperature of Iceland. There is a saying among the Danes that there is mild weather in Iceland when it is cold in Europe, and vice versâ: an observation probably true in the main, though not borne out by fact in the year 1862, when I was in the island. The summer was cold throughout Europe, and also in Iceland: however, there was a difference in one particular; June and July were months of incessant rain throughout England, France, and Germany, whereas I had only three rainy days during the whole of my tour.

Thunder and lightning are rare in Iceland, and only occur during the winter. *Meteorological phenomena.*

The Aurora Borealis is very splendid as soon as the darkness of winter sets in, lighting up the gloomy skies with its glorious scarlet streamers. Other phenomena are the *Hrovarelldur*, or electric flames, which appear about metallic objects, such as buttons, or stream from the head, like the glories of the saints. *Rosabaugur*, or storm-rings, form about the moon, and mock suns, called *hjásólar*, are frequently seen, sometimes to the number of nine at once. Meteors, termed *vigahnöttur*, and shooting-stars, *stjörnurhap*, are often observed. Olafsen and Povelsen give a singular account of a circumstance which took place in the summer of 1754. They say that, on a morning when the weather was serene, though the sky was rather cloudy and a slight wind prevailed, there was seen at Eyrarbakki a black cloud coming from the mountains in the north-east, and descend-

ing obliquely through the air towards Eyrarbakki. The nearer
it approached the smaller it became, and it darted along with the
rapidity of a hawk. This cloud, which then appeared round,
flew towards a spot where several persons had assembled, as well
strangers as natives, for the purposes of commerce; and on
passing rapidly before them, it touched the jaw of a middle-aged
man, causing him such pain that he instantly became raving
mad and threw himself into the sea! Those who were near
him prevented him from drowning; but he remained insane,
uttered all sorts of extravagant expressions, and made many
forcible attempts to free himself from those who held him. They
wrapped his head in flannel, and held him down for some time
upon the bed; after two days, the madness abated, but he was
not restored to his senses till the expiration of a fortnight.
Another account of this phenomenon states that the persons in
the company did not perceive the cloud till it came up with them,
but simply heard a hissing in the air while it passed: those,
however, who were farther off observed its rapid course, and saw
it sink and disappear on the sea-coast. The cheek of the man
who was touched turned of a deep black and blue colour, which
gradually disappeared as he recovered.

There can be little doubt that this "cloud" must have been
some material flung from Hekla, which erupted that very year.
A fireball of a terrific nature was cast up during the eruption of
Kötlugjá, in 1755.

Hurricanes and whirlwinds traverse Iceland with great rapidity
and with enormous violence. One is mentioned in the Gisla
Surssonar Saga, which tore the roof off a hall. Olafsen and
Povelsen saw a similar whirlwind detach from the shore of
Reykja-fjord a large block of stone to which a ship's cable was
attached, and whirl it into the sea.

Natural
history.

Of wild animals, foxes are the most plentiful. Reindeer were
introduced into the island in 1770. Thirteen head were then
brought from Norway. Of these, ten died during the voyage;
but the remaining three increased rapidly, so that at present
there are considerable herds in the unpopulated districts of the
island, especially in the rolling mountain deserts of the north-
east. In winter, the reindeer are hunted down by the natives,
for the sake of their flesh and horns. They have never been
domesticated, as the country is too uneven and intersected by
rivers to render sledging practicable; and, as they devour the
esculent lichen, which is an Icelandic staple of food, the reindeer
are looked upon with very little favour.

Bears come over with the drift-ice from Greenland, but in no great numbers; only from ten to fourteen are killed during each winter.

There are seven families and thirty-four species of mammals in Iceland; but of these, twenty-four are water creatures. Of Cetacea alone, there are thirteen varieties.

Two quadrupeds appear to be indigenous: the fox, and the somewhat problematical Icelandic mouse; the rest have been brought over since the colonization of the island.

The feathered creation is most fully represented in the country. There are six families, and about 90 species of birds. Of these, fifty-four species are water-fowl.

No reptiles have ever been discovered in Iceland; no frogs croak in the marshes, no snakes or blind-worms wriggle through the coppice, nor do lizards dart among the rocks.

The fish which frequent the numerous lakes, rivers, and fjords are little known. Faber mentions forty-nine varieties, of which seven are fresh-water fish. The list is manifestly incomplete.

The most valuable property of an Icelander consists in his cattle; he possesses cows, horses, and sheep. The Sagas speak of flocks of geese and droves of swine as having formed part of a farmer's wealth; but all the geese are now wild, and there are no swine in the island. The ass is also quite unknown. The dog is of the Esquimaux type, with ruff around its neck, head like a fox, and tail curled over its back. It is of great use to the farmer in keeping his flocks together, and defending his tún or home-meadow, from the inroads of cattle. National industry.

The swan and eider-duck are the only birds turned to any account. The former is shot for its quills; the latter is preserved, by law, for its down. A severe penalty is inflicted on all who kill this bird; and it becomes so tame in the breeding season, as to build and lay on the roofs and in the windows of the farmhouses. The birds strip their breasts to line their rudely-constructed nests, and the down thus pulled off is removed by the peasants. The duck at once pulls off a fresh supply, and this is again removed. If the third lining of down be carried off, the bird will desert its nest, and never return.

Fisheries are prosecuted with great activity around the coast, chiefly by the French, who keep a man-of-war constantly in the Faxa-fjord, to protect the interests of their fishermen. The natives secure enough of the produce of the sea to supply themselves with food for the winter, but never engage in the fisheries with a view to traffic. Cod, haddock, skate, and halibut abound on the coast, and the lakes are filled with trout and char of great size and delicious flavour. Seals and whales are killed for their oil; and shark's flesh is eaten, after having been buried for some months, to free it from its peculiarly rank taste. The rivers teem with salmon, which are caught in creels. The fish are split and wind-dried, then stacked in an outhouse for consumption.

No grain is cultivated, but a species of wild corn, growing on the sand-flats by the sea, is much prized. It is reaped with a sickle, and, after having been thoroughly dried, is threshed. The straw is used for thatching, and as food for the cattle; the corn is baked, then ground, and made into thin cakes or wafers, which, when powdered with cinnamon, are very delicious. The meal is also kneaded into lumps of dough, which are eaten in milk or used with butter. The only places where this corn grows in any quantities are the Myrdals and Skeitharár sandur. The amount grown is only sufficient to render it an article of delicacy, and not a staple of food. The food of an Icelander consists of stockfish, rye cakes, boiled trout, Icelandic lichen, rice, rancid butter, and *skyr*, or curd. The drink is coffee, milk, and corn-brandy. Potatoes and carrots are cultivated in the gardens of the larger farms, but not to any extent. The former have never suffered from the disease which has ravaged the potato-fields of Europe. The only dressing given to the gardens is the ash from the peat and sheep's-dung fires of the kitchen.*

Learning.   Every native reads and writes well; he occasionally understands Latin. The clergy are uniformly well educated, reading Greek and Hebrew, and being sufficiently proficient in Latin to converse in it fluently. They are sometimes acquainted with Danish and German. Indeed, there is a talent for languages observable amongst the generality of Icelanders. The rector of the Latin school at Reykjavik is a master of eight languages, and my guide knew three or four very respectably.

The number of books possessed by the farmers and priests is small, but they borrow of each other and copy the volumes lent them. Their libraries consist generally of Sagas. The following is a catalogue of the books belonging to a farmer in the Vatnsdal :—

1. Bible.
2. Prayerbook.
3. Sermons of Vidallin, late Bishop of Reykjavik.
4. Book of Icelandic plants and their properties, by Hjaltallin.

---

* Further and fuller particulars concerning Icelandic food are given in chap. iv.

5. *Saga Thithriks Konungs af Bern*, ed. Unger.   1853.
6. *Islendinga Sögur*, second vol. only.   1829-30.
7. *Njáls Saga*, ed. Ol. Olavius.   1772.
8. *Vatns dæla Saga*, ed. Sveinn Skúlason.   1858.
9. *Bragtha-Mágus Saga*, ed. Gunnlaug Thortharson.   1858.
10. *Rímur af Gunnari á Hlitharenda.*   1860.
11. *Hrafnkels Saga*, in MS.
12. *Asmundar Saga*, in MS.
13. *Króka-Refs Saga*, copied from the printed edition of Marcusson.
14. Latin moral maxims, in MS.
15. *Lovsamling for Island*, ed. Stephensen.   1853-55.
Besides these there were some old numbers of newspapers and odd parts of the Transactions of Althing.

Iceland is subject to Denmark. It is now divided into three **Government.** amts instead of four, as in olden times.   These amts are subdivided into twenty-three sysla—eight in the South Amt, eight in the West Amt, and seven in the South and East Amt. They are again subdivided into 169 hreppar.

Over two of the amts is placed an amt-man, who is subject to the governor-general, under whose special jurisdiction is the third amt. The former resides near Akureyri, at Fredriksgaf, the latter at Reykjavík.

Each sysel is presided over by a syselman, to whom are answerable the magistrates of the hreppar.   Other officials are the landvogt or sheriff, who controls the financial arrangements of the country, a justice, and two assistants, before whom go all criminal cases.   Natives cannot go to law with each other until their complaints have come before certain umpires, of whom are the bishop and dean of Reykjavík, *ex officio.*

The interests of the people are invested in Althingmen, or members of the national parliament, which sits at Reykjavík. This assembly is administrative, not legislative.

The syselmen are bound to give out proclamations and notices, also to forward to head-quarters registers of births, deaths, and marriages, which they receive from the magistrates of the hreppar.   The governor-general, the landvogt, the amt-men, the chief justice, and the syselmen, are appointed by the Danish crown, but the rest of the officials receive their nomination from the governor.

The ecclesiastical division of Iceland was formerly into two bishoprics, those of Hólar and Skalholt. At present, there is but one bishop, whose cathedral is at the capital. The island is portioned into archdeaconries (prófasta-kalla) and parishes (presta-kalla). The clergy are appointed by the crown, subject to the consent of the bishop.

The island is further divided into medical districts, of which there are six: one medical officer is stationed at Reykjavík, a second in the Vatnsdal, a third in Akureyri, a fourth in the west, a fifth in the south, and the sixth in the Westmann Islands.

Population. The population of Iceland is 68,000, scattered thinly through the fjords and along the rivers, only gathered into settlements at Reykjavík, which contains 1,400 souls, and Akureyri 800. Smaller villages, clusters of poor cottages, around two or three merchants' stores, are at Isafjord and Eskifjord. Elsewhere the people are widely separated, an arrangement necessitated by the scantiness of pasture for their cattle. The places marked on the map must not be taken for towns and villages; with the exception of the localities specified above, they stand for single houses; and any one who wishes to know the number of farms in Iceland, may ascertain it with precision, by counting the specks on Gunnlaugson's map.

The usual age for Icelanders to marry is from twenty-five to thirty.

In 1858, 8 men committed suicide, 65 were drowned, 17 perished by other accidents, and 1,939 died of disease.

Death is most common among children. In the same year died 489 children between the ages of one and five; 68 between the ages of five and ten. The most healthy period of life is from fifteen to twenty, during which only eleven died.

Fifteen old people lived to ages between ninety and ninety-five, and five between ninety-five and a hundred.

In 1858, there were 487 marriages; the ages of the parties were as follow:—

Age.	Bridegroom.	Bride.
Under 20	0	25
20 — 25	107	174
25 — 30	221	182
30 — 35	86	55
35 — 40	34	18
40 — 45	10	14
45 — 50	8	11
50 — 55	12	7
55 — 60	2	1
60 — 65	5	0
65 — 70	2	0
	487	487

The number of children born in a year is about 2,940. The proportion of illegitimate children to those born in wedlock is 15 per cent. Of 2,987 children, only 48 were born of mothers under twenty, of these 28 were legitimate, and 25 illegitimate; 458 had mothers from twenty to twenty-five; 933, of which 764 were legitimate, 169 base-born, were born of mothers between twenty-five and thirty; the mothers of 708 new-born children were from thirty to thirty-five; those of 549 from thirty-five to forty; those of 221 from forty to forty-five; those, finally, of 25 from forty-five to fifty.

About 2,020 people die in a year, so that the annual increase of population is 920. Fewer deaths occur in February, which is the coldest month, and they are most frequent during the warmest month—July. In February, about 128 people die, and in July 205.

The diseases most prevalent in Iceland are those of the skin, occasioned by want of cleanliness and proper nourishment. Diarrhœa is very prevalent in the spring. Typhus and small-pox have swept the country. Leprosy is by no means un-common: it is very prevalent in Grimsey and the Westmann Islands, taking the form of elephantiasis. In Catholic times there were hospitals in the island for the reception of the poor sufferers; but after the Reformation, charity declined, and the hospitals fell into disuse. Consumption is unknown. Reyk-javík would be a good place for patients in the first stage, as the air is remarkably pure, and charged with ozone. A young man in, as I fancied, the last stage of consumption, came out with me in the steamer *Arcturus*. I left him at Reykjavík, expecting never again to see him alive. On my return to the capital, I was astonished to find him almost completely restored: he had regained his flesh, recovered his colour, and lost his wearing cough.

The natives of Iceland are tall and slender, remarkable for the brightness of their complexion, and the profusion of their hair. This is generally light brown, but occasionally red or black. The hair of the children is white, but it darkens with age. The eyes are, in almost all cases, blue or grey: those of the women are bright and beautiful. The girls have graceful figures, which appear to advantage, as they hold themselves very upright, both when walking and sitting. Their features are not regular, but soft and pleasing. In character, the people are phlegmatic, conservative to a fault, and desperately indolent. They have a peculiar knack of doing what has to be done in the clumsiest manner imaginable. When, for instance, it is requi-site that a box should be corded, a native looks at it for a few minutes to discover how it can be most inconveniently and uncouthly tied up; he then slowly sets himself to work on it,

Diseases.

Characteristics of the people.

after the fashion he has excogitated. The Icelanders may possibly employ themselves during the winter, but they certainly do nothing during the summer. I have not had the felicity of seeing a native do any real work. To accomplish a task, he takes as many days as an English labourer would take hours.

Trade.
The trade of Iceland is carried on exclusively by the Danes. Once in the year the Icelanders journey either to Reykjavik, Akureyri, or to one of the merchant stations, to barter their wool and eider-down for rye-meal, crockery, coffee, and timber. Each merchant possesses a vessel which cruises around the coast during the summer, running up the fjords, and anchoring at a station. The peasants come to the ship in boats, bringing with them their articles for barter, and purchasing at the merchant's store, which is supplied with shop counter on the quarter-deck or in the hold.

History of Iceland.
Iceland was accidentally discovered by Naddothr, a Norwegian viking or pirate, in 860, whilst on a voyage to the Faroe Isles. Naddothr satisfied himself, by a casual survey from the top of a hill on the east coast, that the land was too dreary to be inviting, and he called it Snjá-land. Four years later, a Swede, hight Garthr, circumnavigated the island, and gave it the appellation of Garthar-holm. Flokki, another adventurer, was the third on the track; he explored the south and west of the island, and called it by the name which it now bears.

This discovery took place just before the period when Harald Hárfagr made himself paramount in Norway, by crushing all the petty kings, and reducing the nobles to submission. Those who resisted such treatment fled the country, and directed their course to the new land. Ingolfr and Hjörleifr were the first to sail, and landed in Iceland in the year 870. They were followed by large bands of emigrants, who rapidly colonized the fjords and vales of the dreary island, bringing with them their thralls, cattle, household goods, and traditions. These settlers did not, however, find the island totally uninhabited; a few Culdee anchorites had already planted themselves on the coasts, and it is probable that the Irish fishermen were already acquainted with the island. The *Islendinga bók* or *Scheda of Ari Fróth*, written about the year 1120, and the earliest monument of Icelandic literature, says that, at the time of the influx from Norway, the island was inhabited by Christian people whom the new comers called Papar; these eventually deserted the island, as they did not wish to live among heathens. They left Irish books, bells, and croziers behind them, whence it was concluded that the

Papar were natives of Ireland. The same story is repeated in the *Landnama bók*, and it is strongly corroborated by the *ex parte* statement of an Irish monk, Dilcuil, who wrote about the year 825. He says that he had spoken with priests who had visited the remote island of Thule, that it lay far away in the north, and that between it and Britain was a cluster of islets (evidently the Faroes), some only separated by narrow straits; that these islets were thronged with countless sheep and sea-birds, and that they had been inhabited for upwards of a hundred years by Irish hermits, but that these had at last been driven away by Norwegian rovers. The details which the monk gives of the solstice, the length of the summer day and winter night, leave no room for doubt that, by Thule, he referred to Iceland. In corroboration of these testimonies we find several names of places in the island, bearing Irish names, such as Patreksfjord, Papey, and Papyli, and Erlendr-ey.

After the settlement by the Norse, the island was visited by Scotch and Irish, who occasionally chose it for a place of abode. Gaelic and Erse prisoners were taken by the vikings, and brought as thralls to the new country, so that a certain infusion of foreign blood remains in Iceland. This accounts for the introduction of such names as Njál, Kormak, Kjartan, and Erlendr, or "the Irishman."

For sixty years there was a continual influx of settlers, and, at the beginning of the tenth century, the country was as fully peopled as it has ever since been.

In 930 a code of laws was adopted, by which the new nation was to be governed. An annual meeting of the bonders was fixed for midsummer, on the extensive plain of Thingvalla, at which all were expected to attend for the purposes of litigation, adjustment of quarrels, and general legislation. This gathering was called Althing, and continued to be held regularly until 1800, when it was abolished. In 1845 it was restored: it sits, however, no longer in the midst of the glories of Thingvalla, but in a whitewashed room at Reykjavík.

Christianity was accepted as the national religion in 1000, at this annual assembly. The island was afterwards divided into two bishoprics, those of Hólar and Skalholt. The bishops were elected by Althing, and even the saints were canonized by popular acclamation. Clerical celibacy was never enforced, and few of the distinctive dogmas of Rome, or her "pious opinions, which are not articles of faith," received much support. With the introduction of the Church, came the knowledge of Latin letters, and, from the twelfth century to the fourteenth, most

Conversion.

of the Sagas orally existing were committed to writing, and copies multiplied. Monasteries, hospitals, and schools, were established, and learning was cultivated with great assiduity.

Of this, the golden age of Icelandic history, Adam of Bremen, in his *Hist. Ecclesiastica*, cap. 243, speaks as follows: "Thus, spending in simplicity a holy life, since they seek for nothing beyond what nature yields, the Icelanders can cheerfully say with the Apostle, 'Having food and raiment, let us be therewith content.' For they have their mountains for towns, and springs for delights. Happy, I say, the race whose poverty no one envies, and happiest in this, that they have now all received Christianity. There are many remarkable points among their customs, especially charity, from which it comes that, with them, all things are common both to strangers as well as to natives. For a king they have their bishop, and to his nod all the people attend: whatever he has laid down, whether from God or from Scripture, or from the customs of other nations, that they have for law."

Reformation. In the sixteenth century, the Reformation was forced upon the people by the monarch of the united kingdoms of Denmark and Norway; its progress was everywhere marked by blood, and even the Lutheran historian, Finn Jónsson, is unable to veil completely the atrocities which were committed. The venerable Bishop of Hólar, Jón Arnason, the last Catholic prelate, received the crown of martyrdom along with his two sons, uttering with his dying breath, "Lord! into Thy hands I commend my spirit." The clergy either conformed or were ejected from their benefices; and the Crown seized on most of the Church property, which it sold for its own aggrandizement. Since 1551 the Lutheran religion has been established. A mission has been planted at Reykjavík by Romanists, but has not hitherto succeeded, the laws of Iceland forbidding the erection of any place of worship not of the established religion.

Discovery of Greenland. An event of considerable importance in the history of the island was the discovery of Greenland by Eirik the Red. This new country soon became a flourishing settlement, with a cathedral, two towns, and 190 farms. Shortly after this followed the discovery of America, by Leif and others, who pushed as far south as Massachusetts. But these pioneers opened a way which was not traversed by colonists, and the discovery led to no practical results.

Subjection to Norway. In 1261, by a decree of Althing, the sovereignty of Iceland was made over to Hakon, King of Norway, with the understanding that no tribute should be exacted, that trade should be encouraged, and that natives of the island should be suffered to acquire honours and civil offices in Norway itself.

In 1380, when the crown of Norway was annexed to that of Denmark, the island was transferred without opposition to the Danish Government. Subjection to Denmark.

Previous to the Calmar union, Queen Margaret had made the trade with Iceland a royal monopoly, carried on only by vessels belonging to or licensed by the crown ; this monopoly was kept up by her successors, and, after the union, by the Danish Government to a very late period, on account of its lucrative nature, and only abolished in 1776.

This injured the trade of Iceland to a very great extent, and would have been more severely felt, but for the facility of evading its restrictions. English merchant vessels continually trafficked with Iceland, bringing meal and clothing in exchange for fish. In 1413, one of the first acts of Henry V. was to send letters to Iceland, with five ships, relative to the opening of a market for English merchandise.

In 1402, the black death was brought into the country by a shipwrecked vessel, and for three years it devastated the island, sweeping off nearly two-thirds of the population. The contagion had first been carried to Norway by an English boat. At Bergen, according to Torfœus, who follows Pontanus, the plague broke out in its most frightful form, with vomiting of blood ; and throughout the whole country, spared not more than a third of the inhabitants. The sailors found no refuge in their ships ; and vessels were often seen driving about on the ocean, and drifting on shore, whose crews had perished to the last man. Black death.

About the same period, the coasts of Iceland were infested with English pirate vessels fitted out at Hull, Lynn, and elsewhere. Farms were plundered, churches despoiled, and whole families carried off to be sold into slavery. In 1512, the pirates even had the audacity to capture and murder the governor of the island. In 1614, one of these corsairs, named John, came into the harbour of the Westmann Islands, sacked the village, and carried off all the ornaments and valuables of the church. On his return to England, King James I. caught him, and, after having punished him for his offences, restored the pillage ; but the church furniture was destined to be again lost, for in 1627 a Turkish or Algerine privateer, after having cruised along the south coast, plundering the farms and churches, anchored in the harbour of Heimey, and again robbed the church. The corsairs burned every building in the island Piratical inroads.

and carried away with them 400 persons in fetters, whom they conveyed to Algiers and sold into captivity. One of the unfortunates, a Lutheran pastor, named Olafr Egidsson, escaped two years later, and wrote an account of the sufferings they had undergone. In 1686, the Danish Crown ransomed the surviving Icelanders, but only 37 of the 400 were alive, and regained their native island.

In 1808, an English privateer, under the command of a certain Captain Gilpin, made a descent on the island, landed an armed force at Reykjavik, broke open the public chest, and carried off upwards of thirty thousand rix-dollars.

Icelandic revolution. In 1809 took place one of the most extraordinary circumstances on record, which has been dignified by the name of the Icelandic Revolution.[*]

An individual of the name of Jörgensen was at that date prisoner of war in England. This man was a Dane, of a restless and adventuresome spirit. At an early age he had served as an apprentice on board a British collier; after which he had entered our navy and served in it till twenty-five years old, when he returned to his native country. He was speedily appointed to the command of a privateer of twenty-eight guns, the *Admiral Juul*, in which he fell in with two English ships of war, and, after a brief engagement, was obliged to strike his colours. On landing in England, he was paroled, and remained in London till December, 1808. Meeting there with a wealthy merchant named Phelps, he urged him to speculate in a trade with Iceland, where, according to Jörgensen's account, a large amount of tallow was lying ready for exportation.

A vessel, the *Clarence*, was fitted out and freighted with meal, potatoes, rum, sugar, and coffee. .

M. Jörgensen embarked on board the *Clarence*, together with Mr. Savigniac, an Englishman employed as supercargo, and sailed in December, thereby breaking his parole. In January, 1809, the ships anchored off Reykjavik, but permission to land the cargo was peremptorily refused, although it was acknowledged that the island was in extreme want of the various articles on board. Messrs. Jörgensen and Savigniac at once proceeded to capture a Danish brig which had just arrived freighted with provisions, an act which so alarmed the Danish officials of Reykjavik, that they consented to communications being opened between Mr. Savigniac and the natives ; at the same time, how-

---

[*] For the details of this I am indebted to Mr. Hooker's *Tour* and Sir George Mackenzie's *Iceland*.

ever, they issued a proclamation, which was forwarded over the island, threatening any one with death who ventured to trade with the English.

The goods were now brought on shore, and Mr. Savigniac remained in charge of them, whilst Jörgensen returned to England with the *Clarence* in ballast, having, in the first place, restored the captured brig to its owners.

In June arrived Count Tramp, the governor, who had been absent during these transactions. In the same month, the *Rover*, a British war sloop, commanded by Captain Nott, arrived off Reykjavik, and Count Tramp entered into a convention with him, by which it was stipulated that British subjects should be permitted to carry on a free trade with Iceland during the war, but that they should, at the same time, be subject to Danish laws. This agreement was signed, and the governor undertook to have it printed and circulated freely. This he did not do; the old proclamation threatening death still remained up, and was probably still circulated. Five days after the drawing up of the convention, Mr. Phelps himself arrived in the *Margaret and Anne*, a fine ship, carrying ten guns. Jörgensen was also on board. Savigniac at once proceeded to the vessel and laid the state of affairs before Mr. Phelps. The merchant waited several days, expecting to see the proclamation of free trade with English vessels posted up; but as the former notice threatening death was still unrepealed, and there seemed to be no intention on the part of the governor to issue the convention drawn up between himself and Captain Nott, Mr. Phelps gave orders to Captain Liston, the master of the *Margaret and Anne*, to seize the person of the governor and detain him as prisoner, directing him also to make a prize of the *Orion*, a brig belonging to Count Tramp, provided with a British licence, which, however, it had forfeited.

Mr. Liston, in pursuance of these directions, landed twelve of his crew with arms, and made a prisoner of the governor, without any resistance having been offered. Jörgensen, a thorough adventurer, ready for any emergency, at once assumed the government of Iceland, declared himself Protector, and commenced the exercise of his power by issuing proclamations, which announced that Danish authority was at an end in Iceland, that the ancient constitution of the island should be restored, that all Danish property on the island should be confiscated for public use, and that there should be free trade. Jörgensen announced also that Iceland should have her own flag, which was to be three stockfish, split, proper, on a ground azure.

The bishop and clergy in a synod accepted the new government: Jörgensen made a journey to the north, and was received with open arms everywhere, and several natives were enrolled as soldiers. The new army, which was to resist all warlike inroads on the coast, consisted of eight men armed with old fowling pieces and swords, in green uniforms, and mounted on sturdy ponies. The soldiers scoured the country, intimidating the Danes, securing confiscated property, and rousing the enthusiasm of the natives.

Delighted with his army, Jörgensen assumed the title of his Excellency, the Protector of Iceland, Commander-in-Chief by sea and land, and posted up a proclamation, on the 11th July, which shows the impudence and assurance of the man. Its first article ran:—"We, Jörgen Jörgensen, have taken upon ourselves the government of the country until a regular constitution is established, with the power *to make war and conclude peace* with foreign potentates." The second article announced that, having been unanimously elected by the enthusiastic soldiery, he had taken upon himself the conduct of the military department. By the third article a new flag was appointed for Iceland, and Jörgensen swore to defend its honour with his heart's blood. By the sixth, all factious officials were banished to the Westmann Islands. The eighth announced that an ambassador would be sent to his British Majesty; and the sixteenth forbade all irreverence towards the sacred person of Jörgen Jörgensen.

The harbour of Reykjavík was next put in a state of defence by order of his Excellency, Mr. Phelps executing this order with great alacrity, assisted by a crowd of natives. A battery, denominated Fort Phelps, was speedily formed, and mounted with six guns which were dug out of the sand, where they had lain for a hundred and forty years, and which were perfectly useless. The public money chest, which contained 2,700 rix-dollars, was seized by the new governor, and a Danish vessel which came into the harbour was also captured by his orders.

But the end of Jörgensen's power was nearer than he had expected. The *Talbot* sloop of war, Captain Jones, arrived in Havnafjord, and to him the Danish merchants applied. The vessel was at once brought round into Reykjavík harbour, and among the first objects that met the captain's eye was the deep blue flag with the three white stockfish waving over the courthouse. Immediately upon his arrival, Count Tramp, who had now been detained for nine weeks a prisoner on board the *Margaret and Anne*, demanded an interview with him. This was granted, and Captain Jones, after having examined into the *pros* and *cons*, concluded that he was officially bound to interfere in a business in which the honour of his country was implicated. He gave orders that the new Icelandic flag should be taken down, the battery destroyed, and that Count Tramp, Mr. Phelps, and Jörgensen, should come to England, that the whole case might be laid before his Majesty's Government and thoroughly investigated. These conditions were complied with; Mr. Phelps, together with Count Tramp and Lieutenant Stewart, of the *Talbot*, sailed in the *Margaret and Anne*, while Jörgensen was placed on board the *Orion*. On the third day of voyage, the former ship was set on fire by the Danish prisoners, but the passengers and crew were saved by the *Orion*.

The result of the investigation and representations made in England was that an Order in Council was issued, of which the following is a copy :—

" At the

Court at the Queen's Palace, February 7th, 1810.

Present :
The King's Most Excellent Majesty in Council.

" Whereas it has been humbly represented to his Majesty that the islands of Ferroe and Iceland, and also certain settlements on the coast of Greenland, parts of the dominions of Denmark, have, since the commencement of the war between Great Britain and Denmark, been deprived of all intercourse with Denmark, and the inhabitants of those islands and settlements are, in consequence of the want of their accustomed supplies, reduced to extreme misery, being without many of the necessaries, and most of the conveniences, of life :

" His Majesty, being moved by compassion for the sufferings of these defenceless people, has, by and with the advice of his Privy Council, thought fit to declare his royal will and pleasure, and it is hereby declared and ordered, that the said islands of Ferroe and Iceland, and the settlements on the coast of Greenland, and the inhabitants thereof, and the property therein, shall be exempt from the attack and hostility of his Majesty's forces and subjects, and that the ships belonging to the inhabitants of such islands and settlements, and all goods being of the growth, produce, or manufacture of the said islands or settlements, on board the ships belonging to such inhabitants, engaged in a direct trade between such islands and settlements respectively and the ports of London or Leith, shall not be liable to seizure and confiscation as prizes.

" His Majesty is further pleased to order, with the advice

aforesaid, that the people of all the said islands and settlements
be considered, when resident in his Majesty's dominions, as
stranger friends, under the safeguard of his Majesty's royal
peace, and entitled to the protection of the laws of the realm,
and in no case treated as alien enemies.

"His Majesty is further pleased to order, with the advice
aforesaid, that the ships of the United Kingdom, navigated
according to law, be permitted to repair to the said islands and
settlements, and to trade with the inhabitants thereof.

"And his Majesty is further pleased to order, with the
advice aforesaid, that all his Majesty's cruisers and all other his
subjects be inhibited from committing any acts of depredation
or violence against the persons, ships, and goods of any of the
inhabitants of the said islands and settlements, and against any
property in the said islands and settlements respectively.

"And the Right Honourable the Lords Commissioners of his
Majesty's Treasury, his Majesty's Principal Secretaries of State,
the Lords Commissioners of the Admiralty, and the Judge of the
High Court of the Admiralty, and the Judges of the Courts of
the Vice-Admiralty, are to take the necessary measures herein,
as to them shall respectively appertain.

(Signed)   " W. Fawkener."

Free trade.   Of late, free trade has been granted to Iceland, and undoubtedly it will be a great boon to it. The heavy restriction which
formerly existed checked the life-blood of the people, and was
probably the main cause of the entire loss of the old Greenland
settlement, which died out from want of supplies, so completely,
that the place thereof knew it no more, and only the ruined
churches and the tombstones remain to tell that former generations had peopled those lone shores.

Iceland has withered under the same paralyzing influence,
and the whole character of the people has been deteriorated by
the grinding want of the necessaries of life, so that there is now
none of the energy and enterprise among them which were the
distinguishing features of the early population.

Eminent men.   Iceland has produced some few great men in modern times,
whilst its ancient palmy days teemed with heroes, lawyers, poets,
and historians. In grace of diction, and charm of putting a
scene vividly before one, working up the details, and bringing the
actors and speakers forward, by a few bold strokes which leave
nothing to be desired, the old historians, or saga writers of
Iceland, are inimitable. Of later times, Finn Jónsson, or Fin-
nœus Johanœus, as he delighted to call himself, the accom-

plished writer of the ecclesiastical history of his native isle, is one of the great lights. To Arnas Magnœus we owe the preservation of the Saga MSS. This truly great man was born in 1631, and, at the age of thirty-one, became professor of philosophy in the Copenhagen University. A few years later he was made professor of northern antiquities, in which capacity he made frequent trips to his native isle, to gather the MSS. which were dispersed through the farms. After having collected together a large library, he had the misfortune of seeing it consumed by fire in 1728. Still many sagas remained in his own private library, and others were afterwards sent him from Iceland, so that, at his death, two years later, he was able to bequeath to the university library a considerable store of valuable MSS.

Another distinguished Icelander is Thormod Forfœus, author of *The Series of Dynasties and Kings of Denmark* and the *History of Norway*.

Last in the list comes the sculptor Thorwaldsen, whose grandfather was an Icelander.

I close this introductory chapter with a list of the principal *Travels.* travels in Iceland.

Olafsson og Pálsson, *Reise igjennem Island* (Sorœ, 1772).

Von Troil, *Letters on Iceland* (2nd ed., 1780).

Hooker's *Tour in Iceland* (1st ed., 1811 ; 2nd ed., 1813).

Sir George Mackenzie's *Travels in Iceland* (2nd ed., 1812).

Henderson's *Iceland* (1st ed., Edinburgh, 1818; 2nd ed., 1819).

Barrow's *Visit to Iceland*. 1835.

Ida Pfeiffer, *Reise nach dem Skand, Norden.* 1845.

Sartorius von Waltershausen, *Physisch-Geographische Skitze von Island.* Göttingen, 1847.

Pliny Miles, *Rambles in Iceland.* 1854.

Chambers' *Tracings of Iceland.* 1855.

Lord Dufferin, *Letters from High Latitudes.* 1857.

Captain Forbes' *Iceland.* 1860.

Mr. Clarke's " Notes on Iceland," in *Vacation Tourists.* 1860.

Rev. F. Metcalfe's *Oxonian in Iceland.* 1861.

Mr. Holland's " Tour in Iceland," in *Alpine Club Volume for* 1861.

Drs. Preyer und Zirkel, *Reise nach Island.* 1862.

Mr. Symmington's *Pen and Pencil Sketches.* 1862.

Carl Vogt, *Nordenfahrt von Dr. Berna.* 1863.

Besides these, there is the large and expensive work by Gaimard, published in 1840 by order of King Louis Philippe, and costing about 21*l.* It is to be regretted that some English tourists have transferred to their volumes the grossly exaggerated and unreal pictures of Gaimard's great work.

The great majority of the tourists mentioned above have only visited the scenes in the neighbourhood of Reykjavík, such as the Geysir and Hekla, and perhaps extended their range to Snœfell's Jökull.

ICELAND:

ITS SCENES AND SAGAS.

CHAPTER I.

COASTING.

First sight of Iceland—Hjörleif's Head—Dyrhólar—The Westmann Islands
—Life of the Islanders—The Yankee's Story—My first Saga: " The
Red Rovers "—Cape Reykjanes—The Mealsack—The Great Auk.

"BRITTISHERS! I take you all to witness that I aire the first to
catch sight of Iceland!"

This, uttered in a shrill nasal twang, woke me on the
morning of June 15th.

"How so? You have not been on deck."

"I've looked through the port-hole though. Ay! I guess
you seem pretty well streaked, young man; now, if—you—
please, your five shillings."

Confound the Yankee! On his own proposition we had
come to an agreement on the previous evening, that every
passenger should pay the sum now demanded to the first who
saw Iceland.

I had made up my mind to be the first on deck, had
resisted the solicitations of Martin to have another drop of
whisky-toddy, had slept through Mr. Briggs' jokes, had knocked

1

under for once in an argument with the Yankee on the rights of the American war, so that I might retire early to my berth, —and yet Mr. Abraham Blank was before me.

I hurried on deck, but the Yankee, satisfied with having won his bet, turned round for another snooze.

The morning was beautiful; full in front, Hjörleif's Head, a bluff of dark rock, seven hundred feet high, stood up proudly from a low, repulsive plain of volcanic dust. We could distinguish the smoke from a little farm on the top, but the house itself was invisible, either sunk in a depression, or with the slope of its turf roof towards us, grass-grown like the soil around.

Behind the headland rose the white dome of Kötlugjá, now calm and hushed, looking down on the devastation it has made; for these extensive flats were once grassy meadows, but Katla buried them deep in sand and pumice. Between Katla and Myrdals Jökull, the adjoining heap of snow, is a grim chasm, hitherto never reached by man.

Hjörleif's Head is named after one of the first settlers in Iceland. Hjörleif and Ingolf sailed from Norway together, but when the latter came on shore under the Oræfa, the other pushed west and established himself at this point. Having built two farms, he set his Irish slaves to plough the sandy soil with his one ox. The rogues were idly disposed and soon tired of their task; they then slew the ox, and came back to their master with a story that a bear had carried off the creature. The Paddies seem then to have thought that having perpetrated one evil deed they might just as well make a pair of them, so they killed their master, and laid his bones by the bones of his ox: then, fearing lest Ingolf should take it into his head to look up his foster-brother, and haul them all over the coals for their morning's work, the scamps loaded the boats with everything that they could carry off, and fled to the Westmann Islands, which are called after them. So landlord stalking is no new invention among the Irish!

A glorious panorama was gradually unfolding before us as the steamer advanced. The long stretch of Myrdals Jökull,

pure and undinted in the morning sun; the black burnt crags,
so steep that snow will not veil them, but only festoon their
heads, now mellowed in the fresh beam of dawn; the tremulous
blue sea flickering along the sand flats, and frothing about the
Needles of Portland, flashing white through the mighty natural
arch of Dyrhólar; a pale grey glacier from Myrdals, bristling
with silvery spikes, and discharging its melted ice in broad
streams, seaming the flats in all directions: these are the
leading features of the scene, and the minor details are striking
also; the great gull on poised wing, vibrating overhead; the
kittiwakes dipping in the wavelets; the gannet descending with
a rush into the water to spike a luckless fish he has seen from
aloft; a shark asleep on the surface, rolling with the swell,
his dorsal fin standing above the water like a ploughshare;
and, out at sea, a whale blowing off a fountain of spray.

Dyrhólar is an arch drilled by the waves through the
blackest of rocks. The gap is 200 ft. across, the thickness
of the crag, 75 ft., and the height above the water, from
80 to 90 ft.; so that, weather permitting, the little steamer,
Arcturus, might thread the needle eye, without much danger.

We sight the Westmann Islands, a cluster of turf-capped
columns, thronged by sea-birds, which have the majority of
the rocks to themselves, as only one of them, Heimey or Home
Isle, is inhabited. This is the largest of a group of fifteen,
and is as big as all the rest put together. Its length is three
miles and its breadth hardly two, the distance from the mainland is seven miles, but communication between them is cut
off entirely for months, and it is only when a cascade near
Holt reaches the sea in an unbroken silver thread, as it does
to-day, that a boat can venture across the straits. When the
wind is at all violent, the stream is taken and tossed in spray
high above the cliffs over which it shoots.

The islands form a syssel or county by themselves, and
have a magistrate, parson and doctor resident on Heimey.
There is one church near the harbour, which has been completely sacked of all its plate and old vestments by the pirates
of the seventeenth century. One of the skerries, Geldingasker,

1—2

is so abrupt, that sheep, brought to the foot in boats, are hoisted
on to it by ropes. Another, Hellarey, has caves in it which
serve as folds during boisterous weather, when the winds
are violent enough to sweep the unprotected live-stock over
the cliffs. Bjarney, which fronts the little harbour, is a highly
picturesque pile of rock; the name signifies Bear Island, but
when Bruin made it his resort and managed to clamber up its
perpendicular sides, I have been unable to discover. As the
Arcturus, our steamer, approaches the neat and tight little
harbour of Heimey, we see the lava which has flowed from
Heima Klettir, a volcanic cone, nine hundred and sixteen feet
high, connected with the island by a narrow ridge of crag.
The Westmann Islands get their name, as already stated, from
the Irish scoundrels who slew Hjörleif. It is a satisfaction to
know that Ingolf, the dead man's foster-brother, killed the
rascals one day when they were taking their siesta after dinner.
By the way, is the expression dead man's foster-brother quite
admissible? I use it on the following authority. Shortly
before leaving England I was visiting a clergyman who had
just returned from a funeral; when there was a tap at the
door, and the servant looked in. "Well, Mary, am I wanted?"
asked the parson. "Please your reverence, there's the
corpse's brother wants to speak with your honour."

The Westmann islander must lead a melancholy life, cut
off from the world, living on the bleakest spot imaginable,
with an ever-boiling sea below him, its throb and roar always
sounding in his ear, the wind sweeping over the unsheltered
turf, searing it as with a hot iron, as it rolls from the snow-
fields of Eyjafjalla. His life is one of daily peril, battling
with the sea in slight fishing boats, or slung over the precipices, collecting eggs, his life depending on the hair rope his
own hands have woven, and these Icelandic hair ropes are frail
enough, Heaven knows! Offer him a stout English cable and
he will shake his head; his father, and his father's father used
rope such as his, why should he try one of other texture?
His home is like a rabbit-burrow, under turf mounds, redolent
with the disgusting odour of the fulmar petrel (*Procellaria*

glacialis), whose strong flesh is his food, its bones his fuel, and its feathers his bedding. Nay, more! the young birds have a wick run down their throats, and they are so charged with oil that they serve as lamps.

The fulmar petrel has to be caught with caution. The fowler must drop on it suddenly, and by a sharp grasp of the throat secure it before the bird has time to spit the oil from its beak, which it will do, when alarmed, knowing it to be the treasure for which it is hunted.

As the feathers can never be purified, the offensiveness of the fowler's home may be guessed; to a stranger, the atmosphere is perfectly unendurable. One or two cottages in Reykjavik are furnished with feather-beds of fulmar down, and none but natives will enter the hovels.

Children do not live, unless removed to the mainland; the coarse flesh of the sea-birds, and the early age at which they are weaned, produce a fearful disease amongst them, called gjuklofi, which is quite incurable. Violent cramps and spasms contort the body and continue at intervals, increasing in violence, till death ensues.

The history of this cluster of islands is marked by misfortune, as may be gathered from the Icelandic annals. Shipwreck succeeds shipwreck; half the crew of a merchant vessel rise in the night and murder the other half, as the ship lies at anchor in the tiny harbour; the crags become a refuge for outlaws. "Lopt in the islands biting puffin bones, Soemund on the moor munching berries," says an old saw, commemorating two famous exiles. Finally, pirates enter the harbour, rob, burn, and carry into captivity, as has already been mentioned in the introduction.

A boat started from among the rocks as we came to a standstill just off the harbour, and brought letters from the post ship; the sysselman, the doctor and the parson were in the bows, and eight lusty fishermen rowed. The post-bag was handed up the side, and one letter, the only one that year, was flung to them from the deck. "Some newspapers!" in an imploring voice from the little cockle-shell which danced

at our side, and we tossed down all t'
could scrape together in the cabin. '
loose, we waved caps, and shouted
captain opened the bag, he found
dozen letters, of which only two h
is just the way of these out-of-th'
" They write to any one they kn'
years, and always forget the '
break the seals and discover th

We steamed on again. '
shoulder of Eyjafjalla, 5,4?
and, as we steam past the
threads the scenes of th'
catch the white cap of '
of Trihyrningr, and the
around him.

As evening set i'
low about Eyrarbak]
There was a cry o'
deck to have a '
sending up spou'
There must ha'
from his Whit
the ungainly

The Ya'
" I calcul'
U-nited f
  " M
derisio
on h
fish'

Pl. 1.—Fig. 1

Almannagjá.   Hrafnabörg.   Thingvalla Church.   Thingvalla Lake.   Hengill.

THINGVELLIR.

Fig. 2

Eyjafjalla.

Heimaklettur.   Helena.   Eyjafjalla.

COAST NEAR THE WESTMANN ISLANDS.

at our side, and we tossed down all the Danish papers we
could scrape together in the cabin. The painter was flung
loose, we waved caps, and shouted "Adieu!" When the
captain opened the bag, he found that it contained half-a-
dozen letters, of which only two had any direction. "That
is just the way of these out-of-the-way people," laughed he.
"They write to any one they know, or have known in former
years, and always forget the directions, so that I have to
break the seals and discover the destination of their letters."

We steamed on again. To the right, the glorious silver
shoulder of Eyjafjalla, 5,432 feet high, shines in the sun,
and, as we steam past the mouths of the Markarfljot, which
threads the scenes of the glorious story of Burnt Njal, we
catch the white cap of Tindarfjalla Jökull, the brown horns
of Trihyrningr, and the distant Hekla with his mantle of snow
around him.

As evening set in, we lost sight of the shore, which is
low about Eyrarbakki, and we stood direct for Cape Reykjanes.
There was a cry of "Whales!" and we ran to the hurricane
deck to have a view of the monsters rolling and tumbling,
sending up spouts of foam, diving, and rising to spout again.
There must have been a shoal of twenty; Martin sent a shot
from his Whitworth rifle amongst them, but without disturbing
the ungainly creatures in their frolic.

The Yankee eyed them with gravity, and then observed:
"I calculate we could make better whales than them in the
U-nited States."

"Make them! Nonsense! stuff!" from Mr. Briggs, in
derision. Mr. Briggs was a rather portly gentleman, who was
on his way to Iceland with his friend Martin for a little
fishing and shooting. "Make whales! Fiddlesticks' ends!"

"Yes, make them," retorted the Yankee, bridling up.
"There was an intimate friend of mine made the great
Leviathan or sea-sarpint; and a pee-owerful sight better it
was than anything in natur', I reckon."

Mr. Briggs gave a long whistle, expressive of astonishment,
dismay, or perhaps doubt.

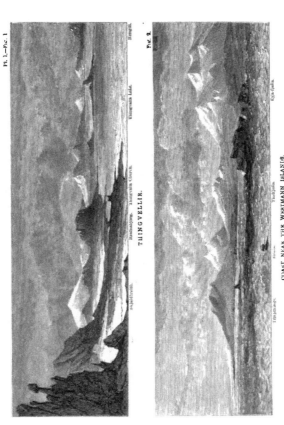

Pl. 1.—Fig. 1

Hengill. Thingvalla Lake. Hrafna Gjörä. Meeneshiag. Sulpelbruith.

THINGVELLIR.

Fig. 2.

Eyja-fjalla. Vindgjalla. Herran. Ejiö-jökvärr.

COAST NEAR THE WESTMANN ISLANDS.

"You may whistle, Britisher; but if you'll only listen, I'll tell you the story."

"Wait a moment," said Martin; "let us order some grog to be brought up here to us, then we shall get a bucket each, and sit round the funnel, chattering till we see Reykjanes."

"Well," quoth Mr. Blank, when we had provided ourselves with seats; "this is what my brother did."

"Brother!" exclaimed Mr. Briggs. "'Twas friend just now."

"We're all brothers in the States; and that's one of the great advantages of a free and enlightened Constitution, it makes every man equal, and——&c., whilst you in the old countries, with your kings and queens, your dukes and duchesses——"

"That's enough," said Mr. Briggs; "now for the story."

"Well, I'll just tell you what *he* did—that's my brother. You see he'd a tarnation long bill to meet, and he'd just got about nothing at all to meet it with. 'So,' says he, 'wise men live upon fools, and New York is full of 'em,' that is, of Britishers and natives of Ireland, who come over in shoals: a genu-ine Yankee is no fool, I can promise you. So he buys a score or two of old molasses' casks, strings 'em together on a cable, covers 'em with tarpaulin, and anchors the whole, a bit out to sea just off Long Island. Then he hires a steamer, and puts an account in the papers, stating that the sea-sarpint had been seen playing off Long Island on the top of the waters, and a-dancing there just like a baby on a spring sofa. Next day he advertised that his steamer was going in search of the cratur, and he charged a dollar and a half for any one who would come with him, food and liquor extra. Sure enough, he got a ship's load, and they all had a be-utiful sight of the sarpint as nat'ral as life. The news spread, and my brother made a dozen trips; and he'd have made more, but that some whalers harpooned the sarpint and found out what 'twas made of. I estimate my brother walked into those sight-seers pretty con-siderable. What d'ye think of that—ay?"

"Why, 'twas a swindle," answered Mr. Briggs.

"So it was, pretty con-siderable, I reckon."

"Padre!" said Mr. Briggs to me,—he called me Padre because I wore a warm Franciscan cloak and hood, which he himself had lent me, as I had not brought sufficient warm things with me, whilst he was overstocked. "Padre! you have promised us a story from some of the musty old Icelandic Sagas; you had better tell us one now, as there is nothing to see and nothing to do."

"I shall be glad to comply with your request," I answered, "as I wish much to introduce you to my hero Grettir, and it is necessary that you should know something about him before visiting the scenes of his great deeds."

"Tell us first where do you find any records of him?"

"In the *Gretla*, a Saga composed, or rather, I should say, committed to writing, in the thirteenth century."

The Red Robbers [*]
(A.D. 1012).

ONE morning after a night of storm on the coast of Norway, the servants ran into the hall of a wealthy bonder, named Thorfin, to inform him that, during the night, a ship had been wrecked off the coast, and that the crew and passengers were congregated on a neighbouring sandy holm, signalling for help. Up started the bonder, and hastened to the strand; he ran out a large punt from his boat-house, and jumping in with his thralls, rowed lustily to the rescue. The shipwrecked people belonged to a merchant vessel from Iceland, which had been driven among the breakers, during the darkness, and had gone to pieces; yet not before a portion of the lading had been brought ashore.

Among the shivering beings gathered on the sand strip was Grettir, the son of an Icelandic chief who lived at Bjarg in the middle-frith; he was then a young man, tall and muscular, with large blue eyes, bushy hair, and a freckled face.

Thorfin received the half-frozen wretches on board his

---

[*] *Gretla*, chaps. 17—20.

boat and rowed them to the mainland, after which, he returned to the holm, and brought off the wares. In the meantime, the good housewife had been lighting fires, preparing beds, routing out dry suits, and making hot ale, ready for the sufferers ; and, right kindly they were treated, you may be sure.

Well, the chapmen stayed a week at the farm, whilst their goods were being dried, and till the women of the party were sufficiently recovered from cold and exposure to continue their journey to Drontheim, whither the whole party were bound : after which they left Thorfin, with many thanks for his courtesy and kindness. Grettir, however, remained, not at the request of the bonder, who did not much like him, but to suit his own convenience. Indeed, he stayed somewhat longer than Thorfin cared to keep him, considering what a fellow Grettir was, never joining in conversation, unwilling to lend a helping hand in any work, a great stay-at-home, crouching over the fire all day, and withal eating voraciously. Thorfin was much out of doors, and, as he was a sociable man, he often requested Grettir to accompany him, either into the forest, or about his farm, but could get no further answer than an impatient shake of the head and a grunt. Now the bonder was a fellow with a right merry heart and a kind one, and one too that loved seeing all around cheerful. With such a disposition, it is no wonder that the morose and indolent Grettir found no favour.

Yule drew nigh, and Thorfin busked him to depart, with a number of his freedmen, to keep high festival at one of his farms, distant a good day's journey. His wife was unable to accompany him, as the eldest daughter was ill, and wanted careful nursing ; and Grettir was not invited, as his sullenness would have acted as a damper to the joviality of the banquet.

The farmer started for his farm in Slysfjord some days before Yule, accompanied by his thirty freedmen, expecting to meet a goodly throng of guests, whom he had invited from all quarters.

Norway had for some time been in a disordered condition, from the mischief caused by numerous Berserkirs and Corsairs

who roved over the country, challenging bonders to mortal
combat, for their homes, their wives and families.   If a bonder
declined to fight, as the law stood, his all was forfeited to the
challenger; and if he fought and was worsted, he lost his life
as well.   With the advice of Thorfin, Earl Erik Hakon's
son put down these holm-bouts, and outlawed those whose
custom it had been to make a business of them, going round
the country and riding rough-shod over the peaceful bonders.

Among the worst of these, were two brothers, well known
for their wickedness, Thorir wi' the Paunch, and Bad Ögmund.
They were said to be stronger built than most, and to care for
no man under the sun.   They robbed wherever they went,
burned farms over the heads of the sleeping inmates, and
with the points of their spears drove the shrieking wretches
back into the flames.   When these pirates wrought themselves
up into their Berserkir rages, they howled like wolves, foamed
at the mouth, their strength was increased to that of Trolls,*
and they rushed about, demon-possessed, murdering and
destroying every living being that came in their way.   Thorfin
had been the prime instigator of their outlawry through the
length and breadth of Norway; and, as may well be con-
jectured, the brothers bore him no good-will, and only waited
for an opportunity of wreaking their vengeance upon him.

The eve of Yule was bright and sunny, and the sick girl
was so far recovered as to walk out and take the air, leaning
on her mother's arm.

Grettir spent the whole day out of doors, in none of the
sweetest of tempers, at being excluded from the festivities of
the season, and left to keep house with the women and eight
dunderheaded churls.   He fed his discontent by sitting on a
headland watching the boats glide past, as parties went to convivial gatherings at the houses of their friends.   The deep
blue sea was speckled with white sails, as though countless
gulls were playing on the waters.   Now a stately dragon-ship

---

* Trolls are mountain gnomes or demons, generally of prodigious size and
power.

rolled past, her fearful carved head glittering with gold and colour, her sails spread like wings before the breeze, and her banks of oars flashing in the sun, then dipping into the sea : now a wherry rowed by, laden with cakes and ale, and the boatmen's song rang merrily through the crisp air.

The day began to draw in, but still the red sparks from little vessels, fleeting by in the dusk, showed that all guests had not yet reached their destination.

Grettir was on the point of returning to the farm, when the strange proceedings of a craft, at no great distance, attracted his attention. He noticed that she stole along in the shadows of the islets, and darted with velocity across the open-water straits between them ; she hugged the shore wherever she could, moved in a zigzag course, and suddenly came flying with quick oar-sweeps towards the bay which Grettir was overlooking. In the twilight, he could make out thus much of her, that she floated low in the water, that she was built for speed, and that her sides were hung with shields. As she stranded, the rowers jumped on the beach. Grettir counted them, and found that they were twelve, armed men, too ! They broke into Thorfin's boat-house and dragged forth his great punt, in which thirty men were wont to sit, pushed it out into deep water, and drew their own boat under cover, and pulled her up on the rollers.

Mischief was a-brewing—that was plain as a pikestaff ! So Grettir descended the hill, and sauntered up to the band, with his hands in his pockets, kicking the pebbles before him and humming a tune with the utmost nonchalance. " May I ask who is the leader of this party ? " quoth he.

" Ah ! ah ! I'm the man," responded as ill-looking a fellow as nature could well turn out of her laboratory ; " why, I am ! Thorir wi' the Paunch, and here's my brother Ögmund with all his rascals. I reckon the Bonder Thorfin knows our names. Don't you think so, brother ? And we have a little account to settle with him. Pray is he at home ? "

" Upon my word, you are lucky fellows," spoke Grettir ; " coming here in the very nick of time, if you are the men

I take you for. The bonder is from home, with all his freedmen, and won't be back till after Yule; his wife and daughter, however, are at the farm. Now's your time, if you have old scores to wipe off; for there is everything you can possibly want at the house: silver, good clothes, ale, and provisions in the greatest profusion."

Thorir held his tongue whilst Grettir talked; afterwards he turned to his brother Ögmund and said: "This is just what I expected, is it not? Now we can serve Thorfin out in thorough earnest, for having made us outlaws. What a chatterbox this fellow is! There's no need of pumping to get anything one wants to know out of him."

"Every man is master of his own tongue," retorted Grettir. "Now come along with me, and I will do the best I can for you."

The rovers thanked him, and accepted the invitation; so Grettir, taking Thorir by the hand, led him towards the farm, talking the whole way as hard as his tongue could wag. The housewife happened at the moment to be in the hall, putting up the hangings, and preparing for the Yule banquet; and hearing Grettir speaking with such volubility, she stood still in astonishment, and asked whom he was greeting so cordially.

"It is quite the correct thing to receive guests well, is it not, mother?" asked Grettir; "and here are Thorir o' the Paunch, Bad Ögmund, and ten others, who have kindly come to join us in our Yule carousal, which is delightful, for without them our party would have been wofully scanty."

"Oh, Grettir! what have you done!" cried the poor woman. "You have brought hither the greatest ruffians in Norway. I would have given anything that they had never come. This is the way in which you return the good Thorfin has done you, in rescuing you from shipwreck, in taking you into his house and caring for you through the winter, as though you were one of his freedmen; and when you had not a farthing in your pocket to bless yourself withal!"

"Stop this abuse!" growled the young man. "There's time enough for that sort of thing another day. Now come, and take off the wet clothes from the guests."

"You need not scream before you are hurt, my good woman," quoth Thorir; "you will want all your words for to-morrow, when I shall carry you and your daughter away with me, and you will have to say good-by to home for many a day. What think you of that?"

"Capital!" roared Grettir. "That is capital."

On hearing this the housewife and her daughter fled to the women's apartment, crying and wringing their hands with despair.

"Well," said Grettir; "as the women won't attend on you, I suppose that I must; so be good enough to hand me over anything you want to have dried, such as your wet clothes and weapons."

"You're different from every one else in the house," spoke Thorir. "I almost think that you would make a boon-companion."

"As you please," answered the young man. "Only, I tell you I don't behave like this to all folk."

Then the freebooters gave him up their weapons; he wiped the salt-water from them, and laid them aside in a warm, dry spot. Next, he removed their wet garments, and brought them dry suits which he routed out of the clothes-chests belonging to Thorfin and his freedmen.

By this time it was quite night. Grettir brought in logs, raked up the fire, and made a noble blaze.

"Now, my men," quoth he; "sit at table and drink; for, i' faith, you must be thirsty after all the rowing you have done in the day."

"We are ready," said they; "only we don't know where to find the cellars."

"Will you let me fetch ale for you, or will you help yourselves?"

"Oh, go after it yourself, by all means," answered they.

So Grettir brought the strongest ale and poured out for

very naturally, mistrusted his intentions, and had besides secreted herself, from fear of the pirates.

"Come, answer!" shouted Grettir; "I have captured the whole twelve, and all that is wanting is a supply of weapons. Call up the thralls and arm them; quick! there is not a moment to be lost."

"There are weapons enough here," answered the poor woman, emerging from her hiding-place. "But, Grettir, I have no faith in you!"

"Faith or no faith," exclaimed Grettir; "I must have weapons at once. Where are the churls? Here, Kolbein! Svein! Gamli! Hrolf! Confound the rascals, where have they skulked to?"

"It will be a mercy of God if anything can be done!" said the housewife; "for we are in a sorry plight, to be sure. Now, look here. Over Thorfin's bed hangs an enormous barbed spear. You will find there also helmet and cuirass, also a beautiful cutlass. No lack of weapons, if you have only the pluck to use them!"

Grettir seized the casque and spear, girded on the sword, and dashed into the yard, begging the woman to send the churls after him. She called the eight men, and bade them arm at once and follow. Four of them obeyed—rushing to the weapons and scrambling for them, but the other four ran clean away.

I must tell you that, in the meantime, the Berserkirs had rather wondered at Grettir's disappearance, and from wondering had fallen to suspecting that all was not right. Then they sprang to the door, tried it, and found it locked from without. It was too massive for them to break open, so they tore down the partition of boards between the store and the office. The Berserkir rage came on them, and they ground their teeth, frothed at the mouth, and burst forth with the howl of demoniacs through the office door, upon the landing at the head of the steps, just as Grettir came to the foot.

Thorir and Ögmund were together. In the fitful gleams of the moon they seemed like fiends, as they scrambled forth armed with splinters of deal, their eyes glaring with frenzy, and great foam-flakes bespattering their breasts and dropping on the stones at their feet. The brothers plunged down the narrow stair with a yell which rang through the still snow-clad forest for miles. Grettir planted the spear in the ground and caught Thorir on its point. The sharp double-edged blade, three feet in length, sliced into him and came out beneath his shoulders, then tore into Ögmund's breast a span deep. The yew shaft bent like a bow, and flipped from the ground the stone, against which the butt had been planted. The wretched men crashed to the bottom of the stair, tried to rise, staggered, and fell again. Grettir planted his foot on them, and wrenched the blade from their wounds, drew the cutlass and smote down another rover as he broke through the door. Other Berserkirs poured out, and Grettir drove at them with spear, or hewed at them with sword; he slew another as the churls came up. They were late, for they had been squabbling over the weapons, and now that they were come they were nearly useless, as they only made onslaughts when the backs of the robbers were towards them, but the moment that the vikings turned on them, they bounded away and skulked behind the walls.

The pirates showed desperate fight, armed with chips of plank, or sticks pulled from some pine-faggots which lay in the homestead. They warded off Grettir's blows, and fled from corner to corner, pursued by their indefatigable foe. In the wildness and agony of despair they could not find the gate, but bounded over the wall of the yard, and ran towards the boat-house with Grettir at their heels. They plunged in and possessed themselves of the oars; Grettir followed into the gloom, and smote right and left. The bewildered wretches climbed into the boat, some strove to push her into the water, whilst others battled in the darkness with their unseen enemy; but some pulled one way, some another, and the blows from the oars fell on friend as well as foe, so that the panic became more complete.

In the meantime the thralls had quietly returned to the

2

farm, quite satisfied when they saw the robbers take to their
heels, and no entreaties of the house-wife could induce them to
follow Grettir; the four churls had had quite enough of fighting; true, they had killed no one, but then they had seen some
men killed. Grettir sprang into the boat, and stepped from
bench to bench driving aft the terrified vikings. As the boat-
house was open to the air on the side which faced the sea,
whilst the farther end was closed with a door, Grettir was in
shadow, whilst the black figures of the rowers cut sharply
against the moonlight, so that he could see where to strike,
whilst his own body was undistinguishable.

One stroke from an oar reached him on the shoulder, and
for the moment paralyzed his left arm; he killed two more
vikings, and then the remaining four burst forth, and, separating
into pairs, fled in different directions. Grettir followed the
couple which was nearest, and tracked them to a neighbouring
farm, where they dashed into a granary and hid among the
straw. Unfortunately for them, most of the wheat had been
threshed out, so that only a few bundles remained. Grettir
shut and bolted the door behind him, then chased the poor
wretches like rats from corner to corner, till he had cut them
both down. Then he pulled the corpses to the door and cast
them outside.

In the meanwhile the sky had become overcast with a thick
snow fog which rolled up from the sea, so that Grettir, on
coming out, saw that it would be hopeless attempting to pursue
the two remaining Berserkirs. Besides, his arm pained him,
his strength was failing him, and there stole over him an over-
powering sense of weariness after his protracted exertions.
The housewife had placed a lamp in the window of a loft, so
that Grettir, seeing the light, was able to find his way back
through the snow-storm without difficulty. When he came to
the door she met him, and, extending both her hands, gave
him a cordial welcome. "You have, indeed, shown great
valour!" quoth she; "you have saved me and my household
from insult and ruin. To you, and you alone, are we indebted."
"I am not much altered from what I was last evening, yet

now you sing quite a different strain; then you abused me
most grossly," grumbled the young man.

"Ah! but we little knew your metal then. Come, be a
welcome guest within, and tarry till my husband returns.
Thanks are all that I can render you, but, be assured, Thorfin
will not rest content till he has rewarded this deed of yours
munificently."

Grettir replied that he cared little for a reward, but that he
gladly availed himself of her invitation : " And now I hope you
may sleep without much fear of Berserkirs." Grettir drank
little, but lay down fully armed for a sound and well-earned
sleep.

On the following morning, as soon as day broke, a party
was formed to search for the two remaining vikings who had
escaped from Grettir in the darkness. The snow had fallen
so thickly during the night that the ground was covered, and
all traces were obliterated, so that the search proved ineffectual
till dusk, when the men were discovered under a rock, dead
from cold and loss of blood. The bodies were removed to the
shore and buried under a cairn between tides.* Then all
returned to the farm in high glee, and Grettir chanted the
following verse :—

> " Twelve war-flame branches are buried
> Low by the loud-resounding ;
> Unasked, sent I them, singly,
> To speedy death ; O ye gold-sallows,
> Well born ! bear me all witness !
> What is wrought mightier ? tell me,
> If ye wot,—this being little."†

---

* Burial between tides was looked upon as disgrace ; hence the Gula
Thing's law commands : " Every dead man is to be taken to church and
buried in consecrated ground ; except vile evildoers, betrayers of their masters,
inveterate murderers, breakers of promised peace, thieves and suicides. Those
men, who have been guilty of the aforesaid crimes, shall be buried within
reach of the tides, where the water licks the green turf."

† I give this verse nearly literally, as a specimen of the curious style of
Icelandic poetry of the period. War-flame is a periphrasis for a sword ; branch,
or grove, for man ; consequently, war-flame branch is a swordsman. Gold-
sallow is similarly a periphrasis for woman. Loud-resounding, for sea.

"There are not many men like you, certainly," answered the lady. "At all events in this generation."

Then she seated him on the high stool of honour, and treated him with every distinction.

So passed the time, till the return of the bonder.

It was not till the Yule festivities were well over, that Thorfin busked him for return; then, after having dismissed his guests with presents, he and his freedmen started for home, before news had reached him of what had taken place during his absence. The first startling circumstance was the appearance of his great punt, stranded. Thorfin bade his men row to land with all speed, as he suspected that this could not be the result of accident. The bonder was the first, in his anxiety, to leap ashore, and run to the boat-house. There he saw a ship hauled up, on the rollers, and, at the second glance, he knew it to be that of the vikings. His cry of dismay brought the rest around him: he pointed to the vessel and said, "The Red-rovers have made an attack on my farm. I would give house and lands that they had never come."

"What cause is there for fearing that a hostile visit has been paid?" asked some of his men.

"I know whose boat this is," answered the former. "It belongs to Thorir o' the Paunch, and Bad Ögmund, the two wickedest and most brutal of all the Norwegian pirates. No effectual resistance can have been offered, I fear, as the farm was deserted by all fighting men, except, perhaps, that Icelander, but I put no trust in him whatsoever."

The freedmen now consulted with the farmer as to what steps should be taken, supposing that the house were occupied by pirates.

All this while, Grettir was at home, and he was to blame for leaving Thorfin in uncertainty and alarm. He had seen the master's boat round the headland and enter the bay, but he would neither go himself to meet him on the strand, nor suffer the thralls to do so.

"I do not care even though the bonder be a little distracted at what he sees," said the young man.

"Have you any objection to my going to the shore?" asked the wife.

"None in the least; you are mistress of your own actions."

Then she with her daughter ran to meet her husband, and greeted him with a bright smile on her face. He was delighted at seeing her, and said, kissing her forehead, "God be praised, sweetheart, that you and my child are safe and sound! but tell me how matters have stood during my absence, for, from the look of affairs, I do not think that you can have been left quite undisturbed."

"No more have we," she replied. "We have been in grievous danger of loss and dishonour; but the shipwrecked man, whom you have sheltered, has been our helper and guardian."

Thorfin said: "Sit by me on this rock, and tell me of what has taken place."

Then they took each other's hands and sat together on a stone; the freedmen gathered round, and she told plainly and truthfully the story of the Rovers and Grettir's gallant conduct. When she spoke of the manner in which the young Icelander had decoyed them into the storehouse and fastened them in, all the freedmen raised a shout of joy, and when her tale was ended, their exultant cries rang so loud that Grettir heard them in the farmhouse.

Thorfin spoke no word to interrupt the thread of his wife's recital, but the workings of his heart were clearly legible on his countenance. After she had ceased, he sat still and rapt in thought; no one ventured to disturb him. Presently he looked up, and said, "The old saying proves to be true— 'Despair of no man!' Where is Grettir?"

"At home," answered the wife. "He is a strange man, and would not come to meet you."

"Then let me go to him," said the farmer, rising and walking towards the house, followed by his men.

When he saw Grettir, he sprang to him, and thanked him in the fairest words for the heroism he had displayed.

" This I say to you," spoke Thorfin, "which few would say to their dearest friends,—that I hope one day you may need support, so as to prove how earnestly and joyfully I will strain every nerve to assist you; for, assuredly, I never can repay you for what you have done in my behalf, till you are brought into great straits yourself. Abide with me as long as you list, and you shall be held in highest esteem by me and my followers."

Grettir thanked him heartily, and spent the rest of the winter at his house. The story of his exploit was noised throughout Norway, and it was especially praised on the spots where the Berserkirs had given any trouble.

"Now, Padre," said my friend Mr. Briggs; "tell me whether Thorfin ever had the opportunity he so coveted."

"He had indeed, and right nobly did he fulfil his promise. I cannot tell you that part of the story, as it is too long; moreover, if I be not mistaken, yon wild fringe of coast ahead of us is Cape Reykjanes."

I was right; the ragged line of bristling lava spikes, on which not a trace of herbage was visible, was the "Smoking Cape," as its name signifies: rightly is it so termed, for not far out to sea—just where some columns and spits of black stone project from the waves—volcanic eruptions have taken place, eleven times since the colonization of the island.

One jagged prong of rock rising from the waves at the farthest seaward point of Reykjanes is called the Kerl, pronounced Kedtle, and means "Old man."

Whilst I had been telling my saga, the wind had risen to a gale, so that now the waves rolled against the headland with fury, and shivered into drifts of spray, which filled the air with a haze of flying brine, called by the sailors "Spoondrift." The battle of the surge around the Kerl was magnificent; columns of foam shot high above him and fell in a fierce shower on his ragged cap. Certainly the old man stood up gallantly before the repeated shocks of the billows, hurled at him with all the swell of the ocean unbroken from the coasts of Labrador. Through gaps in the lava fringe, great spouts of foam were driven with a roar like that of artillery, and up the shelving sides the surge rushed and then retired, leaving a mass of white rivulets, which poured down to the flood from which they had come.

To our left we could distinguish the Mealsack, a columnar rock, in shape resembling the Bass. It is just eight miles from Reykjanes, and stands 200 feet above the sea. Its sides are perfectly perpendicular; it is about 150 feet in diameter, and is nearly circular. The head is not horizontal, but dips towards the north-west; it is white with the excrement of sea birds, which throng it in myriads. The Mealsack has never been scaled and never will. The great Auk (*Alca impennis*) is said to breed at its foot, and among some neighbouring skerries. The sea is rarely sufficiently calm for a boat to venture near the rocks. The great Auk has not been seen since 1844; it was not obtained in the Arctic expeditions, and Mr. Audubon could gain little authentic information regarding it in Labrador.

In ancient times these birds were common in Iceland, and many skerries are named after them. Those near Reykjanes were famous for them, and their bones are said still to be found heaped on the rocks of that cape; but probably the repeated eruptions in the sea have obliged them to desert their favourite haunt. This Auk used also to be found on S. Kilda, according to an old description of that islet, by a Mr. Martin, who visited it in 1697, and who says, " The sea-fowls are, first, the gairfowl, being the stateliest, as well as the largest, of all the fowls here, and above the size of a solan goose, of a black colour, red about the eyes, a large white spot under each eye, a long broad bill; stands stately, his whole body erected, his wings short; he flyeth not at all; lays his eggs upon the bare rock, which, if taken away, he lays no more for that year: he is palmipes, or web-footed, and has the hatching-spot upon his breast, *i. e.* a bare spot, from which the feathers have fallen off with the heat in hatching; his egg is twice as big as that

of a solan goose, and is variously spotted, black, green, and
dark; he comes, without regard to any wind, appears the 1st
of May, and goes away about the middle of June."

The bird was becoming scarce in the Faroe Isles in the end
of the last century, according to Landt, who lived there from
1791–98. The bird cannot fly, but is a powerful swimmer;
its wings are about four inches in length, and the length of the
body nearly three feet. The wings are used for swimming
under water, and are without produced feathers. The feet are
thrown very far back, so that the Auk sits remarkably upright;
it can only shuffle in an erect position, balancing itself by its
flaps. The bill, four inches long, is jet black with transverse
furrows, the grooves white; the culmen of the upper mandible
is considerably arched, and so also is the gape; the lower
mandible has its outline marked by two curves meeting at an
angle.

Between the beak and eye is a large oval patch of white;
the breast, belly, and under parts are white; the head, throat,
back, wings, and tail are of a glossy black with a metallic lustre.
The feet are also black; the tips of the secondaries are white,
forming a band along the wing.

A hundred pounds has been offered in Denmark for a
living bird, and fifty pounds for a dead specimen, and
one would fetch, in England, a very considerable sum;
yet, since the date mentioned above, no penguin or great
Auk has been seen and captured. Enthusiastic naturalists
may perhaps be stirred to going a voyage of discovery when
I tell them that an Icelandic record of the beginning of the
seventeenth century speaks of a whole boat-load of these birds
having been taken from certain skerries, called in the Sagas,
Gunnbjarnar-eyjar, situate somewhere between Iceland and
Greenland. (*Grönland's Hist. Mindesm.* i. 124.) Where
these skerries are I cannot say with any precision, but a paper
Codex in the Copenhagen Royal Library—quoted by Thorfœus
(in his *Grœnlandia*, 73), states that " Greenland was colonized
by Erik the Red, but it was first discovered by a man hight
Gunnbjörn. After him is named the Gunnbjarnarsker. This

is six weeks' sail from the penguin-skerries beyond Reykjanes, and one has to sail due south from it twelve weeks to reach Garth in Greenland, which is the Bishop's seat."

In the British Museum there are two specimens only of the great Auk, one being the individual obtained by Mr. Bullock, from Papa Westra; the other was purchased at a sale which took place in Holland, about 1861. A specimen is also to be seen in the Newcastle Museum.

In old times the Auk must have been common in the Isles of Denmark, for among the *Kioekkeen-moeddinger*, or refuse of the kitchens of the early Danish inhabitants, a considerable number of the bones of this bird have been discovered.

The only spots which the bird can now frequent are the islets in Breithi-fjord, the north-western friths and skerries, or Grimsey; but, in all probability, it is extinct.

CHAPTER II.

REYKJAVÍK.

Arrival—Description of Reykjavík—" A very antient fish-like smell!"—The
School—The Residence—The Cathedral—Thorwaldsen—Icelandic Silverwork—The Merchant Stores—Mr. Briggs and the Porters—Dinner—I
visit the Rector—A would-be Suicide—The Catholic Mission—The
Origin of Iceland.

ON the following morning, June the 16th, we were in the
bay of Reykjavík, the wind blowing from the south-west in
stormy gusts, laden with vapour, which clung to the high
grounds and condensed into showers.

The captain took me ashore in his yawl, whilst the others
were making up their minds how to transport themselves and
their multifarious luggage to land.

I made for the inn, and secured rooms. Dinner was
promised in an hour's time, and the landlord, Jörgensen,
engaged to board us at the rate of one specie dollar (4s. 6d.)
a piece, per diem, exclusive of wine and spirits.

Reykjavík is a jumble of wooden shanties, pitched down
wherever the builder listed. Some of the houses are painted
white, the majority black, one has broken out in green shutters,
another is daubed over with orange. The roofs are also of
wood, and coloured black or grey. The town lies between
the sea and a fresh-water lake full of reeds and wild-fowl ;
it is in the shape of a rude parallelogram, facing the sea on
one side, showing its back to the lake on the other ; the other
sides rise up the slopes of hills from three to four hundred

feet high, the one crowned by a windmill, the other by the
Roman Catholic mission.

. . Near the lake is a square, or market-place, covered with
turf, the cathedral forming the most conspicuous object in it.
At right angles is the French consulate and apothecary's shop
combined.

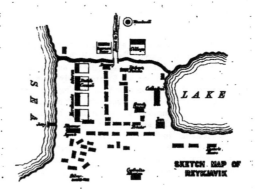

There are but two streets, and these are hardly worthy
of the name. One leads from the jetty to the inn, and
is called the Athalstræti, or High Street; in it live the agent
for the steamer and the printer. The second starts from
this street, and terminates at a bridge crossing a brook,
which flows from the lake into the sea. In this thoroughfare
live the sheriff (Landvogt) and Professor Pjétur Pjéturson,
head of the theological seminary. The sea-front is occupied
by a line of merchant stores. The moment that the main
thoroughfares are quitted, the stench emitted from the smaller
houses becomes insupportable. Decayed fish, offal, filth of
every description, is tossed anywhere for the rain to wash
away, or for the passer-by to trample into the ground.

The fuel made use of is dry sea-weed, fish-bones, and any refuse which can be coaxed to smoulder or puffed into a blaze; so that the smoke, as may well be imagined, is anything but grateful to the olfactory nerves.

An Icelander seems to have no sense of smell; perhaps it is well that he has none, for there is no possibility of gratifying that sense, whilst there is every opportunity of mortifying it. The enormous amount of snuff consumed is one cause of this deadness in the perception of scent. Nature has made a mistake in forming Icelanders' faces; she should have inverted their noses, so as to facilitate their plugging them with tobacco. Queen Charlotte, it is said, was wont to lay a train down her white satin sleeve, and snuff it up with a sweep of the nose; an Icelandic lady of a certain age does much the same sort of thing, laying her train in spirals from the wrist to the knuckles. But as " my nose knows not the titillating sensation which her nose knows," I am keenly alive to all the unpleasantness of the Reykjavikian slums, and I hasten to the market-place.

Yon tall house with men on the roof mending it, and recruiting exhausted nature by a pull at the snuff-horn after driving home each nail, is the Latin school (Lœrtha skóla). There are about forty-six boys now in it, which, considering that it is the only educational establishment in the island, is a scanty percentage. This is to be accounted for by the jealous fear which parents feel, lest their offspring should be corrupted by the grandeur and dissipation of the forty or fifty decent shanties which form the capital; lest, also, they should become too fond of the cleanliness of a Danish household, to return with enthusiasm to the ancestral dirt of the parental piggery. For Reykjavik is a Danish town essentially; Danish dress, Danish fashions, Danish neatness, Danish cookery, prevail.

The course of study at the college comprises a thorough grammar - school education in the classic tongues, with instruction in English, French, German, Danish, music, mathematics, geography, and history. But sufficient attention is hardly paid to mathematics; the consequence is, a very perceptible incapacity among educated Icelanders for following a train of reasoning or appreciating the force of an argument. At the school much care is taken to keep up the purity of the Icelandic tongue, which has suffered great corruption in and around Reykjavík.

The college has not always stood where it does now. The first Christian school was founded in 999 by Hallr at Haukadalr, near the Geysir. Sœmund the Wise, the supposed collector of the elder Edda, established a second at Oddi. A more systematic attempt was made in 1057, by Bishop Isleif, at Skalholt, and in 1105, S. Jón Ogmundsson, Bishop of Hólar, formed a school in connection with the cathedral. The boys must have been pretty well taught, for we are told in the Saint's Saga, that the master builder, then engaged on the construction of the cathedral church, learned Latin by overhearing the lads reciting their lessons. After the Reformation the school languished, so that in 1789, King Christian VII. of Denmark had entirely to reorganize it, and to remove it to Reykjavík. It was subsequently transported to Bessastathr, where there was a printing-press, but in 1846 it was brought back to the capital. The food of the lads, as prescribed by law, is as follows:—

On Sunday morning, before the students go to church, they have some breakfast with butter. At midday:—1st course, stock-fish and butter; 2nd course, meat broth; but if this cannot be got, pease with meat. In the evening:—1st, stock-fish and butter; 2nd, barley-water grout, with milk and butter.

Monday, midday:—Stock-fish and butter; meal grout, with milk. Evening:—Stock-fish and butter; curd with cold milk.

Tuesday, midday:—Stock-fish and butter; pease and meat. Evening:—Stock-fish and butter; cold cods-sound.

Wednesday, midday:—Stock-fish and butter; meat soup. Evening:—Stock-fish and butter; warm fish.

Thursday, midday:—Stock-fish and butter; pease and meat. Evening:—Stock-fish and butter; cold or warm sausages.

Friday, midday:—Stock-fish and butter; meal grout, or buckwheat porridge. Evening:—Stock-fish and butter; haddock and flounder.

Saturday, midday:—Stock-fish and butter; warm sausages. Evening:—Stock-fish and butter; curd with milk.

"They are also to have with each meal some bread or kager (a flat rye and bran cake), or ship's biscuit; also 'brose,' as much as the pupil may want. At Christmas and Easter, they are to be specially indulged in sausages, sheep's head, and tripe."

The pupils no longer feed in the college, though they sleep in it. The food, as above specified, represents pretty fairly the diet of a well-to-do Icelander. From it will be seen how important a staple in the country, the split and wind-dried stock-fish proves to be.

Beside the stream flowing from the lake, and near the bridge, is the bishop's palace, a wooden booth, tarred all over, much like a settler's cabin in a Canadian clearing; I saw flowers in the window, and little girls peeping out between the pots.

A stone building, whitewashed during the reign of Jörgensen, having a turf patch before the door, and a little tree, twelve feet high, trained against the wall, is the governor's house. It was originally designed as a prison, but Icelanders are too lazy to become great criminals, so it remained untenanted for some time, and was then adapted to its present use.

It is the third stone edifice in Iceland; the first being Hólar Church, the second, Fredriksborg, the seat of the northern governor, and of these three buildings every native is proud. The "Residence" boasts no architectural merit; it might do for a shed in a dockyard.

The governor is postman, and the letters for England may be left with him; but, as the postal arrangements are not very effective, many of the Danish merchants prefer sending their more important despatches by the captain, or one of the passengers, of the *Arcturus*. I have received letters myself, from Iceland, through private hands, whilst those sent me through the Post-office never reached their destination. On leaving Reykjavík I was loaded with letters which I was begged to post in England, some for Denmark, others for Germany and France, so unsafe is the Government Post-office considered to be.

Now let us push down the street, avoiding the drunken man who lies wallowing on the ground, sobbing as though his heart would break, because his equally drunken adversary will not turn his back and stand steady, to let him have a comfortable kick.

: Stride across the gutter which traverses the road, and we are in High Street.

As the boat has not yet left the steamer with my comrades and their luggage, I saunter to the cathedral, a large stuccoed edifice, consisting of nave, chancel, east sacristy and south tower rising out of the nave (the church points north and south). This tower is perhaps the most hideous erection which head of man could devise or hand execute. The sides curve inward and are capped by a saddle-back roof, tarred, and surmounted by a vane never at rest. The interior of the church is no better. There are galleries for the sake of contributing additional ugliness, I presume, as they can be of no manner of use, as the whole population of Reykjavik could be accommodated on the floor. It would hold three or four hundred people easily, and the ordinary Sunday congregation in summer—I know nothing of winter—is from fifty to eighty, out of a population of one thousand four hundred souls!

In the lower gallery is a small organ played by the music-master of the Latin school.

The chancel is raised three steps above the nave, and is unfurnished with stalls, as the Icelandic established religion has but one service, entitled the "Mass," which is performed at the altar. There is no form for matins, evensong, or for the hour services. The roof of the chancel is painted blue, with gold stars. There is no east window, and the place is occupied by a *baldacchino* enclosing a painting of the Resurrection, feeble in design and bad in colour, belonging to the worst French sentimental school. The windows of the church are roundheaded and placed high in the walls. The brass chandeliers are fit only for a gin-palace. Thorwaldsen's font is in the chancel, a present to the isle from which his family came. It is beautiful as regards sculptured detail, but bad in

general design, the *motif* being a Pagan altar. It is a squat
rectangular block of marble, with a wreath around the diminutive bowl. On the front is a representation of our Lord's
Baptism; on the right, Christ blessing little children; on the
left, the Virgin and Child; whilst the back is occupied by a
festoon and a cluster of fat cherubs, supporting the legend:
" Opus hoc Romæ fecit, et Islandiæ, terræ sibi gentiliacæ, pie-
tatis causâ, donavit Albertus Thorvaldsen, anno M.DCCC.XXVII."
When will the stupid affectation of putting Latin inscriptions
in churches cease ? The living vernacular is so infinitely
superior to the stilted dead Latin. A word or two about
Thorwaldsen, and I shall return to the cathedral.

Albert Thorwaldsen was born on the high seas, November
the 19th, 1770. His father was a ship-builder, and his
grandfather the Icelandic parish priest of Miklibær. He
was early sent to school at Copenhagen, where, whilst quite
a child, he got into a little trouble, was arrested by the police
and conveyed to the guard-house, but dismissed with a repri-
mand. At the age of eleven, he was placed in the school
of the Society of Arts, and obtained several prizes and medals.
He afterwards left Denmark to stay in Italy and study the
remains of classic art; there his statue of Jason attracted
the attention of Mr. Hope, who gave him an order for its
completion. His star now rose, and numerous orders poured
in upon him from all sides, so many of which were from
the government of Denmark, that he determined on returning
to Copenhagen. The king at once placed a frigate at his
disposal to convey him and his works to Denmark; and, on
his arrival, showed him every mark of consideration. Rooms
in the Charlottenborg royal palace were allotted him, and
there he spent the remainder of his days.

Thorwaldsen was much attached to a Miss Mackenzie,
but never married her. He died on the 24th March, 1844,
and was buried with great pomp in the churchyard of the
Fruenkirke at Copenhagen.

In character, he was open and straightforward, simple
in his habits, and of a kindly generous disposition. He is

said to have written once to an Italian prince of blood royal, who owed him for a statue, and had kept him waiting for his money a considerable time : " May it please your royal highness—I *must* have my money."

Another story told of him is as follows :—King Christian VIII. visited him in his studio one day, and invited him to dinner on the following day. Thorwaldsen looked towards his servant, and asked whether he had any previous engagement. " Yes," replied the man, " to Oersted." " Sire," said Thorwaldsen, " I must decline your invitation, for I have promised to be with Oersted to-morrow, and it is his birthday, so I cannot disappoint him."

After this digression I must return to the cathedral. I asked for the vestments, and was shown the chasubles of crimson and violet velvet, with gold crosses on the backs. The mediæval vesica shape probably never prevailed in Iceland ; still the ancient form must have been very different from that now in vogue throughout the island, which is quite as hideous as the modern Roman shape. Both have been curtailed over the shoulders, and end in an ungainly curve, but these northern vestments have been clipped in front, so that they fall no lower than the breast, their width in front being sometimes as little as six inches ; a singularly uncouth mutilation.

A magnificent ancient cope, presented by Julius II. to the last Catholic bishop of Hólar, and thence removed at the suppression of that ancient see, attracted my notice. The embroidery has been remounted on crimson velvet; at the back is a delicately worked representation of our Lord enthroned in glory, between SS. Mary and John Baptist ; below are angels gathering the dead from their graves. The Orphreys are worked with equal tenderness ; they consist of a series of tabernacles enclosing saints, of whom I could distinguish the four evangelists, and SS. Margaret and Barbara.

On one side of the choir is a monstrous pulpit, on the other the pews of the governor and bishop. The prelate, it seems, takes but little part in the church service ; the dean

3

celebrates and preaches, whilst the bishop is seated in a
comfortable pew, amongst his female relatives, not even in
his cathedral church parted for one moment from the society
of his cherished spouse and accomplished daughters.

Between the ceiling and the roof of the church is the
public library, consisting of 8,000 volumes, chiefly in Danish
and Icelandic. It was founded in 1821, and for a subscription
of a dollar per annum, the inhabitants of Reykjavik may take
advantage of its privileges.

Running my eye along the shelves, I noticed a great
preponderance of Lutheran divinity, with dust thick on the
volumes, and all the Sagas very much thumbed. The library
contains most of the Latin and Greek authors, though not
the best editions. No manuscripts of any age are to be found
there, but I believe that the school library contains transcripts
of the last century from some of the unpublished Sagas.

Leaving the cathedral, I made my way to the silversmith's,
ascended the hill south of the town, climbed a wall, and
came upon the house, without discovering any path leading
to or from it. The silver bracelets and brooches of Icelandic
manufacture are strikingly like Maltese work, and are, for
the most part, very beautiful. There is no great variety in
design, and they are modelled from old examples in the
national style of art.

A right principle prevents them from becoming vulgar.
A firm outline is sought, and within its restraints tendrils
of twisted wire gracefully wheel and curl. It is only when
the outline is broken and frittered away, in conformity with
foreign bad taste, that its beauty evaporates. Dispersed through
the island, there is a large amount of old silver worked into
filigree buttons, clasps and pins, selling by weight, the ornaments being flung into one scale, and dollars into the other,
till the scale turns.

These silver decorations consist mainly of buttons and
belts, the buttons being balls, from the size of a marble to
that of a crab-apple, wrought with pierced flower-work, and
made to jingle by a leaf or a letter $\overline{A}$ being suspended from

them. Why this same letter should be repeated over the whole country as a pendant to the silver balls, I cannot understand, and I could find no native who had any explanation to give of the fact. Some of these buttons are of great age; the late priest of Mosfell had some with figures of the Madonna on them, indicating a date prior to the Reformation. Many families have silver spoons; I have seen a spoon with the date 1520 engraved on it; but none are remarkable for their beauty, and they are less carefully wrought than the horn ladles common to every byre.

Having made a few purchases at the silversmith's, I returned to the strand, which is lined by merchants' stores, each surmounted by the Danish flag,—red charged with a white cross—the only bit of bright colour in Reykjavik.

Phaugh! a whiff of fish comes down with the blast: there is a heap of stock-fish before the door of each shop, with a tarpaulin flung over it.

The town is full of idle men, who follow the stranger whithersoever he goes—provided he does not walk too fast for them. They hang about the stores as thickly and stupidly as flies round a sugar-barrel; they stream into the shops after me, throng so closely round me that I can hardly move, listen to what I say, eye me from head to foot, ask the price of every article of clothing I have on; bid for my knicker-bockers, which, of course, I cannot spare; feel my stockings, and laugh to scorn their loose texture; criticize my purchases, want to examine my purse, but I object, and by so doing, hurt the feelings of half-a-dozen; they pull out of my hand the comforter and sou'-wester I have just bought, and would proceed to try the latter on their own heads, only I snatch it from them. Then they tell the merchant that he has charged too high for the muffler, and put too low a figure on the sou'-wester. They make advances towards familiarity, shaking hands, asking my name, then my father's name, then they inquire who was my mother; they offer me a pinch of snuff, or rather a pull at their snuff-horns, which are like powder-flasks, and are applied to the nostril, the head thrown back,

and the snuff poured in, till the nose is pretty well choked.
One man, very dirty and very drunk, insists on having a kiss
—the national salutation ; and, when the merchant explains
that such is not the English custom, he kisses all the natives
in the shop, and embraces the merchant across the counter.

Finding that the store is suddenly emptied of these
loiterers, and hearing an unusual hubbub on the beach, I
suspect that the boat containing my companions has arrived,
so I rush down to the jetty.

I found the Yankee and Mr. Briggs in a high state of excitement, charging up the landing-place with fishing-rods fixed,
discomfiting shoals of men and boys, who had laid hands on
boxes, bags, bundles, tent-cases, and shouldered them with
the purpose of conveying them to the inn.

" This will never do ! " gasped Mr. Briggs, who was
thoroughly exasperated ; " these ugly scoundrels will not leave
me alone. I will *not* pay five-and-twenty men, where one
will suffice. Confound the fellows ! they will not keep their
hands off. Padre ! you speak a little Icelandic, just say
something strong and stinging to them, will you ! "

I vainly protested in broken Icelandic, and with the most
frightful pronunciation : I very much doubt whether I made
myself understood at all, as my knowledge of the language
was entirely from books.

" I've been cursing them in French, German, and English ;
but it is just throwing so many naughty words away, they roll
off them like drops of water from a duck's back, or dust shot
from a gull's feathers. Martin there ! Bless the man if
he has not become perfectly bewildered and useless, instead of
holding hard to a tent-case, and clinging to a whisky bottle ! "

Whilst Mr. Briggs was speaking, all the luggage was
whipped up, and there was no help for it, but to surround the
company of porters, and prevent any from decamping with
some of the traps. Mr. Briggs disposed his forces well, he sent
Martin on ahead with the thermometer tubes, he placed the
Yankee on one side, made me take the other, he himself
brought up the rear of this human drove.

"Look sharp after them, Padre!" he vociferated; "there are some ugly rabbit-burrows to your left. Yankee! that fellow by you looks as though he would bolt with my dressing-case. You've a corkscrew in your hand—drive it into a fleshy part, if the rogue attempts to dodge. Martin! straight along, don't turn to your left. Ah, ha! Yankee! leave the spirit-flask alone now, and keep your eye on the luggage!"

On reaching the inn, the five-and-twenty men, not to mention the boys, clamoured for payment.

"Which was the bell-ewe, Padre!" asked Mr. Briggs; "the fellow with the green box, ay? Well, I secured him on the wharf; him I shall pay—not another rogue of them."

My portly friend singled out his man and demanded the charge. "And, whilst I'm paying him, run in to the land-lord, and find out what 'Good-by,' and 'Don't you wish you may get it?' are in Icelandic."

"Voer-thu sœl!" I learned from the landlord, Jörgensen, served all sorts of purposes. Literally it means; "May you be blessed!" I explained to Mr. Briggs.

"But!" he burst forth; "I don't want these thieves to be blessed, in any way; I want, I particularly desiderate, their being just the very other thing." Then, standing on the doorstep, he met all claims with "Voer-thu——" pointing with vehement jerks of the forefinger towards the centre of the globe.

Having at last escaped from his tormentors, Mr. Briggs adjourned, much exhausted, to the large room where dinner was laid.

We commenced with the favourite Danish soup of raisins floating in sago, sweetened, and coloured with claret. Eider-duck's eggs followed, hard boiled; then came stewed mutton, with the gravy in a butter-boat by itself; and we wound up with suur-kryder, a jelly of cloves and fruit, something like damson-cheese.

Having had coffee, we were in condition to pay visits in the town. I went first to the Latin school, where I inquired for the rector from the porter Olavur, who was Mr. Metcalfe's

guide in 1860, but whom he dismissed for bad conduct. I
was ushered up a broad flight of steps and along a passage
with dormitories on one side, into which I looked through
square window-panes inserted in the doors, and saw that they
were clean and airy.

The rector's apartments were at the end of the corridor;
I was shown into a well-furnished room, with fauteuils, sofas,
curtained windows, and French engravings on the walls.   On
the table was a wreath of immortelles, surrounding a China
Cupid seated on his haunches, and drawing his bow at a
venture.

In a corner was a large highly-polished stove, the pome-
granate at the top, crowned by a delicately poised life-sized
bust of a fine intellectual head, modelled in plaster from life,
and painted flesh-colour, with hale rosy cheeks.   The eyes
were closed; eyebrows and lashes, hair, whiskers, were all
of a light brown tint.

This singular cast was taken from the rector's head when
he was younger, and it is to be hoped that it will always
remain with the college, of which he has been such a distinguished ornament, as a remembrance of one who has done
more than most men for the advancement of learning in his
native isle.   The rector can speak fluently English, French,
German, Danish and Latin; his knowledge of Greek and
Hebrew is also considerable.   "He has but a single fault!"
said my guide, one day.   "Having spent a great part of his
life in France and Germany, he does not think Iceland the
most glorious locality in the universe!"   And small blame
to him! I thought.

I brought my visit to a speedy termination, as I found that
examinations were in progress, and that the rector could ill
spare time to be with me.

On descending the street, Mr. Briggs rushed up to me
with an exclamation of—"Oh, Padre! There is not the
slightest use in coming to Iceland!"

"How so?"

"Why, it is just like everywhere else! I have been

looking in at one of the stores, and what do you think I saw? Crinolines, real crinolines, man! hanging up for sale. Crinolines! is it not horrible! we are not beyond the range of fashion yet! Oh, for Jan Meyen!"

Another horrible circumstance which distressed Mr. Briggs and made him despair of getting beyond the reach of modern civilization, was the fact that there was a photographer in Reykjavík; yet I think he became quite reconciled to the idea, when he heard the story of the man's career, which was peculiar enough in its way. The photographer had been so unsuccessful in his line at Copenhagen, that, in a fit of despondency, he had cut his throat, then rushed to his landlady to show her what he had done, and fallen a bleeding victim at the feet of one who had relentlessly dunned him for arrears. She, with great presence of mind, tied her apron round his neck and called a policeman, who in turn called a cab, and conveyed the would-be suicide to the hospital, where his throat was promptly sewed up, and he soon recovered. The circumstance had, however, sent such a thrill of commiseration through the benevolent hearts of the Copenhageners, that a subscription was raised in his behalf, and with the proceeds he was shipped to Reykjavik, where, it was presumed, he would drive a flourishing trade, there being no competition. I should not say that he was by any means a first-rate artist. I saw a photograph which he took in 1861 of Messrs. Shepherd and Holland, with their guides; one of the party having moved his eyes, it became necessary that they should be painted in. This was effected with a little brown and white paint, quite *en règle*. But the curious part of the circumstance was, that, by the middle of the summer of 1862, every portion of the photograph had vanished except those eyes, which held their ground and stared out of a blank surface.

I spent the rest of the day in purchasing horses and securing guides. I bought five nice little ponies at prices varying from 24 to 38 dollars (2l. 15s.—4l. 7s.) A good horse which has been fed on hay during the winter always sells at a

comparatively high figure, whilst those which have fished for
themselves, subsisting on sea-weed, moss, and anything they
can pick up, are fit for very little work during the summer,
and are worthless for a long journey. I proposed buying
three more horses farther up the country, one to serve as a
riding pony, to make up a complement of four riding and four
baggage horses—eight in all. It is necessary for a traveller,
who wishes to get over much ground in the short summer, to
be provided with an extra set of horses, so that he, his guide,
and each of the baggage loads, may change ponies twice or
thrice in the day. Guides were more difficult to procure, as
the only good one, Olavur Steingrímsson, was absent with
Messrs. Shepherd, Upcher, and Fowler, who had left Reykjavík in May. Another guide, Zöga, who speaks English,
goes no farther than to the Geysir, or at farthest to Hekla.
However, at last, my companions and myself were suited
with three. Gúthmundr and Magnús were to be the servants
of Mr. Briggs, the Yankee and Martin, whilst a theological
student, with a licence to preach in his pocket, was to accompany and attend on me, at two dollars and a half per diem,
and I had to find him horses, lodging and food. All three
guides proved to be worthy, honest fellows, but I made a
mistake in hiring a student, who, as I might have expected,
would not do the rough work which an ordinary guide would
accomplish. He has his failings, as we all have, but his are,
I believe, on the surface, and overlay a true and upright heart.
His faults will appear prominently enough in the following
pages, as they afforded me some amusement, though they
also, at times, caused considerable irritation, which I found a
difficulty in suppressing; his name shall be, through my
book, Grímr Arnason.

We drew up an agreement with our guides, as well as we
could, none of the party being lawyers, and I have no doubt
that it is full of flaws. I give it as a guide to tourists; it
is quite sufficient for all practical purposes in Iceland.

"It is hereby agreed upon between A B and C D, that in
consideration of the sum of —— per diem, paid to the said

C D, he will honestly and faithfully serve A B to the best of his ability, as guide and general servant, during his journey in Iceland; such service to be from the 18th day of June. But if at any time C D shall misconduct himself, or in case of his inability or incompetence to perform his duties, A B shall have power to dismiss C D from his service, upon paying him six days' wages from the day of discharge." It is of essential importance that some such agreement should be made, as the Reykjavik guides are, many of them, very incompetent, and sad drunkards.

Before leaving Reykjavik I paid a visit to the Catholic Mission, and found it a snug farm; it was once surrounded by a tún or home-meadow, which has been spoiled by the priest in his endeavours to introduce French husbandry. I found the missioner walking up and down to the leeward of his house, wrapped in a warm cloak. He invited me indoors, and I had a long conversation with him relative to the ecclesiastical condition of the island. After this I asked for a few practical hints for my journey.

"First and foremost," said the priest, "take with you plenty of small change, you can get none in the country, and there is no reason why you should give a dollar when half will do. You will have to pay for everything you want, and rightly too. The Icelander has a hard struggle to keep himself and family alive; and food is expensive. He has often to take a journey of many days to the capital, that he may provide himself with a year's stock of the necessaries of life—perhaps he loses some of his horses on the way, and, as likely as not, the goods he has purchased are damaged in crossing the rivers. It would be a wrong thing for a visitor, who comes to the country for pleasure, to prey upon the scanty supplies of these poor people without sufficiently remunerating them—yet this has been done! The Icelanders are inclined to be hospitable, but they cannot afford to follow their inclinations."

"Is no reliance to be placed on the statement of travellers respecting the Icelanders, that they decline to receive payment?"

" The farmers will always make a charge, but the priests will now and then refuse money. They will, however, often allow themselves to be persuaded to accept a present, such as an illustrated book, a Latin author, or a silk handkerchief for their wives. You will find, also, that where their *amour propre* would be hurt by accepting dollars, yet they will take an English half-sovereign to make into a ring for a favourite daughter."

" What must I reckon upon as my daily expenditure ? "

" A guinea, exclusive of what you pay your guide. It may not always amount to so much, but it will sometimes exceed. Of course it is not to be expected that your Reykjavík guide can know every corner of the island, and, in crossing passes with which he is unacquainted, you must hire an extra man, who, knowing that you cannot do without him, will demand a fancy price."

" I suppose," said I, " that the country and scenery are most magnificent."

" Magnificent indeed ! " answered the abbé; " there is the magnificence of Satan imprinted deep in the face of this land. Did you ever hear the Danish account of the origin of Iceland ? "

" Never," I replied.

" Well, then ; after the creation, Satan was rather taken aback, and he thought within himself, ' I'll see now what *I* can do ! ' So he toiled at creation, and lo ! he turned out Iceland. This myth gives you a notion of the place : all is horrible and gloomy. You are reminded again and again of the scenes in Dante's Inferno. This land is magnificent too ! for there still lingers majesty about the handiwork of the fallen angel."

CHAPTER III.

MÓSFELL.

Leave Reykjavík—Rich Soil—Leiruvogr—Accidents—Camp out—A lazy
Guide—Arctic Midnight—Rank Breakfast—The Story of Sonartorrek
—Alpine Road—Icelandic Dogs—A Farmhouse—Filth—On the Heithi
—Birds—Icelandic Indifference—Arrival at Thingvöllum—A Quarrel.

OFF at last! Farewell comfort, ease, good food, snug beds!
Welcome hard riding, rain and cold, scanty diet, and the
ground for a couch!

At eight o'clock on the evening of June 19th, we left
Reykjavík for the interior and north. The sun had come out
right gloriously from his pavilion of clouds, and furled them up
in long white folds on the mountain tops.

The baggage train preceded us by half an hour, that we
might have the pleasure of galloping out of the town, cracking
our whips, whilst Bob, Martin's Newfoundland dog, careered
at our heels.

We caught up the advanced party on the brow of the first
hill, and our pace was brought to a jog-trot in conformity
with that of the sumpter animals.

We were a merry party, making as much noise as boys
breaking loose from school. The guides smacked their long
Icelandic whips and vociferated to the horses; the American
sang "Yankee-doodle," with the fervour of a patriot; whilst
Grímr trolled forth a "kvoethi" or lay, to a dismal chant,
and the loose pots and pans hammered out a harmony from
the back of the pony, which trotted along with them on its

saddle. Martin discharged his fowling-piece at gulls and skuas, which were flying in numbers over head, quite beyond range; and Mr. Briggs occasionally yelled with pain, as the corners of boxes or the prongs of fishing-rods violently impinged, at various points, on his person, whenever a pack-horse charged past him on the narrow track, and brought its load into collision with every thing and person which obstructed its course. On one occasion my portly friend was nearly annihilated by a body of erratic ponies, which Gúthmundr and Magnús drove into the path, at the very spot where he was bickering along on his reeking steed.

We formed a cavalcade of twenty-five horses and six men; our luggage consisted of tents, two boxes of comestibles, two cases of brandy, an apparatus for cooking over a spirit-lamp, guns and fishing-tackle. I must not omit from enumeration the great bed—a horse-load in itself—belonging to Mr. Briggs, who had no intention of sleeping uncomfortably, even in Iceland.

The baggage is adapted to an Icelandic horse's back in the clumsiest manner possible; two square sods of turf are laid against the sides of the pony, and a framework of wood in two curves is laid over these and strapped together under the belly. To this frame are affixed eight wooden pins, and the traveller's luggage is attached to these by woollen ropes of little strength, forming a net-work of cord full of knots, which take half-an-hour to be untied. If the burden on one side of the horse be greater than that on the other, round goes the pack-saddle, and bags and boxes strew the road; consequently one of the arts of loading consists in estimating weights and producing equipoise.

For two or three miles the track is cleared of stones, and we jogged along with comfort; but on crossing the Elitha-á, there remained no trace of road-making, and we found it impossible to ride abreast. The sumpter horses, as well as the reserve, were continually breaking away in quest of grass, and the inconvenience was so great that the guides proceeded to lead them *í taumi* (cognate with Eng. team), which consists

of tying the head of one horse to the tail of another, so that a
line is formed, and a refractory beast has no chance of getting
away.   One guide leads the foremost pony by its halter, a
second brings up the rear, and the third oscillates between the
extreme points, alarming each pony individually by his shouts
and whip-cracking, besides keeping his eye on the packages
and straps.

To right and left opened friths, with wild duck rocking on
the crisp wavelets which flowed in before the breeze.

We rode under bluffs of rich mould, ten to twenty feet
high, formed of disintegrated volcanic rock, ragged and mangled
by the spring torrents, which sweep off snags of turf and cakes
of soil into the sea.   The earth seems ready enough to produce
most luxuriant crops were the climate less rigorous.   If we
had such deposits in England, we should quarry them out to
fertilize our fields.   Draining might make the soil produce something even in Iceland, but at present it is coated with only a
scanty crop of grey moss and wiry grass, whose tender shoots
are charred by the snow-water as soon as they begin to spring.

As the night progressed, if I may call that night when the
sun was still in the heavens, the clouds left the mountains, and
the stately pile of Esja stood out purple-mantled, with its
ravines choked with snow.   We had a fine view of it from
above Leiruvogr, a bay studded with islands, on one of which
cows were pasturing, and light blue smoke gave evidence of a
farm.   Beyond, in the remotest distance, shone the sugar-loaf
Snœfels Jökull, nearly seventy miles distant.   To our right,
was a bold scarth of dark rock thronged with ravens, at its
foot a hot spring, near which nestles a small byre, whose tún
or home-field, bright green sprinkled with golden-cups, was
gladdening to the eye wearied with the prevailing grey and
black tints of the landscape.   The only land cultivated in
Iceland is the tún, which is a meadow surrounding the house,
varying in extent according to the number of cows kept on the
farm ; this field is dressed with their dung, and produces the
hay which constitutes the food of the cattle during the winter.

An oyster-catcher sailing near us, was brought down by

the gun and thrust into my saddle-bag, though Grímr shook
his head, and pronounced it execrable eating.

From Leiruvogr, a flat of deep bogs extends to Mósfell,
traversed by a river, the waters of which are rendered lukewarm by the existence of hot springs in and around it. The
name Leiruvogr (the ei is pronounced like ai in laird), signifies the bay of muddy deposit. I wonder whether the Frith
at Plymouth, called the Laira, which is muddy enough in all
conscience, has any etymological connection with this Icelandic bay!

We noticed a mallard feeding in the swamps, so the
Yankee cocked his gun and rode into the river; he discharged
his piece just as the bird observed him, and plunged into a
pool hard by. Unfortunately, the Yankee not only missed
his bird, but lost his seat, for the pony, unaccustomed to
hearing explosions on his back, bounded forward suddenly,
and deposited its rider in a sitting posture, gun in hand, in
the water. This produced a general laugh, led by Mr. Briggs;
but the Yankee had soon an opportunity of turning the tables
on him; for as, a few minutes later, my fat friend was ambling
ahead of us with the utmost composure, his pony came to a
dead halt before an ugly quagmire, and by the impetus with
which he had been moving, threw Mr. Briggs head over heels,
with his arms up to the elbows in the mire, and his feet on
either side of the horse's neck. Mr. Briggs bore our jokes with
the greatest good-humour, protesting that he had not fallen,
but that the ground had risen up and hit him on the head,
which was a very different thing. Our merriment was cut
short by Mósfell (pronounced Mósfedtl) church appearing
above us, perched on a mound, and looking much like a diminutive Noah's ark. The parsonage is close by, and its tún is
crossed by a gill, draining the two ridges which rise above
the church, one of which is crowned by a rock strikingly like a
roast sucking-pig. We unsaddled and hobbled the horses,
unravelled the intricate meshes of cord which bound up the
parcels, rolled out Mr. Briggs' great bed, drew the tent
from its case, and spread it on the ground. Grímr in the

meantime lit his pipe, and put his hands into his pockets.
" Well ! " said I ; " here is the tent to be put up, help me
with it."

The student grumbled at our giving unnecessary trouble,
when the house was close by, in which we could be all
accommodated. The exterior of the parsonage was so uninviting that we were hardly prepared to risk a night in it, and
preferred sleeping under canvas in the pure air of heaven.
As we did not choose to take his advice, Grimr sulked, and
would have nothing to do with the tent, so that Mr. Briggs and
I had to erect it ourselves, spread the floorcloth and lay out
the rugs for beds, whilst my amiable guide watched us from
his seat on one of the boxes. When the canvas was taut,
Grimr, taking the pipe from his mouth and pointing with
the stem over his shoulder, said : " The grey horse is down ;
and it will probably break its back."

" Well then," said I ; " go and help the beast to its
feet."

" It's of no use ; it will fall down again."

Not satisfied with this reason, I made him follow me to
the grey's assistance. The horse had lain down to roll, and
was wedged between heaps of turf, and could not extricate
itself, as its feet were hobbled. We soon brought it to its
legs again, and then Grimr muttered, " The chestnut is running away ; before morning it will be back in Reykjavík."

" Then run after it ! "

" It is of no use ; if I catch it, it will run away again."

So I was obliged to pursue the steed myself, through the
river, getting myself soaked, catch it, and secure the hobbles
which were loose.

" Now I shall go to bed," said Grimr, retiring towards the
farm.

It was a still arctic midnight. The sky was flooded with
light, toning the azure to the tenderest green. Clouds were
transmuted to rose flakes, and mist to a nebulous haze of
flame ; some ragged cloud patches, high above the mountain
peaks, flamed like gold in the furnace, their shadows picked

out with carmine. A crown of rays extending to the zenith,
streamed from behind Esja, which was thrown into grey
shadow. Rock and mountain were distinct, as though seen
through an opera-glass, every crag and furrow was pencilled
with wondrous minuteness, each mountain top cutting against
the sky with intense precision. Though no direct rays of sun
touched the earth, yet the reflected light from above made
everything even clearer than by day, when a slight haze
softens outlines and blends colours.

The most perfect stillness reigned, only broken by the
rippling of the stream over a bank of pebbles, before it
hushed its murmurs in the bogs. In another hour the sun
would be above the horizon, so we hastened to our beds on
the hard ground, using our saddles for pillows, and our horse-
rugs for blankets.

I cannot say that the first night in the tent was pleasant;
the cold was sharp, the ground, where we lay, full of stones,
and I was roused at intervals by the horn of the snipe and
the melancholy strain of the swan, as these birds flew past
our tent, after the sun had broken over the northern hills.

At seven o'clock Gúthmundr and Magnús came to look
after the horses, and Grímr popped his head in at the door,
with the cheerless announcement that the priest had nothing
to give us for breakfast, except a bowl of milk and a few
cakes of sheep's-dung for fuel.

"That is all we want!" exclaimed the Yankee starting up.

In a few minutes Mr. Briggs was hard at work coaxing the
dung into a flame, and the American was busy skinning and
cutting up the oyster-catcher. The fragments of the bird
were flung into a pot and boiled. Mr. Briggs and I tasted
before we served the rations out, and mutually shook our
heads.

"It is rather rank, Padre!" said my friend.

"It is so, most decidedly," I answered; "we must put
something into the pot to subdue the flavour. What shall we
put? I have it! there is a bottle of cayenne gargle I brought
with me in case of sore throats!"

That was the oddest breakfast I have had for many a day, oyster-catcher with gargle sauce! still it stopped our hunger, and we all wished for more; one bird between four being short allowance. The guides had breakfasted on stock-fish in the house.

It was late before we started, as the horses had strayed far in quest of grass. Whilst they were being caught by Gúthmundr and Magnús, I read over some chapters of the *Aigla*, the saga of a mighty warrior and chieftain whose house was this same Mósfell.

One of the most beautiful incidents of this saga I shall relate for the benefit of the reader, premising first of all, that the hero, Egill Skallagrímsson (pronounced Ey-il Skadlagreemsson), had fought with King Athelstan against the Scotch, and had been the slayer of the son of King Eric Bloodaxe, of Norway. He was now an old man of seventy-one; and had retired to rest and talk over his mighty deeds in this wild, grim island home, whose scenery was so fully in keeping with his own character.

Sonartorrek.*
(A.D. 975.)

On a day in the summer of 975, five house-churls belonging to Egill Skallagrímsson rowed to a merchant vessel, stationed in the Borgar fjord, at the mouth of the Hvítá or White River; with them was Böthvar, a son of Egill.

The boat started at high tide, which was in the evening, and remained alongside of the ship for a considerable time, whilst divers articles purchased by Egill were handed down the side, and deposited in her. During the time that the churls were thus employed, a fierce wind had risen, and now rolled the sea before it in tumultuous billows, which, meeting the out-current of the river, after turn of tide, formed eddies which engulphed the boat on her return, and every soul on board perished.

Next morning, the bodies were washed ashore in the

* *Aigla*, chap. 81.

4

fjord; that of Böthvar by Einar's-ness, the others along the
southern strand, and the boat was found on the beach under
the Smoking crags.   On the same day, Egill heard the news.
He mounted his horse, and rode in search of the corpses :
that of his son he found lying uninjured on the shingle, laced
round with sea-tangles.   He lifted it on his knee, brushed the
sodden hair from the young face, placed it in front of his
saddle, mounted himself, and rode with the body of his son
wrapped in his arms to Digranes, where stood the cairn of his
father Skallagrím.   Egill fetched a spade and dug into the
mound; he was occupied the whole of the afternoon at this
work, and in the evening he had reached the wooden chamber
wherein lay the ancient warrior, busked for the last battle at
the "Twilight of the Gods," with casque about his brows,
and sword between his hands.   Egill bore the corpse of his
child into the tomb, and laid it by that of the grandfather,
then filled up the pit he had dug, and restored the cairn to its
former condition.   After this he rode home, and, without
uttering a word, went into the chamber where he was wont to
sleep, bolted the door behind him, and lay down on the bed.
His face was so stern and grave as he entered the house, that
no one ventured to address him.   The old man had gone out
in the morning, dressed in a scarlet fustian tunic, tight fitting
about the body, and fastened with wrought silver buckles at
the sides : he had also worn closely fitting hose.   On his
return, the farm-servants noticed that the kirtle was torn down
the back, and the hose split, by the working of his muscles
when he dug into the tomb.

Hours passed, and Egill did not open the door; he took
neither meat nor drink, and so he lay both day and night.
Folk walked softly through the house, and the wife listened
anxiously on the threshold, but the old man neither spoke nor
moved.   So passed a second day, yet no one dared to interfere with the master in his grief.

On the third morning, as the day broke, A'sgerthr, the
good wife of Egill, ordered one of the freedmen to mount his
horse and ride, as swiftly as possible, west away, to Hjarthar-

holt, and tell her daughter Thorgerthr what had taken place, and ask her advice as to what course had better be pursued.

The messenger reached Hjarthar-holt by noon, and related all that had happened. Thereupon, Thorgerthr let a horse be saddled for her, ordered two servants to ride with her, and before sun-down, was at the house of her parents.

She dismounted at the door, and stepped quietly into the kitchen, where she found her mother. They embraced affectionately, and the daughter, as she kissed A'sgerthr, felt that her cheeks were wet with tears.

" My dear ! " said the housewife, " tell me whether you have eaten your supper, for, if not, I will order food to be brought you immediately."

" Mother mine ! " answered Thorgerthr, in a voice loud enough to be heard throughout the house ; " I have tasted nothing, neither do I intend touching food till I reach the halls of Freyja : * I can do nothing better than follow my father's example, and accompany him and my brother on the long last journey."

Then she stepped to the threshold and called,—" Father, father ! open the door ! I wish that you and I should travel the same road together."

All within was silent for a space, but presently she heard the old man's step coming to the door, the bolt was drawn back, and Egill, pale and haggard, stood before her. She passed him without saying a word ; then he again bolted the door, and returned with a moan to his bed, but kept his eye fixed inquiringly on his daughter's countenance.

She lay down in another bed which was in the room, saying,—" May we soon sup with the gods, father ! "

Egill answered,—" You act rightly, daughter, in choosing to follow your aged father. Great love do you show in thus joining your lot with mine. Who could think that I should care to live, bowed down beneath the burden of my great and bitter sorrow ? "

---

* Odin received heroes after their death ; Freyja took matrons ; and Gefjon virgins.

Then both for a while were silent.

There was a small circular opening in the wall opposite the old man's couch, and, through it, the evening sun sent an orange spot upon the floor.

Not a sound in the room but the breathing of father and daughter, yet from without sounds of life were borne in upon the summer air. The river, at no great distance, rushed monotonously, yet with a pleasant murmur, over its pebbly floor; far off, up the mountain side, a flock of sheep were being driven to fold, and the barking of the dogs was distinctly audible in the little chamber: presently, a flock of swans passed, with their strange, musical scream; and, now and then, the whinny of a horse reached the ears of those who had laid themselves down to die.

Suddenly Egill spoke: "Daughter! I hear you munching something."

"So I am father. It is söl (Alga saccharina)," she replied. "I think that it will do me harm; without something of the kind, I might live too long."

"Does it really shorten life?" asked Egill.

"Oh, that it does. Will you have some?"

"I see no reason against it," answered he.

Then she rose from the bed, stepped over to him, and gave him some of the sea-weed.

As the plant is saturated with brine, both she and her father soon became exceedingly thirsty. They lay still, however, for some time, without either speaking. The sweet air of summer blew in at the little window fresh as from the gates of Paradise. Without, the churls were making hay, and occasionally a few grass blades were borne into the room by the draught. One of the thralls whetted his sickle; a girl at the farther extremity of the tún began a song. Within, the golden spot reached Egill's bed-board and began to slide up it. A mouse stole from behind a chest, and stood on the floor, looking round with bright beady eyes, then darted under one of the beds.

The thirst of the daughter became, at last, so intolerable

that she rose, saying that she must taste one drop of water. Her father raised no objection, so she stepped to the door, opened it and called for water. Her mother came up, and as the girl bent to kiss her, she whispered a word into A'sgerthr's ear. Directly, a large silver-mounted drinking-horn was brought. Thorgerthr closed the door again, and bolted it, took a slender draught and offered the horn to her father.

"Certainly," said he; "that weed has parched my throat with thirst." So he lifted the horn with both hands, and took a long pull.

"Father," said Thorgerthr, "we have both been deceived; we have been drinking milk, not water." As she spoke, the old man clenched his teeth on the horn, and tore a great sherd from it, then flung the vessel wrathfully to the ground. "What is to be done now, father?" asked the daughter. "This our scheme has broken down at a very early stage, and we can no longer think of continuing it. I have a better plan to propose. Let us live sufficiently long for you to compose a beautiful elegy on your son Böthvar, and for me to carve it in runes on oaken staves; after which we can die, if the fancy takes us. I do not think my brother Thorstein quite the man to make much of a poem on our poor Böthvar, and it would be a disgrace to the family that the gallant boy should remain uncommemorated in song. As soon as your elegy is complete, we will hold. a funeral banquet, at which you shall recite it. Now, what think you of my plan, dear father?"

Egill replied that the spirit of song was gone from him, but that he would try his best. Then he sat up in his bed and chanted the following lay, composing, at first, with difficulty, till the fire of poetry kindled in his soul as brightly as it had burned in the days of youth: and the spot of flame from the setting sun, which had been running up the wall, rested on, and glorified, the old man's inspired countenance. His voice, faltering at first, waxed strong and clear, so that it filled the house. This was his song:—

"I tune my tongue but feebly
  To stir the air with song,
From heavy heart but hardly
  I drag the load of wrong.

From frozen brain but thinly
  The soft sweet metres thaw,
From mines of grief but dully
  The golden dole I draw.

My race to death is drawing,
  As drop the forest leaves;
As in the southland garners
  Are gathered golden sheaves.

Sad is the heart that singeth;
  My sorrows rise and swell;
The lips but feebly mutter
  The bitter tale they tell.

A gap in heart's affections—
  For where's my bonny boy?
The cruel sea hath torn in,
  And swept away all joy.

Ran bitterly has tried me!
  For friend on friend I grieve;
And now cold ocean shivers
  The bright chain I did weave.

The bright chain of my weaving!
  Oh, vengeance! would 'twere mine!
But how can these old sinews
  Resist the ruthless brine?

Of much, too much, despoilèd,
  An old man, sitting lone,
With trembling fingers counting
  The gaps in dear old home.

Bereaved of his last treasure,
  The target of his race,
Borne by the valkyrie
  Up to the Blissful Place.

Oh! would my boy had oldened,
  To wield the bright blue blade;
And Odin's hand extended
  On his fair head been laid!

To father he—e'er faithful—
  Held when all else were cold;
The son's warm pulses quivered
  Through these thews waxing old.

Now, through the long night watches
  I restlessly am tossed:
I cannot sleep for thinking
  Of all that I have lost.

Odin! why hast thou riven
  The green bough from its stem,
And ta'en it up to root it
  In homes of gods and men?

Spear-shaker! our old friendship
  I rend for aye away;
I trust thee now no longer,
  Fell leader of the fray!

Upon the grassy headland,
  Where father, children, sleep,
Above the constant throbbing
  Of the ne'er quiet deep,

Stands Death, calmly waiting:
  What! can I dread to die?
Nay, gladly! oh! how gladly,
  Towards her arms I fly!"

Note.—I know that I shall get into dreadful trouble with Icelandic literati for this version of the famous Sonartorrek, or Son's loss. A literal translation would be quite unintelligible to the majority of readers, so I have culled the sense of the poem, and put it together in a popular form. For the sake of those who would wish to know something more about it, I translate the first seven verses. The original poem is in twenty-four.

"It is much burdensome to me to move my tongue, or stir the weight of the song's balance ;[1] now bootless is Odin's theft,[2] now is it not easily dragged from the lurking place of thought. The hidden store of Odin's kin,[3] which was brought from the home of the Jotuns,[4] is hard-drawn from the thought-house,[5]

---

[1] To stir the weight of the song's balance = to compose poetry.

[2] Odin's theft = Poetry.

[3] The hidden store of Odin's kin = Poetry also.

[4] The home of the Jotuns.—The liquor of poetry was in the possession of the brothers Suttung and Baugi in Jotunheim, till Odin robbed them of it.

[5] The thought-house = the head.

(my heavy sorrow is the cause!) When blameless Bragi arose in the dwarf's boat,[6] the wounds of the Jotun's neck[7] resounded down by the kinsman's sea-gates.[8] For my race draws to its close, as the branches of the forest are beaten with fierce blows. He is no happy man who bears the bones of his child's corpse down from his bed. I must first tell of the loss of father and mother. That forest of poetry, leaf-bearing song, bear I out of the hall of words.[9] Sad is the rent, where the wave washed into my family! I know the gap (caused by the death) of my son, when the sea despoiled me,—it stands unfilled and open. Ran[10] has taken cruel advantage of me; I am bereaved of dear friends. Marr[11] cut the bonds of my race, the cord spun by myself."

Now this is sheer nonsense to any one who does not understand the principle upon which Icelandic poetry is constructed: which is to use periphrases on every possible occasion, instead of the plain straightforward word. Without a knowledge of the signification of these far-fetched similes, the poetry is quite unintelligible. This accounts for the fact, that an Icelander of the present day cannot in the least follow the meaning of an old poem. I add explanations of the difficult expressions in the seven verses, literally translated.

Now, it fell out that, as Egill composed, his grief abated, and, when the Lament was complete, he rose from his bed, and, entering the hall, seated himself on the high stool of honour. Then all the house-folk gathered around him, and his wife and daughter sat at his feet. When a silence was made, he lifted his voice and sang the poem; and this Lament he named the Sonartorrek. Afterwards Egill waked his son in the ancient manner with much feasting, and Thorgerthr returned home laden with rich presents her father had bestowed upon her.

When Egill left King Athelstan, he received from him two chests of silver, and these the old man secreted near the house, with the aid of two thralls, whom he slew, that the secret of the treasure might never be revealed. Some

---

[6] When blameless Bragi arose in the dwarf's boat = when immaculate song awoke among the gods.

[7] The wounds of the Jotun's neck = the blood of Ymir = the sea.

[8] The kinsman's sea-gates = the cairn of Skallagrím.

[9] The hall of words = the mouth.

[10] Ran was a sea-goddess, and is put for the sea itself.

[11] Marr was a sea deity, and is put as well for the sea.

suppose that it was buried in the gill aforementioned, because Anglo-Saxon coins have been found in the rubble brought down by the stream; others, that it is concealed near the warm springs, and that the jack-o'-lanthorns which dance there mark the spot where silver is buried; a third conjecture is, that the treasure was deposited in the bogs east of the house. I think the gill the most probable spot; for the old man was aged ninety and perfectly blind at the time, so that he could not have gone far. Now the bogs by the warm springs are too distant, and those east of the house are completely commanded by the farm, and it is unlikely that he should have chosen a place which was overlooked.

I wished much for a spade and mattock, that I might search for antiquities in the tún, or gill, but it is impossible to persuade an Icelander to allow his meadow-land to be upturned. It provides him with but just enough fodder for his cattle, so that every grass-blade is precious in his eyes. Each year dung is thrown over the tún, and any traces of the ancient buildings are gradually buried. It would be most interesting to make excavations on the site of some farm famous in history; but then it must be purchased for the purpose.

But to return to my journal.

We paid to the priest of Mósfell for the keep of the horses, the breakfast of the guides, and the milk and fuel, the sum of a dollar and a half (3s. 4½d.), which was what Grímr decided on as a fair remuneration.

The morning of June 20th was lovely, but cold; a sharp northerly wind was blowing, but we were sheltered from it during a considerable portion of our ride, by the stately serrated ridge to our left. In the best of spirits, though rather empty within, we started for Thingvellir; the horses, partaking in our glee, trotted cheerfully up the moor slopes, and scrambled rapidly over the slants of rubble borne down by spring torrents.

The track led along the brink of a gorge of great wildness and beauty, down which the river thundered in a succession of

leaps from shelf to shelf of basalt, and then, bursting through
a portal of crag, slipped calmly round a tongue of grassland,
on which smoked a little byre. High above stood the purple
battlements of Skúlafell, composed of fluted rocks ranged like
the reeds in a Pan-pipe. A more perfect subject for a picture
can hardly be conceived, yet I made no sketch, as I was far
behind the rest of the party, having stopped to gather specimens of saxifrage, and feared missing my way; but perhaps
the main reason of my not drawing the noble scene before me
was the expectation of returning by the same route, and having
a second opportunity.

Galloping after the party, I overtook it near a pretty
upland farm, about which the dogs were making a deafening
uproar, being alarmed at our appearance, and puzzled with
the looks of Bob the Newfoundlander, whom they seemed to
mistake for a young bear.

They were particularly ill-disposed towards Grímr and
myself for riding through the tún to the farm-door, and would
hardly give over their noisy demonstrations of anger, when
their master came out and ordered them off. The Icelandic
dog (Canis familiaris Islandicus) has been already briefly
described in the Introduction : its head is just like that of a
fox; it is small, has sharp eyes, short legs, a profusion of hair,
a ruff round the neck, a tail curled over the back, and it is
generally of a white, dappled, or tawny colour. In Iceland
the different kinds of dogs are distinguished by different
names. The sheep-dog is fjárhundr ; the hound, veithihundr ;
the dog which can follow scent, rakkr ; the poodle, lubbr ; the
house-dog, bœarhundr ; and the lap-dog, mjóhundr. The farm,
at the door of which Grímr and I reined up, is celebrated for
its breed of dogs. The price of a puppy is about a dollar, but
the traveller had better not purchase one, as it will not live in
England. A skipper, who visited Iceland yearly, informed us
that he had brought a dog with him to Leith on his return
from every cruise, but that he had never been able to rear any,
with the exception of a pup born on the voyage.

The byre we now visited was a good specimen of Icelandic

domestic architecture. From three sides it presented the appearance of a confused cluster of turf mounds. Among these two are conspicuous, one for having a chimney formed of a barrel with both ends knocked out, the other for being longer than all the rest, and for having two or three glass panes inserted at intervals in the turf. The former roof is that of the kitchen, the latter of the *bathstófa*, or sleeping apartment. On the fourth side of the house is the front, consisting of a series of wooden gables between thick turf walls. The woodcuts will explain the construction of an Icelandic house.

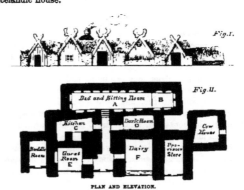

PLAN AND ELEVATION.

Looking at the front of the house, one observes five or more gables made of wood, painted red or black, wedged between turf walls from four to ten feet thick. The apex of the gable is seldom above twelve feet from the ground, very generally only eight, and is adorned with wooden horns, or weathercocks. Under the central gable is the door, around which are crooks upon which the stockings of the family are hung to dry on windy and sunny days. Passing through the door, one enters a long dark passage, too low for a person to

stand upright in it, leading to a ladder which gives access to
the bathstófa, or common eating, working, and sleeping apartment, marked A on the plan. This room is lighted by two or
more glass panes, three inches square, inserted in the roof and
sealed in so as never to be opened for the admission of pure
air. The walls are lined with beds, and the end is divided off
by a wooden mock-partition (never closed by a door) so as to
form a compartment: here the father and mother of the
family sleep, together with such visitors as cannot be accommodated in the guest chamber. In the bathstófa sleep all
the people connected with the farm, two or even four in
a bed, with the head of one at the feet of the other. The
beds are lockers in the wall, lined with wood, and with
wooden partitions between them. They are arranged along
the room much like the berths in a cabin, or the cubilia
in a catacomb. Each is supplied with mattress, feather bed
or quilt, and home-woven counterpane. The Icelanders not
only sleep in this room, but eat in it, making sofas of the
beds, and tables of their knees. In it is spent the long
dark winter, with no fire, and each inmate kept warm by
animal heat alone. The stifling foulness of the atmosphere
can hardly be conceived, and, indeed, is quite unendurable
to English lungs.

Gaimard, in his great work, gave two highly imaginative,
but utterly inaccurate, representations of Icelandic interiors,
with the natives seated around a blazing hearth, reading sagas
or playing the langspiel, a national instrument, and these
illustrations have been, most unfortunately, reproduced in
some English tourists' volumes. The fact is, Icelanders
never have the opportunity of sitting round a fire, and the
only place in the farm where there is one, is the kitchen,
a small cell not capable of accommodating more than three or
four persons at once, and unprovided with seats. Besides, the
fire is made up of sheep's dung, which smoulders without
giving out much heat till it is quickened up temporarily for
roasting coffee with a little willow-root or brush-wood.

The *eld-húsi*, or kitchen (c), is lighted through the barrel which forms the chimney, and is totally unprovided with windows. As the chimney is in the roof ridge, and is not always over the fireplace, the acrid, offensive smoke has to make its way out as best it can, or penetrate every corner of the house, impregnating all articles of clothing with its disgusting odour.

Every house is provided with a dark closet (D), in which are deposited the looms during the summer, together with various articles of lumber. The guest room (E) is the cleanest place in the house, and generally has boarded walls, but not unfrequently they are of turf like those of the other rooms, and I have often cleaned my knife and fork after a meal by driving them into the walls of my dining-room. The floor is sometimes boarded, but very commonly is simply the bare earth into which have been trampled fish skins and bones for many generations. This chamber has in it a bed and a table, also the chests containing the wardrobe of the family, and these serve as seats, for chairs there are none. Over the bed is the library of the house, deposited on a shelf let into the wall. Icelandic farms have only a ground floor, and a cramped attic above the dairy and guest room, in which one cannot stand upright, except precisely under the roof-tree; these attics are used either as store-rooms or as bed-chambers.

To one unpleasant topic I fear that I must allude, so that no traveller may be left unprepared for what inevitably must befall him, if he makes an Icelandic tour, and lodges in the farmhouses.

Man forms but an insignificant item among the countless tenants of a byre. If these same tenants were simply those whom kindly nature has endowed with such remarkable springing capabilities, or even those which infest a seaside lodging and a Welsh inn,—well; but other and more loathsome forms of life teem in the unwholesome recesses of the bathstófa, and it is quite hopeless for the traveller to think of avoiding them if once he enters a farm. Unlike the Icelanders of the genus *homo*, these horrible parasites are endowed with a predilection for novelty, and in a moment scent out the blood of an Englishman, and come in eager hordes, from which he finds no escape till he reaches a boiling spring in which he can plunge his clothes and annihilate his tormentors wholesale. Curiously enough, the natives have a superstitious dread of killing one of these constant companions, and they will remove one which is particularly obnoxious, and lay it gently on the table, without for a moment thinking of depriving it of life.

The ancient Finns made Anteretar, " the wash-tub," their god of health, and Antermen, " the steam of the bath," the preserveress of vigour; alas! the Icelanders have no such gods, or, at all events, never cultivate their worship. I believe they looked upon me as next door to a madman, for plunging into their ice-cold lakes and rivers; but I was compelled so to do, as the amount of water granted me for my ablutions in the morning was scanty. A bowl of milk was placed at my pillow overnight, and when the bowl was emptied in the morning, they brought it in half filled with water for my washing, together with a pocket-handkerchief for a towel, and this was to serve for my guide and myself!

I have already spoken of the tún, or meadow, around the farms; in it stands a turf shed for the hay. Adjoining the house is the cattle-shed, and, enclosed within high turf walls, is the garden, which is manured with the ash from the kitchen fire. In it are grown potatoes, carrots and angelica. It is remarkable, as mentioned in the Introduction, that, notwithstanding the damp of Iceland, the potato disease is quite unknown. May not this be owing to the mode of dressing?

On the trampled ground in front of a house are emptied all the slops, and on it is cast the refuse which cannot be trodden into the floor within doors. There is neither order nor neatness in an Icelandic house : the porch is generally full of clothes, wash-tubs, rakes, turf-cutters, saddles, foxskins, and sacks of lichen, whilst the guest room is littered with brandy glasses, dresses, plates, and whips.

After leaving the byre, we wound round the north side of a small lake, and became so involved in bogs, that it was with great difficulty we got the pack-horses safely through; and, on reaching the sandy banks of a little river which fed the farm, they lay down one after another and rolled, snapping ropes and girths, smashing bottles, and severely trying the Icelandic boxes. After the damage was repaired to the best of our abilities, we ascended the heithi, or moor. High land which can be traversed by horses is called a *heithi*. It is either without vegetation or covered with moss, lichen, Dryas octopetala, and Silene acaulis. These heithies, being exposed to the action of snow-water, are much torn and mangled, the rock being, in many places, quite polished by the streams from the thawed snows, as they slide over them. Mr. Chambers, in passing this same tract of moor from a different direction, saw similar polishings, and at once put them down to glacial action, and the furrows caused by the little rills to the striæ of glacial grooving. I believe him to be mistaken, for the following reasons: the rock is not smoothed except where the streams flow over it, and a slight node of rock three inches high is quite sufficient to divert the striæ and alter the direction of the polished surface. A considerable removal of earth had taken place this spring, and I observed no marks of glacial smoothing on the rock upon which the soil had rested; it was ribbed and curdled like ordinary lava.

These heithies are very awkward things to cross, as the ground is thoroughly broken up. Level turf is excessively rare in Iceland, and is only to be found on low river-flats which cannot be furrowed by the snow; elsewhere, the ground is covered with broken lumps of turf and refuse moss.

The moor abounded in whimbrel and golden plover. My companions shot as many as we could carry, besides ptarmigan in their summer plumage, and snipe. Terns with their jaunty, black caps, snowy breasts, coral legs and beak, skimmed and wavered about us, uttering their strange grating cry. We killed four, but were obliged to throw them away, to make room for birds more suited for the pot.*

On the right of our course lay the lake of Thingvalla in the grey of an overcast night, and beyond it we could distinguish the steam from the Hengill sulphur springs. Clouds had been gathering rapidly over the sky, and hung, big with rain, over the mountain tops. We met an Icelander on horseback, who at once fell into conversation with Grímr; and presently let out that he had got a fine pony for sale.

"Where is he?" I asked.

"At half an hour's ride from Thingvöllum."

"Then you had better bring him over to-morrow morning; and if he is as good as you say, I will purchase him."

"Do you doubt my word?" asked the man. "I shall not take the trouble of bringing the pony over on the chance of your rejecting him after all."

"But you surely do not intend to suggest that I should buy a horse without ever casting an eye on him!" I exclaimed, much amused at the fellow's coolness.

"Pay me the money, and I will send you the horse," said the native. "If you do not trust my word, good-night!" and with a wave of the cap, he galloped away. This trait of indolence is thoroughly characteristic of the Icelanders. At Reykjavik I offered the horse-dealer a hundred dollars in paper, and he refused to take them, though they could be changed for cash four doors off. I mentioned this fact to him, and he stared at me with astonishment, before he replied: "You do not suppose I will take the trouble of going to get them changed? You go, and I will stand here till you return."

We came abruptly out on the brink of the Almannagjá,

---

* I set these down at the time, without hesitation, as the common tern (Sterna hirundo, Lin.) But I see that M. Preyer asserts that the only species found in Iceland is the Arctic tern. Mr. Fowler assures me that he never saw the common tern during the three months that he was in the island. I deeply regret that I did not preserve the skins. The description given above is that entered at the time in my journal, with the dead birds before me.

that famous rift, to see which, Lord Dufferin says, it is worth going the whole world over.

The scene from the edge was most striking. Below us, in the dim light which pierced the cloud canopy, we could see a great plain, bounded by mountains ; the air was perfectly still below, and a column of white smoke rose far off under the mountain roots, where charcoal-burners had kindled their fires. The foundations of the eternal hills were lost in the deepest blue black shadow. The rocky edge on which we stood was 100 feet in abrupt precipice, and from the other side of the rift the ground dipped rapidly to the plain. To our right was the lake, cold and still, one grey island rising out of it. Suddenly the edge of the cloud veil was lifted, and a pale white gush of light ran along the plain, showing us the little church of Thingvalla, the Hill of Laws, the seamed face of the land, and the dusky foliage of the birchwood beyond the region of the rifts. Then we descended the rocky stair in the Almannagjá, cantered along its turfy bottom, broke through the wall on the other side, and after crossing the river Oxerá, rode up to the parsonage and dismounted at the churchyard gate.

The rain began to patter down, and we longed to get under shelter, and fill our hungry stomachs, for we had eaten nothing all day except the oyster-catcher for breakfast. But the priest declined to receive us, as his house was full ; and he apologized for not being able to give us any food, as his supplies were exhausted, and he had not yet been to Reykjavik for more. However, he consented to our birds being cooked in his kitchen, and he pointed us out a spot in the churchyard by an old upright stone, on which is the measure mark of a true ell, where we were to pitch our tents.

Gúthmundr and Magnús started with the horses to the place where they were to feed ; the Yankee took Martin off with him, to give him a lesson in carving, and Grímr, with his pipe in his mouth, conversed with a native who was lounging near.

Mr. Briggs and I spread out the tent.

5

"Well," exclaimed the former, suddenly looking up; "Grímr, are you not going to help us?"

My guide understood and spoke English, so he replied—

"I cannot, I am busy here."

"Yes! busy doing nothing! Come and help to get the tent up."

"You have no right to order me. I have done my work as guide."

"You have done nothing all day but smoke and grumble," retorted Mr. Briggs, getting very wroth. Grímr reddened, and walked off with his friend. Mr. Briggs followed him—"I insist on your helping us to put up the tents."

"It is of no use more than two doing it; and there are you and Mr. Baring-Gould to do it."

"You are an idle, good-for-nothing fellow," burst forth my indignant friend.

Grímr turned to his comrade, and called him to witness the fact that he had been insulted, as he intended demanding legal satisfaction on his return to Reykjavík. I now ran up and quieted the angry men as well as I could, and sent Grímr to fetch some of the boxes; this he consented to do, but, in fact, he brought nothing but a rug and a pack-saddle, and then sauntered off to rejoin his friend.

I was put in a somewhat difficult position; if I dismissed Grímr, I should find no other guide, and a great amount of time would be lost. I determined accordingly on humouring him and letting him have his own way, a plan which succeeded, for I believe that a more upright and trustworthy fellow is not to be found anywhere; but he was very obstinate, and had the disinclination for hard work which characterizes his race.

Mr. Briggs and I erected the two tents ourselves, brought the boxes under cover, made the bed, and slung my hammock, as I had no fancy for another night on the hard ground.

By this time the birds were ready, and were brought to us in a saucepan, and placed on the church steps, where we eat them with biscuits, and then, after a stiff glass of hot toddy, turned into our beds for the night.

Pl. 2.

S. Baring Gould, delt.

Edmund Evans, sc.

ALMANNA-GJÁ.

CHAPTER IV.

THINGVELLIR.

The Plains of the Council—The Almanna-gjá—The Hill of Laws—History of Althing — Lava Broth — Icelandic Food — Butter — Curd —Lichen— Sunday—The Burning of the Hostel : a Saga.

THINGVELLIR,* or the Plains of the Council, have been already described fifteen times, and, if it were not that my journal would be incomplete without some account of this remarkable spot, I should decline giving one. As it is, I shall only put the leading characteristics of the scene before the reader, as briefly and concisely as possible.

The plain forms a rough parallelogram, eight miles long by six broad ; two of the opposite sides being determined by the parallel Allmen and Raven rifts. On the south-west, the plain dips into the Thingvalla lake, and on the north-east, terminates at the base of some low hills, between which open glens lead to the glacier volcano Skjaldbreith.

The Almanna-gjá (pronounce gjá like gee-ow), or Allmen's rift, is a split in the lava extending nearly four miles, to the roots of Armanns fell : the river Öxerá shoots over the north-west verge, and flows for a quarter of a mile through the chasm, then breaks through a gap on the other side, rolls down to the plain, and pours into the lake close to Thingvalla church. The Hrafnagjá (pronounce the f like p), or Raven

---

* Thingvellir is the nominative, the genitive Thingvalla, the dative Thingvöllum. The word is pronounced as though the *ll* were written *dl*.

rift, bounds the plain on the south-east. This is less remarkable than the other chasm, as the height of the walls is less considerable, though the length is somewhat greater. These chasms have been formed by the great plain between them having suddenly sunk, leaving sharp edges in lines at the sides, between which it has been depressed. The plain itself is full of fissures of great depth, half-full of clear water. The sketch on page 69 was taken in one of these after a somewhat difficult descent to a turfy ledge above the water. The bed of the lake is full of similar rents. The greatest height of the north-west side of the Allmen's rift is 130 feet; this is just above the lake; as the rent nears the mountains, the walls are less lofty. The edge is splintered into chimneys, windows, and mushrooms of rock. At the base of the wall runs a belt of sward, forty feet wide, to the point where the river occupies the bed of the chasm. The wall on the south-east side is less elevated than the other, and seldom reaches the height of fifty feet; it is not perpendicular like the other, but stands backward towards the plain.

One of the quaint pillars on the north-west side is called the hanging rock, as culprits were wont to be tied to it and flung over into the horrible abyss.

Below the second fall of the Oxerá, where it tears through the south-east wall, is a pool of blue foaming water, in which women convicted of child-murder were drowned in ancient days. The little island in the river on which duels used to be fought, till they were abolished in 1006, has disappeared, and its site is marked by a patch of mud. The spot is shown where witches were burned; the last who suffered for this crime was Halldórr Finnbogason, in 1685. The last burned in England were Temperance Lloyd, Mary Trembles and Susannah Edwards, at Exeter, in 1682.*

The scenery around Thingvellir is beautiful, but tinged with melancholy, from the deficiency of life and vegetation. Standing on the spot from which I described the landscape

---

* Probably the last execution of this kind in Greenland was that of Kolgrim in 1407.

last night, I will sketch it in its day colours, now that the mountain tops are unveiled. If the reader will take the map and my panorama (Plate I. fig. 2), whilst he reads my description, and compare them together, he will obtain a pretty clear notion of the scene.

VIEW IN NIKOLASA GJA.

To the left is the broad-based Armanns fell, now snow-topped, extending over a considerable area, and backed by the white peaks of Súlur. Turning the eye towards the middle of the plain, we see the opening of an extensive valley, or

rather a portion of the plain, hemmed in between the skirts
of Armanns fell and a fringe of cinder mountains ending in
Hrafna-björg, the Raven's Fortress. A line at the foot of
this range, is the ragged edge of the Raven's rift.

Above the extremity of the valley rises the snowy dome of
Skjaldbreith, soft and beautiful, yet treacherous, for that
white heap poured forth the lava which fills the plain and
forms the lake bed. Skjaldbreith is about fourteen miles
distant from the point at which I stand.

The eye next reaches the lake, and gets an extended view
over it to Burfell and Hengill. The cone in the lake is
Sand-ey, a dry, dusty heap of erupted volcanic matter.

On a height at no great distance from the lake, near the
spot where the Oxerá leaves the Gjá, is the Hill of Laws,
situated between two rifts, the Nikolása and Flossa chasms, so
called because into one plunged a Syselman deeply involved
in lawsuits and not seeing his way out of them, and over the
other leaped a death-doomed criminal. In a line with the
Hill of Laws, just above the mouth of the river, is Thingvalla
church, built of wood and tarred all over.

Thingvellir is interesting, not only on account of its
scenery, but also because of its intimate connection with the
history of Iceland. About sixty years after the colonization
of the island, a code of laws was drawn up for its government,
and for the settlement of disputes, by Ulfljót, a man of royal
descent. His brother, Grímr Goatshoe, after investigating the
whole island, decided upon Thingvellir as the most fitting site
for the grand annual Parliament, or Thing, to meet. It was
eligible on three grounds; first it was the point of junction of
the tracks which crossed the deserts of the interior; secondly,
it was well provided with wood, forage for the horses, and
water; and lastly, it could be got for nothing, as the land was
confiscated, on account of its possessor having committed murder. For his trouble in finding out the best spot for Althing,
the great meeting, Grímr Goatshoe was paid a penny by every
householder in the island, but he handed over the sum as a contribution to the temples of the gods. The time of assembly was

fixed, at first, for the middle of June, but in 999 it was postponed a week later. After Iceland had lost its independence, Althing met still later, the opening being advanced to the 29th June, the festival of SS. Peter and Paul, and subsequently to the 3rd or 5th of July. Althing continued to meet on the old spot till 1800, when it was abolished. On its restoration in 1845, it was removed to Reykjavik. Althing was presided over by a lawgiver, "whose bounden duty it was to recite publicly the whole law within the space to which the tenure of his office was limited. To him, too, all who were in need of a legal opinion, or of information as to what was and was not law, had a right to turn during the meeting of Althing. To him a sort of presidency or precedence at the Althing was conceded; but with a care which marks how jealously the young Republic guarded itself against bestowing too great power on its chief officer, he was expressly excluded from all share in the executive; and his tenure of office was restricted to three years, though he might be re-elected at the end of that period." *

Here ends all I am going to say about natural scenery and historical incident connected with Thingvöllum, being the sixteenth such description existing in print.

We spent Saturday and Sunday at Thingvöllum (I have put this name in the right case, though travellers generally use the genitive, but I shall not treat other names thus). We sketched, fished, and shot—learned a little of the Icelandic culinary art, too, from the smoke-dried servant of Pastor Simon Beck, and taught her a little about English cookery in return.

As the good priest had nothing to give us, and the proceeds of our fishery amounted to one trout, we stood a fair chance of being starved, and the servant-maid expressed her concern for us in vehement terms.

" Bless your heart, old lady !" said Mr. Briggs ; " there is everything needful for making a capital soup scattered broad-

* *Story of Burnt Njal,* vol. i. lvii.

cast over the country." Then turning to me, "It is an old
joke, Padre! indeed, I fear quite a Joe Miller, yet, I'll be
bound, it is new to the Icelanders." Then he called to the
servant to bring him two lava blocks of such a size as would
go into the saucepan, as he intended to make broth out of
them. The woman stared, and pulled out a couple of the
stones which formed the hearth.

"They will do capitally, put them into the pot!"

The woman placed them carefully where he desired.

"Now then," said he, "let them simmer gently; take
care that they are not hard-boiled, nor burned at all; stir,
stir away, never be afraid of stirring, it extracts the
flavour!"

With the utmost gravity, the servant complied, and kept
the spoon revolving in the pot, whilst Mr. Briggs held his
nose over the steam.

"They are doing nicely; have you any salt?"

"Já já!" answered she, and she handed him over a bag-
ful. He poured some in, and let her continue the stirring.

"Any flour in the house?" asked my portly friend.

"Já já!" and she handed him a mug full of rye-meal.

Presently he brought in a couple of tins of Fortnum
and Mason's preserved vegetables, and emptied them in with
the stones.

"It smells good!" said the poor woman with great
solemnity; "and this is hraun-súppi (lava-broth)! Wonderful,
wonderful!"

"We'll give it a little extra relish," observed Mr. Briggs,
throwing in some slices of concentrated soup. "Now, old
woman! what do you say to my cooking?"

The servant put her fingers into the pot, then sucked
them, and put them in again.

"Mikit gott (very good)!" she muttered, in a sad state of
bewilderment; "and all this comes from lava-stones, too!
Wonderful, wonderful!"

"Now for a bowl!" called he, and proceeded to pour
out the fluid contents of the saucepan. The smoke-dried

servant bore the bowl to the guest room, and we did ample justice to the excellent stone-broth.

A few words on the food of Icelanders,—

The meals in an Icelandic house are three,—breakfast, dinner, and supper; but the young lady of the house brings the visitor, before he rises in the morning, a cup of coffee, which is sweetened with sugar-candy, the lump of candy being placed in the mouth, and the beverage sucked past it. How far this indulgence is extended to the household I hardly know.

The main staple of food is stock-fish. Cod is never eaten fresh, it is prepared in a peculiar manner; the spinal bone as far down as to the third vertebra from the vent is removed, an operation which causes the fish to dry speedily. It is also cleansed from all the blood it may contain, and thus acquires a white colour; some experienced fishermen are so particular on this point that they gut the fish the instant they are caught. The cod is then exposed to the wind and dried to the consistency of leather, after which it is eaten with butter.

"The bladder of the fish contains yellow, viscous matter, which forms an agreeable, wholesome, and nourishing dish, and is used instead of isinglass."*

Salmon is cut into flakes and hung in the kitchen to be smoked; it thus acquires an unpleasant smell and taste from the fumes of the sheep's-dung fire. Lake fish are, however, boiled shortly after they are caught, and are very delicious eating. Shark's flesh requires to be buried for six months before it loses its rankness and becomes fit for table.

The only meat eaten in Iceland is mutton, and this generally boiled. The most common way in which it is prepared, is to boil it, then subject it to pressure sufficient to expel all moisture, cut it into junks, lay it by, without salt, and keep it often for twelve months. It is brought to a visitor as a delicacy, but it is with difficulty that he can swallow it, as it is covered with hair and dirt. When we bought a sheep

* Olafsen and Povelsen. I did not taste it myself.

and had it boiled for us, the good people invariably squeezed all the juice from it, and gave us the meat dry and tasteless.

Mutton is, however, sometimes stewed in milk, and is thus very palatable. I only tasted roast mutton twice during my journey, and, on both occasions, at large farms, where the kitchens were provided with more than ordinary conveniences.

Butter is used when either fresh or sour, though the natives prefer it in the latter state.

Whatever care may be taken in salting butter, it is not possible to preserve it good beyond a year, whereas sour butter can be kept for ten or twelve without losing either its goodness or its first acidity. In the old Catholic times, there were large magazines attached to each bishopric, serving as storehouses for sour butter; and in years of scarcity, it was distributed among the poor; but these magazines fell into disuse after the Reformation. This kind of butter is prepared by freeing it from all its milk by repeated churning and washing. When it has been deprived of every trace of milk, it is laid aside; it first becomes mouldy and unsavoury, then loses its yellow colour and becomes white: in six months the butter is sour, and fit for consumption, whereas ordinary butter would become completely rancid. If the sour butter be too old, it loses its acidity and weight, turns rancid, and dries up. If melted in this state, it is found to have lost a very large proportion of its oil.

Besides butter, *skyr*, or curd, is made from milk. The milk is placed in a warm spot near the fire, but not allowed to boil. After it has become lukewarm, rennet is put with it to curdle the milk. It is still left on the hearth till the whey has completely separated from the curd, after which it is strained off and set aside. The curd is much more solid than Devonshire junket, and has a sour taste: Icelandic lichen is sometimes chopped up and put with it, but without improving the flavour. It is eaten out of large bowls, with milk, and is most nutritious. The whey, when left to stand, deposits crystals very much like brown sugar. These are the Icelandic cheese, and are spread on potbröd or kaager; the taste is harsh and acid.

The Icelandic esculent lichen in appearance resembles that which grows on old tree stumps or apple branches, it is black and grey and has a most uninviting look before it is prepared. The preparation consists in soaking it for twenty-four hours and then boiling it in milk; it is very glutinous, and has a pleasant grassy taste.

The drink of the Icelanders is corn-brandy, butter-milk, and *bland,* which is milk and water. Whey is used as a beverage, and also for pickling, but if the whey has not properly fermented, the things immersed in it for preservation will spoil, though when it is good, they acquire a pleasant flavour, and will keep for upwards of a year.

On Saturday evening, the Yankee and Martin started in a leaky boat belonging to the pastor of Thingvalla, hoping to obtain some wild duck, that we might have a good dinner on Sunday and not be driven again to stone-broth, especially as we wanted our preserved meats for the desert interior of the island, where houses are not to be found, and birds are scarce.

In a couple of hours the boat returned half-full of water, with the Yankee pulling, and Martin baling lustily. They had been so far successful, that they had killed three couples of wild duck. Mr. Briggs added to the stock, by shooting some teal, so that our anxiety for the morrow was relieved.

There was no service in the church on Sunday, as a sufficient congregation did not muster. We spent the day in rambling over the plain, descending into fissures, picking flowers, and bathing. After dinner, as the sun shone out brightly and warmly, we lay down on a sheltered slope, away from the piercing wind, whilst I told the story of Grettir's outlawry.

The Burning of the Hostel.[*]

(A.D. 1016.)

THERE was a man named Thorir, who lived at Garth in
Athaldal; he was a mighty Icelandic chief, with numerous
retainers and extended influence. He had two sons, fine promising fellows both of them, and, at the time of my story,
pretty nearly full-grown men. Thorir had spent the summer
in Norway when King Olaf returned from England, and had
got into favour with the king, and also with Bishop Sigurth, as
may be judged by the fact that Thorir, after having built a
ship, asked him to consecrate it, which was a great con-
descension on the part of Thorir.

Thorir left Norway for Iceland; he reached it safely, and
then chopped up his boat, as he was tired of the sea; the two
beaks of the prow he set up over his hall doors, and they were
sure indications of the direction of the wind, for the north wind
piped in one, and in the other wailed the south wind.

As soon as the news reached Iceland that King Olaf was
supreme over the whole of Norway, Thorir considered that
there might be a good opening at court for his two sons, so he
packed them both off late in the autumn, to pay their respects
to the king, and remind him of his old friendship for their
father.

They landed in the south of Norway, and then, getting a
long rowing-boat, they skirted the coast on their way north to
Drontheim. Reaching a fine frith, in which there was shelter
from the gales which began to bluster violently as the winter
drew nigh, the sons of Thorir ran their boat in, and determined on waiting till the storms blew over, in a comfortable
hostel, built some way up the shore for the accommodation of
travellers. Their days they spent in hunting bears among the
mountains, and their nights in merry carousal.

It happened that Grettir was on board a merchantman

---

[*] *Gretla*, chapters 38, 39, 48.

then off the shores of Norway, beating about in the gale seeking safe harbourage.

Late one evening the vessel ran up this same fjord and stranded on the side opposite that on which was the hostel. The night was cold and wintry; heavy storms of snow rolled over the country, whitening the mountains and forming drifts behind the rocks. The men from the ship were worn out and numbed with cold, and they knew not on what part of the coast they had stranded.

When they reached land, they hurried from the shore to seek a sheltered nook where they might pass the night.

It was a wild night! The moon had been clouded over by piles of grey mist, which rolled through the sky, sending out arms of vapour; haggard and ghastly, she seemed to steal over her course swathed in grave-clothes. Now and then some crags caught a straggling gleam and flashed forth, but directly after were again blotted out; then the fjord caught the light and shone like steel till the shadows turned it to lead. An uncertain light flickered down the mountain side over the pine forests, which raved and bent as the wind poured through them.

Suddenly a spark, then a flame, was distinguishable, twinkling among the trees, on the opposite side of the fjord. This was a tantalizing sight for the poor shivering fellows, and they began to wish that some one of their number would swim across and bring over a light. No one, however, offered, and the crew hesitated about pushing the ship off and rowing across, lest they should fall among rocks and injure the vessel.

"In the good old times there must have been men who would have thought nothing of swimming across the frith by night," said Grettir.

"Maybe," answered some of the party; "but it is of no odds to us what men have been, if there are none now up to the mark. Why do you not venture yourself, Grettir? You are as strong and plucky as any of the old heroes. You see what straits we are put to for want of a little fire."

"There is no great difficulty in procuring a light," answered the young Icelander; "but I know that I shall get no thanks for my pains."

"Then you must have an uncommonly poor opinion of us," said the chapmen.

"Well," quoth Grettir, "I will risk it; at the same time, I tell you, I have a presentiment that you will bear me no good-will for what I do."

They pooh-poohed his objections, and assured him that he was the best fellow going.

Then Grettir flung his clothes off, and busked him for swimming. He had on him a fur cape and a pair of *wadmal* breeches; these he hitched up and strapped tightly round his waist with a bark cord. Then catching up an iron pot, he jumped into the sea and swam across. On reaching the farther side, he stood up on the beach, and shook the superfluous water from him; but before long his trowsers froze hard, and the water formed in icicles round the hood of his cape.

Grettir ascended through the pine wood towards the light, and on reaching the hostel from whence it proceeded, he walked straight in without speaking to any one, and striding up to the fire, he stooped and began to rake the embers into his iron pot, and to select a blazing brand which he could carry across in his mouth. The hall was full of revellers, and these revellers were the sons of Thorir and their boat's crew. They were already half intoxicated, and on seeing a tall, wild-looking man enter the hall, half-dressed in fur, and bristling with icicles, they concluded at once that they saw a Troll or mountain demon.

Whereupon every man caught up the first weapon he could lay hold of, and rushed to the attack. Grettir defended himself as best he could, warding off the blows with the flaming log, and eluding the missiles flung at him. In the scuffle the hot embers on the hearth were scattered over the floor, which was strewn with fresh straw and rushes.

In a few moments the hall was filled with flame and smoke,

and Grettir broke through it, escaped to the shore, plunged into the waves, and reached the other side in safety.

He found his companions waiting for him behind a rock, with a pile of dry wood which they had collected during his absence. The cinders were blown upon and twigs applied, till a blaze was produced, and before long the whole party sat rubbing their almost frozen hands over a cheerful fire.

On the following morning the merchants recognized the fjord, and, remembering that on its bank stood the house of refuge which King Olaf had built for weather-bound travellers, they supposed that the light Grettir had procured must have come from it; so they determined on running the boat across and seeing who were then quartered in the hostel.

When they reached the spot, they found nothing but an immense heap of smoking ashes. From under some of the charred timber projected scorched human limbs. The chapmen, in alarm and horror, turned upon Grettir and charged him with having maliciously burned the house with all its inmates.

"There now!" exclaimed Grettir; "I had a presentiment that misfortune would attend my undertaking last night. I wish that I had not taken so much trouble for a set of thankless churls like you."

The ship's crew raked the embers out, and pulled aside the smoking beams in their search for the bodies, that they might give them decent burial. In so doing they came upon some whose features were not completely obliterated, and among these was one of the sons of Thorir. It was at once concluded that the party brought by Grettir to such an untimely end, was that of Thorir's sons, which had sailed shortly before the chapmen. The indignation of the merchants became so vehement, that they drove Grettir with imprecations from their company, and refused to receive him into their vessel for the remainder of the voyage.

Grettir, in sullen wrath, would say no word in self-defence; but, turning on his heel, he stalked proudly into the woods, with his sword by his side, and his battle-axe over his

shoulder, determined on exculpating himself before King
Olaf, and him alone. The vessel reached Drontheim before
him, and the news of the hostel-burning caused universal
indignation.

One day as the king sat at audience in his hall, Grettir
strode in, and going before Olaf, greeted him. The king
eyed him all over, and said—

"Are you Grettir the strong?"

He answered: "Such is my name, and I have come hither,
sire, to get a fair hearing, and rid myself of the charge of
having burned men maliciously. Of that I am guiltless."

Olaf replied: "I sincerely hope that what you say is
true, and that you will have the good fortune to clear yourself of the imputation laid against you."

Grettir said that he was willing to do anything the king
wished, in order to prove his innocence.

"Tell me first," quoth the king, "what is the true
version of the story, that I may know what steps are to be
taken."

Grettir answered by relating all the circumstances; and
he asserted that the men were alive when he left the hostel,
carrying the fire.

The king remained silent for some moments.

"If I might fight some one!" suggested Grettir; "I
should rather like it."

"I have no doubt that you would," replied Olaf. "But
remember you have not a single accuser, but a whole ship's
crew, and you cannot fight them all."

"Why not?" asked the Icelander; "the more the
merrier. Let them come!"

"No, no, Grettir," answered the king. "I cannot allow
such a proceeding to take place. But I will tell you what
you shall do. Go through the fire ordeal."

"What is that?" asked the young man.

"You must lift bars of iron heated till the furnace can
make them no hotter, and walk with bare feet on red-hot
ploughshares."

"I'll do it at once," said Grettir. "Where are the ploughshares?"

"Stop," said the king. "You would be burned to a certainty if you ventured without preparation."

"What preparation?" asked Grettir.

"A week of prayer and fasting," was the reply.

"I do not like fasting," said the young man.

"But you cannot help yourself," answered Olaf.

"I cannot pray," said Grettir; "I never could."

"Then the bishop shall teach you," answered the king, with a smile at the bluntness of the Icelander.

Grettir was removed, and kept in custody by the clergy, who did their best to prepare him for the solemn moment of the ordeal, but they found him a troublesome fellow to manage.

The day came, and Drontheim was thronged with people, who streamed in from all the country round, to see the Icelander, of whom such stories were told. A procession was formed. The king's body-guard marched at the head, followed by the king himself, the bishop, the choir, and the clergy, amongst whom walked Grettir, a head taller than any of the throng, upright, his wild brown hair flying loose in the breeze, his arms folded, and his honest blue eyes wandering over the sea of heads which filled the square before the cathedral doors. The crowd pressed in closer and closer, but without in the slightest degree disconcerting him. Opinions seemed to be divided as to whether he were guilty or not; his dauntless bearing and open, sunny countenance were not those of a truculent Berserkir. Among the mob was a young man of dark complexion, who made a great noise, wrangling and shouldering his way till he reached the procession.

"Look at him!" exclaimed he. "This is the man, who, in cold blood, could burn a house down over helpless victims, and exult at their shrieks of despair; yet now is about to be given a chance of escape, when every one knows that he is a deep-dyed villain!"

6

"But he says that he is guiltless," quoth a man in the crowd.

" Innocent ! " exclaimed the youth. " A plea of innocence
has been set up as an excuse because the king wishes to have
him in his body-guard."

" He should have a chance of clearing his character,"
spoke a person standing near.

" Ay ! but who knows how the irons may be tampered
with by the king and clergy, so that this ruthless murderer
may escape the punishment he deserves ? "

" Young man ! " spoke Grettir, with a voice like thunder,
whilst flame leaped up in his eyes, and his strong limbs
quivered with rage. " Young man, beware ! "

" Beware of what, pray ? " laughed the youth. " Though
you may escape the punishment you have so richly deserved,
yet you shall not escape me."

And, springing up, he thrust his nails into Grettir's face,
so that he brought blood, calling him at the same time, son
of a sea-devil, Troll, and other insulting names. This was
more than the Icelander could bear ; he caught the young
man up, shook him, as a cat shakes a mouse, and flung him
to the ground with such violence that he lay senseless, and
was carried away as if dead.

This act gave rise to a general uproar ; the mob wanted
to lay hands on Grettir ; some threw stones, others assaulted
him with sticks ; but he, planting his back against the church
wall, rolled up his sleeves, and guarded off the blows, shouting
joyously to his assailants to come on.

A flush of honest joy at the prospect of a fight mantled in
his cheeks, and his eyes sparkled with delight. Not a man
came within his reach but was sent reeling back or felled
to the ground.

Grettir caught a stick aimed at him, while it was in the air,
and dealt such blows with it, that he cleared a ring about him,
whilst still, with a voice clear as a bell, he called to the mob
to come on manfully, and not shrink back like cowards.

In the meantime the king and bishop had been waiting
in church ; the processional psalm was ended, the red-hot

ploughshares were laid in the choir and were gradually cooling, yet no Grettir came.

At the same time sounds of uproar entered the church, and the king sent out to know what was the matter. His messenger returned a moment after with a report that, without the cathedral, the Icelander was fighting the whole town.

The king thereupon sprang from his throne, hastened down the nave, and came out of the great western door whilst the conflict was at its height.

"Oh, sire!" exclaimed Grettir; "see how I can fight the rascals!" and at the word, he knocked a man over at the king's feet.

"Hold, hold!" exclaimed Olaf. "What have you done, throwing away the chance of exculpating yourself from the charge laid against you?"

"I am ready now, sire!" answered Grettir, wiping the perspiration and blood from his face, and smoothing down his hair, which was standing on end; "let us go into the church at once; I am longing for the red-hot ploughshares."

He would have pushed past the king had not Olaf prevented him, saying that his opportunity was past, as he was guilty of mortal sin in having killed the young man who had assaulted him, and maimed so many other persons.

"What is to be done?" exclaimed Grettir. "I have undergone all that week of fasting for nothing. Sire! might not I become your hench-man? you will find me stronger than most men."

"True enough," answered the king; "few men have the strength and courage which you possess, but ill luck attends on you. Besides, I dare not keep you by me, as you would continually be getting into hot water. Now, this I decree: you shall be in peace during the winter, but with the return of summer you shall be outlawed, and go to Iceland, where I forewarn you, you shall lay your bones."

Grettir answered, "I should like first to get rid of the charge of the hostel burning, for, honour bright! I never intended to do the mischief."

6—2

"That is likely enough," said the king; "but it is quite impossible now for you to go through the ordeal."

After this Grettir hung about the town for some while, but Olaf paid no further attention to him, so at last he went off to stay the rest of the winter with a kinsman.

On the return of spring, the news of what Grettir had done reached Iceland; and, when they came to the ears of Thorir of Garth, he rode with all his friends and clients to Thing, and brought an action against Grettir for the burning of his sons. Some men thought that the action was illegal, as the defendant was not present to take exception; however, the end of the action was, that Grettir was outlawed through the length and breadth of Iceland. Thorir set a price on his head, and proved the bitterest of Grettir's foes.

Towards the close of the summer, Grettir arrived in a vessel off the mouth of the White River, in Borgar-fjord.

It was a still summer night when the ship dropped anchor. The Skarths-heithi chain was purple, but Baula's sharp cone was steeped in gold, and the distant silver cap of Ok shone in the sun's rays, like a rising moon. The steam rising from the numerous springs in Reykholts-dale was rounded and white in the cool still air. Flights of swans sailed overhead with their harp-like melody. As the gulls dipped in the calm water, every feather of their white wings was reflected. A boat came from shore and was rowed to the ship.

Grettir stood watching it from the bows, leaning on his sword. As the smack touched the side of the ship, "What news?" he called.

"Are you Grettir Asmund's son?" asked a man rising in the boat.

"I am," replied Grettir.

"Then we bear you ill news: your father is dead!"

Another man stood up in the boat, and said, "Grettir, your brother has been murdered!"

"And you," called a third boatman, "have been outlawed through the length and breadth of Iceland!"

It is said that Grettir did not change colour, nor did a muscle in his whole body quiver; but he lifted up his voice and sang,—

> " All at once are showered
> Round me, Rhyme-collector,*
> Tidings sad—my exile,
> Father's loss, and brother's,
> Branching boughs of battle ! †
> Many blue-blade breakers ‡
> Shall bewail my sorrow."§

One night Grettir swam ashore, obtained a horse, and reached the Middle-frith in two days. He arrived at home by night when all were asleep; so, instead of disturbing the household, he went round to the back of the house, opened a private door, stepped into the hall, stole up to his mother's bed and threw his arms round her neck.

She started up and asked who was there. When he told her, she clasped him to her heart, and laid her head, sobbing, on his breast, saying, " Oh, my boy! I am bereaved of my children; Atli, my eldest, is murdered, and you are outlawed; only my baby Illugi remains ! "

Grettir remained at home for some days, till Thorir of Garth learned where he was, and then he was compelled to fly; he was hunted from place to place, and to the last Thorir remained his implacable enemy.

" Thank you, Padre," said Mr. Briggs; " I should be glad to know what was the end of Grettir."

" I will tell you more of my hero when we are on the scenes of his exploits, and visit the place where he died."

---

* Periphrasis for poet.
† Periphrasis for men.
‡ Periphrasis for warriors.
§ Shall suffer for my misfortune.

CHAPTER V.

AMONG THE JÖKULLS.

Leave Thingvellir — Meyjar-skarth—A lovely View —The Big Bed breaks loose—Grettis-tak—A mysterious Vale—Attempts to re-discover it— Ascent of Geitland's Jökull— Brunnir—Dearth of Fuel—Kaldidalr— The Old Woman of Bones—A Vale of Desolation—A Tract of Ash— Kalmanstúnga—Robbers.

I DO not remember a more beautiful morning than that on which we left Thingvellir; not a cloud was in the sky, and the air was fresh and elastic as that of an English spring morning.

Armannsfell had put on a cloak of snow during the night, and a thin sprinkling capped the dark crags of Hrafna-björg.

Our course lay along the foot of the Troll-haunted Armannsfell, through a "skog," or forest of stunted birch and willow, reaching to the saddle. As we pushed among the clumps, we dislodged myriads of grey moths.

The track leads up a long wide valley, bounded on the left by Armannsfell, and on the right by a quaint, jagged cinder-ridge, with teeth and spikes cutting crisply against the sky. (Plate III.) The vale itself is filled with lava poured from the calm, symmetrical Skjaldbreith, or Broad Shield, a jökull wondrously like a knight's maiden shield, as it rises above the black corrugated stone-torrents which gird its base.

The lava is very old, and is consequently much disintegrated, so that the shrubs find snuggeries for their roots in its rifts and shattered vents.

E. Postine Gould, delt.

PART OF THE PLAIN AND FOREST OF THINGVALLA.

Edmund Evans, sc.

London: Published by Smith, Elder, & Co., 65, Cornhill. 1863.

In a few hours we reached a grass-patch, on which cows were grazing, beneath a curious hill, sloping rapidly to the plain from its snuffbox-like cap. This is the Meyjar-skarth, the spot where sports were wont to take place in old historic times, wrestling, goff, and stone-lifting ; whilst the ladies sat on the sides of the hill, watching and backing the players.

After a short halt, our horses scrambled up a narrow gorge to the left, past a stone on which the traveller is expected to inscribe his name. The way is steep and particularly unpleasant, as it lies between ashen grey cliffs, and over heaps of shattered tufa rock.

Tufa is formed of volcanic cinder consolidated into rock, but the fragments are so feebly cemented together, that, with every thaw, masses of crag are dislodged and borne down into the vales. The view from the top of this pass amply repays all the trouble of the ascent. Far away in front is the silver dome of Ok Jökull, with the volcanic cone of Fantofell, or the Scoundrel's Mount, rising, dyed a deep gentian blue, against its matchless white. To the left, the iron grey mountain scarps of Súlur, with here and there a terrace of green moss to relieve its gloom, and a stream flashing over its blackest bluff into the blue still lake at our feet, whose face is only ruffled by three drowsy swans floating in the shadow, like flakes of snow dropped from the mountain ledges. To the right, shoulders of sandrock, striking into the lake and retreating into bays, leaving flat beaches, over which Mr. Briggs and the baggage horses are already careering at a hand gallop.

I leap from my horse and make a water-colour drawing, then scour down the hill in the spoor of the other horses and overtake them at a critical moment.

The sands at the lake-head were so inviting that the mare carrying the great bed had made up her mind for a roll. In effecting this on a hill slope, her girths gave way, the bed broke loose, and I was just in time to see it bounding like a foot-ball down the incline, making straight for the lake, with my portly friend in full pursuit, uttering wails of dismay. Fortunately for him, the bed stopped dead in its course,

wedged between two sand hummocks, just above the water's
edge. By the time that this was rolled up-hill, the horse
with the boxes had turned over for a roll, and was kicking in
the air between the strong trunks, unable to recover himself,
like a cockchafer on its back. As soon as the boxes and bed
were readjusted, and we were fondly hoping that all was ready
for the start, it was discovered that the grey with the
brandy cases, was in full chase of another horse which had
been tranquilly grazing near with a drove sent up the mountains for summer pasture. The guides flew in pursuit, and
brought the palpitating beast to a standstill, fortunately without any of the spirits having been lost.

Our course lay next over a hill-shoulder composed of
angular blocks and chips, tossed down as if to try the horse's
legs, without a patch of moss or a blade of grass to fill the
crannies. From the top we had a glorious view of the silver
peaks of Súlur, which reminded me somewhat of those of the
Finster-Aarhorn. Súlur signifies tent-poles, and the mountain
bears a fanciful resemblance to a tent propped on its poles,
before the guys and braces are made fast.

Hard by the road is a Grettis-tak mentioned in the Saga.
It is a large stone, according to tradition lifted by Grettir, the
great Icelandic hero and outlaw. That he " put " a big
stone is not impossible, but that the block in question was
ever raised by him, is preposterous. From this point, a fine
view is obtained of the stately Geitland's Jökull, with its many
snow-dales and gable scarps. (Plate IV.)   In this glacier
mountain is the mysterious Thorir's dale, which has not
been explored since the time of Grettir (eleventh century),
who discovered it and spent some years in its secluded recesses, shut in on all sides by snow-chains.

Grettir found the soil covered with luxuriant herbage, and
warmed by boiling springs emptying themselves into a rill
which flowed through the valley. The brave outlaw left it only
when wearied out by its solitude; and then, that the entrance to the glen might be found by others, he erected a slab
on the side of Skjaldbreith over against the opening of Thoris-

GEITLANDS-JÖKULL.

... ... & Co 65 Cornhill. 1863.

GEITLANDS—JÖKULL.

London, Published by Smith, Elder & Co. 65 Cornhill, 1863.

dalr, and drilled a hole in the stone, so that, by looking through this, the eye might rest directly on the entrance to the glen. The opening is distinguishable enough from the point; but the slab, though it still stands, has been shaken out of the upright by some of the convulsions of the volcano, so that the hole does not point directly to the vale.

My intention was, to have examined this traditional dale, but the grass was not sufficiently grown to allow of the horses being kept more than a day in the neighbourhood, and it was absolutely necessary that we should have fine weather for the passage of Kaldidalr.

Thorir's vale receives its name from a Troll or mountain being who dwelt there; he is vaguely mentioned in the Gretla, but he is spoken of also in the Bárthar-Saga. In all probability, the Trolls of old Icelandic historical romance were nothing more than ruffians who lived in dens and caves of the earth, robbing bonders and preying on wayfarers. This is remarkably borne out by the fact of one of Grettir's friends, Hallmund, spoken of in the Gretla as a man whose hand was against every man, being named in the Bárthar-Saga as a Troll or evil being.

Several attempts have been made to rediscover Thorisdalr, but all have been unsuccessful. Messrs. Olafsen and Povelsen mention an account of the ascent of Geitland's Jökull by Bjärnarson and Helgi, two Icelandic ecclesiastics, but state that their journal was written in such a confused style that it was difficult to make anything of it. According to this account, they arrived towards evening, in delightful weather, at a large valley situated in the heart of Geitland's Jökull; it was of such depth that they could not distinguish whether it were covered with grass or not, and the descent to it was so steep that they were not able to go down, and consequently they returned. Messrs. Olafsen and Povelsen themselves effected an ascent of the glacier, but without discovering the mysterious vale, probably because they climbed the mountain from the Kalmanstúnga side, whilst the glen lies on that nearest to Skjaldbreith. Their account of the ascent is as follows :—

"On the 9th of August we started for Reykholtsdal on our way for the glacier of Geitland; our object was not so much to discover a region or inhabitants different from those we had quitted, as to observe the glacier with the most scrupulous accuracy, and thus to procure new intelligence relative to the construction of this wonderful edifice of nature. The weather was so fine and the sky so clear, that we had reason to expect that we should accomplish our object according to our wish, but it is necessary to state that in a short time the Jökulls draw towards them the fogs and clouds that are near. On the 10th of August in the morning, the air was calm, but the atmosphere was so loaded with fog that at times the glacier was not visible. About eleven o'clock, however, it cleared up, and we continued our journey from Kalmanstúnga. The high mountains of Iceland rise in gradations, so that on approaching them you discover only the nearest elevation, or that whose summit forms the first projection. On reaching this you perceive a similar height, and so pass over successive elevations till you reach the summit of the ridge. In the glaciers, these projections generally commence in the highest parts, and may be discovered at a distance because they overtop those mountains that do not form Jökulls themselves. We found that it was much farther to the Jökull than we had imagined, and at length we reached a pile of rocks which, without forming steps or gradations at the point where we ascended, were of considerable height and very steep; these rocks extend to a great distance, and appear to make a circumvallation around the glacier, for we perceived their continuance as far as the eye could reach. Between this pile of rocks and the glacier, there is a small plain about a quarter of a mile in width, the soil of which is clay, having neither pebbles nor flakes of ice, because the waters which continually flow from the glacier, carry them off. On advancing a little farther we discovered, to the right, a lake situated at one of the angles of the glacier, the banks of which were formed of ice, and the bed received a portion of the waters which flowed from the moun-

tains. The water was perfectly green, a colour it acquired by
the rays of light that broke against the ice. After many
turnings and windings we found a path by which we could
descend with our horses into the valley. On arriving there,
we met with another embarrassment, as well in crossing a
rivulet discharged from the lake as in passing the muddy soil,
in which our horses often sank up to the chest. In some
parts this soil is very dangerous to travellers, many of whom
have been engulphed and perished in it.

"Our object was so far attained, that we were now on
Geitland, but we found it a very disagreeable place. We discovered a mountain peak rising above the ice, and which, as
well as the other mountains, had been formed by subter-
ranean fires. We led our horses over the masses of ice,
after which we left them, and travelled the remainder of the
way on foot. We had taken the precaution of providing ourselves with sticks armed with strong iron points, and with a
strong rope in case of any of the party falling into a crevass,
or sinking in the snow. We had also a compass which we
regarded as indispensable, as well for guiding us, as to observe
whether at so considerable a height there was any perceptible
deflection of the needle. Thus prepared, we began to escalade the glacier at two o'clock in the afternoon; the air was
loaded with a thick fog which covered the whole mountain;
but, hoping that it would disperse, we continued our dangerous
and troublesome route, though at every instant we had to
pass deep crevasses, one of which was an ell and a half in
width, and the greatest precaution was required in crossing it.

"As we mounted higher, the wind blew much stronger,
and drove larger and more abundant flakes of snow before it:
fortunately we had the wind in our backs, which facilitated
our ascent; but we met at the same time with heaps of snow,
which rendered our progress difficult. Hoping, however, that
the weather would change, we agreed not to return till we
had gained the summit, from which arose a black rock that
we could perceive at intervals.

"At length, after travelling for two hours longer, we found that we had made no additional observations, since we could discover nothing in the distance. A rampart of burnt rock of no considerable height rose above the ice, and at this we paused to rest. The snow-flakes now obscured the air so much that we hardly knew how we should get back: we examined the compass, but without observing any change; and we were prevented by our guides from going towards the north-west, where the mountain is highest and least accessible. The weather continued the same on the Geitland, so that we found it impossible to resist the cold much longer, and deemed it prudent to return.

"Although the sky was very heavy and dark, we discovered, on our return, the entrance to a valley; if the weather had been more favourable, we should doubtless have had the pleasure of investigating it; but we doubt whether we should have found Thorir's dale. As we descended, we found the wind in our face, which threw the snow so much against us, that we could not discover the traces of our ascent, and it therefore only remained for us to take the road which was least steep.

"By this means we again met with ravines and crevasses, which rendered our descent very dangerous, because they were from three to three ells and a half wide, whilst the soil that separated them was very uneven; insomuch that we were obliged often to go out of our way, or to run the risk of being precipitated to the bottom."

Of the existence of Thorir's dale I can have no doubt. The words of the Gretla are simple and explicit; the stone which Grettir set up to mark the entrance of the vale still stands, the Icelanders who live anywhere near are unanimous in their opinion that a vale does lie among the Jökulls in the direction indicated by the Saga, and, indeed, an opening which may lead to it is visible from Skjaldbreith. That the valley produces grass now and is full of hot springs, I do not pretend to assert, as there took place an eruption from Ball Jökull, of

which Geitland's is a spur, in 1716, which may have altered
the character of the dale, destroyed its grass, and choked its
springs.

I unfortunately missed seeing M. Gunnlaugson, the compiler of the great Icelandic map, who could have given me
some information on the subject; but, if it please God to
spare me, on my next visit to Iceland, I shall thoroughly
explore Thorisdalr.

Skirting a desert of new lava which has gushed from that
treacherous Skjaldbreith, we reached a high lake-district,
where Martin shot three northern-divers, and missed several
swans. Mr. Briggs was lucky this day, and my saddle-bags
were filled with the results of his and the Yankee's shooting.

I had been suffering for the last two days from a feverish
attack with cramps, so that I could hardly walk; and, as we
pitched our tents amidst a drizzle, under a snow-patch, with
two cold springs near us, trickling into a tarn thronged with
wild duck and swans, I felt so giddy and ill, that I would
willingly have coiled myself in my rugs and gone to sleep.
But this was impossible: wood had to be got, and the birds to
be cooked. The only fuel which can be procured is willow-
root dug from the soil. No willow grows on this spot now,
but the roots remain for the use of the traveller, and, as this
is the only grass spot between Kalmanstúnga and Meyjar-skarth,
it is a favourite camping-place. Grímr told me that, however much of the roots was dug up, the store never failed.
It would have failed us, however, had I not collected a
provision during the day, and carried it in a bundle before me
on my saddle, tightly strapped together: besides, we had
to send Gúthmundr back some way to a spot where we had
noticed a considerable amount of root which had been laid
bare by the wind.

My companions started in pursuit of swans, whilst I cooked
the supper, and boiled the water for tea and toddy. I strove
hard to make oatmeal cakes, but they were failures; the dog
even would not touch them; the tea also was spoiled by
Grímr having put the pepper into the teapot for safety.

When all was ready, my guide made signals to the sportsmen, and we finished the evening with a capital supper within
the tent, seated on Mr. Briggs' great and comfortable bed,
with boxes for our table, and my hammock for the sideboard,
into which the dirty knives, forks, tin plates, and mugs, were
flung, notwithstanding my earnest remonstrances.

On the following morning I should really have been driven
to ask my companions to halt a day, as my fever had increased,
but that I knew that there was not grass for the horses' sus-
tenance during another twenty-four hours; moreover, I have
a theory that the more one gives way to sickness, the worse it
becomes.

By ten o'clock we were *en route*, my companions walking,
gun in hand, on the look-out for game. I rode on ahead at a
fast trot, that I might make a sketch of the waste district at
the entrance to the dale, but found that I could not steady or
direct my pencil; I was consequently obliged to postpone it
till my return. Plate XIV., which represents this scene, was
taken then; the rising columns of red sand were not visible
at this time, and the phenomenon shall be explained in the
chapter detailing my return journey over Ok.

To the left is Ok, or the yoke, a mountain shaped like a
dish-cover, snow-draped, with Fantofell guarding the entrance
to Kaldidalr. This Jökull is not included in the drawing.
Fantofell obtains its name from the tradition that two rogues,
one from the north, the other from the south, met on its top,
and fought till they had mutually slain each other, like the
Kilkenny cats. An ugly gap—the Kaldidalr—apparently
blocked by a low ridge of saw-like hills, barren and precipitous,
separates Ok from Geitland's Jökull, an outstanding portion
of the great Ball Jökull, from which it is parted by the
mysterious Thorisdalr. In the distance, to the right, is
Hlöthufell, or the Stack, with abrupt flanks; more to the right
is another snow point, very distant, and then the symmetrical
Skjaldbreith. The foreground is sand and shale, quite destitute of vegetation. We turn the flank of the iron-black saw,
and the wind moans up Kaldidalr in my face, fierce and biting.

The scene of desolation is quite indescribable : a vast trench between walls of rock and heaps of snow ; the crags of great height and flat-topped, with bare precipices of green ice and snow resting on them, ready to topple over in avalanches with the least disturbing cause, and bury us under their ruins ; here and there a cone of snow, which has thus shot to the bottom and has not yet begun to melt ; now a smooth sweep of undinted whiteness rising to the Jökull top, or barred with black steps of rock glazed with frozen streams. Not a bird, nor insect, not a sound. I stood—

> " Alone, for other creature in this place,
> Living or lifeless, to be found was none."
> *Paradise Lost.*

In the foreground a cairn of rib and leg bones of horses, which have died of starvation in the pass, with a patch of turf about it as large as a horse-walk in a threshing-mill, the grass grey not green, and that the last sign of vegetation we are to see for many hours.

The bed of the vale has not even the flash or tinkle of a rivulet to relieve its hushed monotony. The snow melts, and is absorbed into the spongy ground. Shoulder on shoulder of snow, buttress on buttress of rock, swell on swell of avalanche rubble for us to toil over ; here and there the skeleton of a poor horse which has fallen lame and died before it could reach herbage. It was indeed an awe-inspiring scene among these Jökulls locked in everlasting stillness, folded in a white veil never to be raised till the crack of doom.

> " The palaces of Nature, whose vast walls
> Have pinnacled in clouds their snowy scalps,
> And throned eternity in icy halls
> Of cold sublimity, where forms and falls
> The avalanche—the thunderbolt of snow ! "
> *Childe Harold.*

Two parties had traversed the dale this year before us, and one had left his best horse hopelessly lamed ; and the other— the postman—had lost his way, and had been nearly driven to

cut the throats of some of his horses to spare them a lingering
death by famine.

I waited at the cairn, wrapped in my Franciscan cloak, till
the rest came up.

" This is the Beinakerling (pronounced Bayna-kedling),
or old woman of bones," said Grímr.   " Every traveller is
bound to write a message of ' God speed ' for the next person
who traverses this pass, and secure it in one of the bones of
the heap."

We complied with the custom, and, after drinking a
bumper to the Queen and to the " Old Folks at Home,"
dashed into the scene of desolation before us, in pursuit of the
sumpter horses now crawling over a neck of rubble a mile
ahead.   Half way through this wilderness is a dark headland
of tufa, the Hádœgrafell, pronounced How-daigra-fedtl, or
Half-day mountain ; it flanks a noble and picturesque trachyte
Jökull, whose pale ashen hue contrasts with the blackness of
the tufa around.   This is Thorishöfthi, and is believed to
mask the mysterious glen.

The sky gradually became overcast, and we were afraid of
the clouds descending upon the snow and enveloping us, but
we were fortunate.   A wildly beautiful scene opened on us
now—the glorious heap of Eireks Jökull, an isolated rounded
head of snow supported on abrupt scarps, and looking something like a bride-cake; beyond this a blue horizon with
water-specks flashing on it, the Arnarvatn-heithi with its
network of countless lakes, over which our course was to lie
in a few days.   Still onward we pushed over soft earth, and
through sludgy snow, whose crust had broken through in
several places, and disclosed ugly pits ready to engulph us
should the snow not support our weight; up a desperate
stair of rock with blocks of glistening obsidian and cakes of
amygdaloid, strewn on either side and under foot.   Still more
snow as we scrambled over a spur of Ok glacier, and then
with a shout of joy we hail a wintry flake of turf; our horses
break into a canter, the dog leaps about us joyously barking,
and the pipe of the plover relieves the ear which has tired

with a stillness so oppressive, that few of us had been in
spirits to speak, during the many hours in the cold dale.

But we were not at the end of our journey yet; we had
two hours more fast riding and two rivers to cross, one of the
hue of milk and water from the amount of unmelted snow it
swept along with it. This was separated from the other river
by a monotonous tract of volcanic sand and cinder, sprinkled
with a minute rhododendron.

At eleven o'clock we reached Kalmanstúnga, and partook
of an excellent supper off rice-milk, stirred with the instrument used in poking the fire, and lake trout. I was in
especial glee, as my fever had left me suddenly in Kaldidalr.

The next day was so rainy that we were obliged to remain
at Kalmanstúnga. Mr. Martin was glad of the opportunity
for skinning his birds and preparing them for the taxi-
dermist.

On the second day, June 15th, we started for Little
Arnarvatn, intending to visit Surtshellir on the way. This
cavern has been so frequently visited and described, that I
have no heart for writing a fresh account of it. It has been
investigated by Olafsen and Povelsen, by Henderson, by Capt.
Forbes, by M. Preyer and Dr. Zirkel, and by Mr. Holland.
Suffice it to say, that its interest has been much over-
rated. It consists of a chain of air bubbles in the lava, the
top of two of which have fallen in; out of these branch
tunnels, one of which served long ago as hall and cubicle for
a robber gang, another as a receptacle for the bones of cattle
stolen from neighbouring farmers. These bones still remain
in great numbers.

The band was destroyed through the treachery of a young
man of the party, who led the armed bonders upon the robbers as they lay asleep in the sun on the side of a turfy split
in the lava, some way off.

All the rogues were killed except Eirek, who, having had
one foot cut off, escaped by running like a wheel with hands
and foot, just in the manner of street urchins, till he reached
the jökull, which he climbed, and then vanished among its

7

snows. Many years after, a ship came into the nearest fjord, commanded by a one-footed merchant. The cheap rate at which the goods were sold attracted the young man, among others, to the vessel. Scarcely was he on board, than the one-footed merchant shouted for the anchor to be raised and the sails to be set. The ship rolled out to sea, and neither youth nor merchant were seen or heard of again.

CHAPTER VI.

THE EAGLE TARNS.

Eiríks Jökull—Taxation—The Sheep's Disease—A Swan's-nest—Lava—
Lose our Way—Camping-out—Glorious Scene—Cooking Arrangements
—Rare Skua—Great Northern Diver—Storm on the Heithi—Gunnars-
sonarvatn—The Great Eagle Tarn—Desert—Hólma-kvísl.

WE were now approaching that desert tract which Captain
Forbes says is only to be traversed if a sufficient supply of
hay is taken for the horses. This is not quite correct; but
grass is certainly scarce, and we were warned that it would be
impossible for us to halt at the Great Eagle tarn (Arnarvatn,
pronounced Atnarvat), as there was no herbage there for the
horses.

We were, therefore, now bound for the Lesser Eagle lake,
where the farmer of Kalmanstúnga assured us we should find
enough for our poor brutes to crop. But as no one knew the
way thither through the labyrinth of lakes, except himself,
we were obliged to engage him as guide for a couple of dollars per diem. Messrs. Shepherd, Upcher and Fowler, had
attempted to traverse the road by the great lake earlier in the
year, but had been compelled to give it up, and cross to
Efrinupr by the Wolf lake.

The road, a mere track, ascended continuously; we had to
scramble over old curdled and plaited lava, sprinkled with the
pale lemon-coloured stars of the Dryas octopetala. In many
places the molten stone seemed to have been poured as treacle
from a spoon, and then to have suddenly congealed.

7—2

The day is very lovely; to the right are the snow heaps of Geitland and Ball jökulls, and the mighty dome of Eiriks jökull, an undinted tract of eternal ice and snow, heaved up on strange ribbed and buttressed flanks, down the gullies of which slip wreaths of silver, hardly beginning to melt in a sun without warmth. One purple crag stands forth as a headland, gashed at the top, snow-powdered; it is Eiriks gnypa, which the brave robber scaled though one of his feet was amputated, and at the summit checked the flow of blood by freezing the stump in ice. The wind is from the north, its moisture condenses on the cold head of the glacier mountains, and a thin veil of mist forms and is puffed away, forms again to be again blown aside, and as the vapour curls away from the snow, it is absorbed and vanishes.

Behind us rises the slim cone of Strútr, with a white cap on, for yesterday's rain in the vales was snow on the mountains. Far away in the west I can discern a similar cone, but grey and snowless—it is Baula, whose sides are too precipitous for the snow to lie on. To our right the Northlinga fljot tumbles between lava walls. There is a red-breasted merganser (*Mergus serrator*), floating on yon pool. Bang! Martin's gun goes off. Bob dashes into the water, and the bird is thrust into my saddle-bag. The farmer calls the duck " Lilla Töpond." Whimbrel and golden-plover pipe and wail in all directions; as we have to find our own provision for the next few days, the Yankee and Martin blaze away, and the saddle-bags are soon as full as they can well be stuffed.

" Pray, to whom does this waste belong ?" I asked, as my pony scrambled alongside of the farmer's grey. " Can it possibly be worth anything to anybody ? "

" I farm it," he replied; " but it belongs to an old lady near Reykjavík. I rent near upon twenty square miles."

" And what do you pay for it ? "

" Seventy-eight dollars (about 9l. 10s.) "

" Dear at the money ! " I exclaimed.

" Not so dear," the farmer answered : " for I get good

land round Kalmanstúnga, my house and tún, a 'forest' for firewood, besides this wilderness."

"Which you would as soon be without."

"Far from it. The trout and char from the lakes supply me and my family with food."

"What taxes have you to pay?" I asked.

"My rates amount to forty dollars; but, because of the sheep disease and bad year, there is an extra tax of twenty-five dollars. To the King of Denmark I pay eight, and that I grudge him."

"What are included in the rates?" I inquired.

"My dues to the Althing man (M.P.), amounting to some twelve or fourteen dollars, Easter offerings to the priest, church-rates, &c. I grudge none of them—not a mark! I am proud to give money to the man who represents my interests at Althing; the priests are our brothers and cousins, so we don't mind giving them a trifle; but as for the eight dollars to the king, every one of them is like a drop of blood wrung from my heart."

"How comes it that you have extra rates because of the sheep disease?" I asked.

"You must know," answered the farmer, clearing his throat and preparing for a long story, "one of the blessed things Denmark has done for us has been the introduction of scab among our sheep. Our sheep, according to the Danes, wanted their stock improving; so they introduced some foreign brutes, and at once a terrible malady spread among the flocks, from Reykjavík as a centre. Sheep died all through the south, and scab was appearing in the north, when the farmers of the north unanimously agreed on slaughtering every infected sheep, and on making these jökulls and deserts a boundary beyond which the disease was not to penetrate. A line of demarcation was drawn; every sheep coming north of this was forthwith killed; all the flocks along the friths which form chains of communication between north and south, were slaughtered, and their owners remunerated by the ratepayers. Now, the sheep in the north are quite well, whilst the scab reappears yearly in the south."

The story of the sheep grievance was checked by an exclamation of "Swans! swans!" from Martin.

To our left, a hundred yards off, was a small tarn with reedy marge; on it sailed majestically two noble birds, every feather mirrored in the still blue water. Bang! went Mr. Briggs' gun. With a strange musical scream, the two bright birds rose from the water and flew to some lake north of our route. "We shall see plenty more," said the farmer; "but only two in each small sheet of water. Swans are not sociable beings, and will not suffer a second couple to occupy the same tarn."

"The nest is sure to be close by," said Mr. Briggs, rolling from his saddle.

We left our horses and searched the rim of the pool. Before long we came upon the nest, a heap of mud, rush and willow roots, about a foot and a half high, with a depression at the top lined with feathers; in this were four greenish-white eggs. We left them, as there was no chance of our being able to carry them unbroken through the day, and our guide assured us that we should find plenty more at little Arnarvatn. Mr. Briggs had probably fired as the male bird was returning to relieve the female by supplying her place on the eggs.

Our route lay now through more desolate country. We traversed long tracts of mud and stone, utterly bare of vegetation, but strewn at intervals with white dead roots of dwarf willow. Here and there, in a depression of the heithi,* bloomed a little moss campion; the grass of Parnassus was in bud, though not yet in flower; but the purple butterwort, in full blossom, shook its beautiful head with every icy puff that swept the waste.

We skirted the great lava flood, which has gushed from Eiriks jökull, has climbed the heithi, and now lies in a long

---

* A heithi, as already explained, is barren, or moss-grown hilly country, which can be traversed by horses.

black ridge on the mud desert. We could trace the sweep
of every billow, now intruding on a lake, then shrinking
before a shoulder of trachyte, here tumbling in cakes down
a hollow, there throwing feelers round a sandy knoll, though
too exhausted to meet beyond it.

Lava is a rock in ruin, never picturesque, always horrible;
for during its flow, gases generated in its fiery womb have
exploded, shivering its whole mass, tilting the sides of these
domes into the air with their jagged edges exposed, and
blowing snags and splinters into cairn-like heaps all around.
In the centre of a lava stream, the surface is more even, but
the edge is always shattered and bristling. Blow up West-
minster Palace with gunpowder, and an Icelandic pony will
trot over the ruins; but the skirt of a lava-flood is an insur-
mountable barrier even to him.

We soon sighted other swans, but my companion failed
to shoot any, as the baldness of the land about the lakes and
pools made it impossible to get under cover whilst approaching,
and the birds were very timid. Mr. Briggs and I rode on ahead,
following the spoor of other horses, and it was full an hour
before we found that this was leading us in a wrong direction.
We were threading a network of lakes. The great map of
Gunnlaugsson was at fault, the Fiskivötn (fish lakes) were
marked on it evidently somewhat at haphazard, and incorrectly.
The river traced on the map as connecting the lakes nowhere
exists, but the tarns lie land-locked in every dell and hollow
of the heithi, surrounded by stony barren hills. Little
Arnarvatn is not named on the map, so the compass was
unavailing, we knew not the direction in which to steer.

We were obliged to retrace our steps, and many a weary
mile it cost us—Mr. Briggs at intervals discharging shots as
signals of distress—when far off to the north we descried a
moving speck on the summit of one of the heithi sweeps.

"The farmer!" exclaimed Mr. Briggs, adjusting his opera-
glass; "I can distinguish his grey."

With a feeling of considerable relief we scrambled in that
direction over rock and swamp, past pool and tarn, till we met

the man galloping towards us, he having caught sight of us at
the same time that we had noticed him.

He was in a great state of excitement: " You should not
have left me ! " he exclaimed; " there is no track where we
are going ; no one knows these lakes except myself ; you might
easily be lost here, and I should never find you again. It was
fortunate for you that you kept to the road ! "

" Where are all the rest of the party ? "

" Miles away to the north ; by this time we might have
been at little Arnarvatn, if I had not been obliged to return
for you."

When we reached the caravan, Gúthmundr in distress
assured the farmer that notwithstanding his entreaties both
Martin and the Yankee had strayed, having started in search of
swans. The poor fellow, with an Icelandic recommendation to
the troll to fetch them, started in pursuit, but this time he was
not long in finding the runaways, as the sound of their guns
directed him to the lake over the next hill, where they were
wasting shot on swans which kept out of range.

It was seven o'clock when we reached the lake, a beautiful
sheet of water surrounded by boggy hills covered with ash-
grey moss, and with here and there a patch of snow, the
tricklings of which burned up the scanty vegetation. I could
see no grass anywhere, but our guide assured us that there
was some a mile and a half up the lake, enough at least with
willow sprouts and angelica shoots to last the horses for a
couple of days. We fixed on a little tongue of land projecting
into the lake as our camping spot ; a nodule of rock on it, built
round with turf, served for a fireplace to the fishermen when
sent there by the farmer. As we arrived, a man and boy who
had been ordered thither yesterday, drew their nets, and we
secured enough char to last us for supper and breakfast. As
the fishermen had collected willow roots sufficient for a fire, we
soon kindled one and made the kettle boil. We drank a cup
of tea all round, and then Magnús, Gúthmundr, and the farmer
started with the horses for the lake head, my companions took
their rods and guns, and Grímr alone remained with me, pre-

ferring to lounge about with his hands in his pockets whilst I put up the tents and cooked the supper. The first of these undertakings was not particularly easy, as the ground was nowhere even; like the surface of every heithi, if moss grows on it at all, it was covered with heaps of grey moss so large as effectually to prevent one from sleeping on it. In one spot, and one only, was there a level patch, and that was barely large enough for both tents to stand upon, and was moreover the bottom of a hollow into which water would be sure to flow with the first storm.

"When it rains this shall be bog," quoth Grímr, eyeing me as I heaved up the poles, and strained the canvas; "it look as though it rain to-night." Then thinking it incumbent on him to do something, he volunteered to put up the little pennant which adorned the tent top.

"That," said I, "takes no trouble. Will you, however, kindly drive the pegs home, and stretch the guys?"

"I see no of the hammer!"

"Not unless you look for it certainly." I had to hunt for the mallet, and then he leisurely drove in the pins where the moss was softest and where the hammering would cost him least trouble. I had to pull them all up again and drive them in anew.

"Now, Grímr," I called; "please to unroll the great bed."

"I see no of it!"

"Because it is behind you; turn round and you will see it."

The theological student stooping for a moment unlaced the tarpaulin case, and then sauntered off to light his pipe at the fire. I spread the waterproof floor, slung my hammock, made the beds in the bigger tent, dragged the boxes under cover into the lesser, and when this was done it was high time to prepare our supper.

The night was glorious. The desert was hushed into a death-like stillness, broken only by the note of the snipe as it whirred by. Far off the great snow cupola of Eiríks jökull was flushed the tenderest rose-tint by the setting sun—it was

eleven o'clock, and overhead a few barred clouds burned against
the green sky.   Strútr was enveloped in a fog which stole up
towards the jökull, touched the plum-coloured crags of Eiriks
gnypa, clung to them, pushed farther, threw a gauzy sash
athwart the glacier mountain, and began stealthily to veil its
sides.   Not a moment was to be lost; I caught up brushes
and colours, and running to a spot which commanded our
camp, as well as the mountain, I made a hasty sketch before
the cloud obliterated all.   Grímr pointed to the fog, and said,
" It shall be bad weather to-night ! "

" Prophet of woe ! " I exclaimed ; " find some other spot
for the tents."

" There shall be none other," he answered, and I believe
he was right.

" Now for the cooking ! "   I turned up my cuffs, drew out
my long hunting-knife, and ranged the birds before me—four
ptarmigan, three whimbrel, and six golden plover.   A dish for
a king, indeed !   I skinned the whimbrel, most of the plover,
and three ptarmigan ; disemboweled them, washed them in the
lake, broke the backs of the big birds, and tossed them into
the pot.   Meanwhile the preacher guide, at my particular
request, had undertaken to skin and clean the remaining
ptarmigan and plover.   I heard his dejected sighs drawn
frequently and heavily over the work, but I took no notice of
them.   Presently he brought the ptarmigan, and flung it in
with the rest of the birds.

" Come, Grímr ! " I said, " you must give me some sticks,
and blow the fire ; I will stir the pot, and make a savoury mess."

The dry willow-roots blazed up merrily, and then died out,
so that the work of keeping a brisk fire about the pot absorbed
my attention so completely, that I did not examine Grímr's
handiwork before consigning his ptarmigan to the pot.   A
capital mess it promised to be : I put in a tin of preserved
vegetables, a few slices of portable soup, salt, pepper, some
Brighton sauce, and sprinkled the whole with garlic powder ;
the meal-bag stood temptingly by ; I ventured on a bold stroke,
and poured into the saucepan several spoonfuls of oatmeal.

" Grímr ! " I cried sharply, as a horrible suspicion flashed across me; ".what is the matter with that ptarmigan at the top of the stew?—the fellow you skinned—it is swollen out and looks so fat, that——Why, Grímr l you never——"

" No, I did not take the insides out. I do not know how."

" So you have let me boil it thus, and never said a word, though you saw what I was about. The soup is spoiled."

" I will take of the ptarmigan out and clean him now."

So the matter was settled. But for this untoward accident the stew would have been perfect.

" Fire off a shot as a signal to Mr. Briggs and the others that all is ready," said I.

" I will," answered Grímr. " But wait, there is of a skúa, a rare one, I think."

I saw a dark bird with sharp-pointed wings wheeling near. Grímr lifted his gun, pulled the trigger, and the skúa fell fluttering at his feet.

" I have only once seen this kjói before," said he. " It is a beautiful bird."

It was so indeed. At the time I did not know of its extreme rarity, and I was ignorant of its name; but on reference to M. Preyer's index I find that he saw the skin of a similar skúa, and named it Lestris Thuliaca; his account of it tallies so closely with the notes I made, that I cannot doubt that the specimens are identical, and that the bird is a variety of the Lestris parasitica. The beak, legs, and webbed feet are perfectly black, the plumage grey, with the exception of these spots: 1st, many of the lesser wing covert plumes are white, so also are the scapulars; on the underside of the wings are white flecks, and the whole wing edge of the greater wing coverts are speckled with white. 2nd, on the belly, between the legs, is a V in white feathers, the angle pointing towards the head. 3rd, the throat beneath the beak is white. The quills of the primaries are of a yellowish white; of the tail feathers, white below and black above; the extremities, however, are black; the quills of the other feathers, excepting those on the white flecks, are grey. The plumage of the

specimen shot by M. Preyer was brownish grey, in that shot
by Grímr the colour was more of an iron grey.

	Inch.	Line.
Length of beak	1	1
Length of leg . . .	1	5
From beak to tail . .	17	8

I now proceeded to cook the lake char (*Salmo alpinus*),
which the natives call silungur; their delicate salmon-
coloured flesh is delicious. Having split and stewed them
with a little meal, all was ready by the time that my comrades
returned.

" Mr. Briggs—what luck ? "

" Not much. I have been trying the fly for trout, but
they will not rise. I believe that they are so unaccustomed to
such things as flies here, that the only chance of catching
them is with minnow or spoon."

" Have you shot anything ? "

" No," from Martin ; " but we came upon a northern
diver's nest. Here are the two eggs I took from it—nest it
can hardly be called though, for the bird seems to have laid
haphazard on mud and stone."

He held out to me two olive-brown eggs sprinkled with
grey and brown spots—length, 3 in. 5½ lines ; width, 2 in.
2½ lines. The bird had escaped, but we could see it now and
then sailing on the water out of range.

The diver is a noble bird ; its dark plumage has a metallic
lustre ; the head and neck are black or green, according to
the light in which they are seen ; one broad white collar sur-
rounds the neck, beneath the chin is a thread of white like the
commencement of a second collar ; the black of the body is
flecked with white, as though the bird were dressed in magnificent black lace over white. The eye is of a blood-red
colour. The bird swims with great celerity, and it is hopeless attempting to come up with it in a boat ; it rarely
lands, as its short legs thrown beyond the point of equilibrium
in the body almost preclude its walking, yet are calculated to
give great propelling force in the water. It can remain below

the surface for a considerable time, and when it rises, if
alarmed, it will keep its body submerged, the dark head alone
showing.   As one comes suddenly on the diver in a lone tarn,
its harsh loud cry, like the howl of a wolf mixed with jeering
bursts of laughter, or the screams of a man in distress, is
sufficiently startling.   These Eagle lakes teem with wild fowl.

GREAT NORTHERN DIVER.

I saw several red-throated divers.   Ptarmigan scud among the
tufts of grey moss, and breed in the low willow-beds ;
phalaropes perch on the highest rocks ; whimbrel, with a

dreary cry, cower over their young before the wheeling falcon; and the plover pipes sadly on the stony hills.

My cookery was highly relished; the fish were done to a turn; the game would have been better, but for the carelessness of Grímr.

After a cup of hot toddy, we went to bed; Grímr, the farmer, Gúthmundr, and Magnús, were put with the boxes into my tent, 7 ft. by 5½ ft., and the rest of us huddled into the larger one. It was high time for us to get under shelter, for a dense fog had covered the lake, and the wind was beginning to sob over the waste, and bluster round our canvas, shaking the tent sides ominously.

About five in the morning my hammock vibrated like a pendulum; and I woke to find that the gale had increased to a storm, and threatened to upset tent and hammock on the sleepers below. The pegs I knew were not driven into firm soil, and the guys were loose. The sides of the tent bellied in, and flapped like sails.

"Mr. Briggs!" I called.

"Ay, what?"

"Oh! you are awake!" I said. "Do, there's a good fellow, take the mallet, drive in the pins all round, and tighten the cords."

"It is raining as it rains nowhere but in Iceland, driving horizontally great splashes of water, and the wind is blowing like——"

"Some one must go, but I cannot," said I, "for if I were to alter my balance, the whole tent would collapse."

"Well, let us make the Yankee go. Mr. Blank!"

But the American was too "wide awake" to be otherwise than fast asleep at that moment.

"Throw a boot at him!" I suggested.

"I can't," answered Mr. Briggs, "for the salt is in one and the pepper in the other; but here is your fishing stocking with the leg of mutton in it, or better, an old preserved-meat tin;" and that whizzed below me and struck the Yankee on the back. A loud snore proclaimed him to be invincibly asleep. We called again, but there was no awaking him. It would be useless asking Martin to turn out, we knew; so Mr. Briggs, with a groan, crawled from his snug bed and opening the tent flap, exposed himself to the full brunt of the gale.

I heard him outside cursing Iceland, hammering at a peg, tugging at a guy, and tightening it. Snap! went one of the braces.

"Ha!" exclaimed Mr. Briggs from without. "The extra guy rope I bought at Reykjavik, Icelandic manufacture!"

Poor Mr. Briggs! he must be drenched by this time, and very cold, I thought, and so he was—he came in with his night gear dripping, his teeth chattering, and the rain trickling off his fat body.

Next morning—I mean four hours later, when we rose— he was complaining of rheumatism down his back, not, he assured me, from having got wet through outside, but from the water having flowed into his bed from without, the tent being in a hollow. I saw now that the flooring was under water, as Grímr had prognosticated, and that the great bed had absorbed it like a sponge, no wonder that Mr. Briggs was rheumatic!

The weather cleared up, and the sun showed through a watery veil; so I determined on pushing on at once to Grímstúnga, at the head of Vatnsdalr. Mr. Briggs agreed to accompany me, as he did not relish the notion of another night on the heithi.

"We shall have a long ride," I said; "twelve hours in the saddle."

"Anything rather than sleeping in a puddle," he answered; "so the sooner we start the better."

It was not, however, till twelve o'clock, that we were under way. Mr. Martin and the Yankee remained for the fishing, and Gúthmundr was to take charge of our luggage. Thus, Mr. Briggs, Grímr and I, could ride without encumbrances, and with a change of horses, to get over the ground as speedily as possible.

The farmer of Kalmanstúnga guided us as far as Big Arnarvatn, where we were to fall in with the road. We passed several lakes, and noticed that in some places, snow patches, instead of discharging their thawed water by a stream into the tarns, decanted them down circular funnel-shaped openings in the mould. The beautiful angelica leaf (*Archangelica officinalis*) starred the black soil, by the margin of the lake : the stalk is delicious eating when full-grown, and is much used by the Icelanders. The black Iceland lichen everywhere; who first thought of using it for food, I wonder!

Presently a large sheet of water opened before us; it is not marked on Gunnlaugsson's map; the farmer called it Gunnarssonarvatn.

"And pray who were the sons of Gunnar?" I asked.

"Two young men, who, about a thousand years ago, came hither from Kalmanstúnga, fishing. They never returned, and when their father went to seek them, he found them seated by this lake, quite dead, with a plate of fish between them. It is supposed that the char and trout here are poisonous, and net has never been flung in the lake since."

After three quarters of an hour's fast riding, we were at the Big Arnarvatn, the most desolate spot imaginable. The lake is very large, it winds about among the hills, so that it is quite impossible to catch sight of the whole sheet at once : its waters looked chill and milky with undissolved snow; the high ridges of hill all round were perfectly barren, and swept abruptly to the marge, strewn with iron-grey and black rock in ragged splinters. Discoloured snow-blotches filled every cranny of the scarps, and the only green spot visible was the mossy headland opposite, where Grettir the outlaw, in 1019, had planted his cottage. Poor Grettir! a sad place of exile indeed! Twice had he there to do battle for his life against hired assassins, and yonder is a cleft in which he and his brave friend Hallmund defended themselves against fearful odds.

To the west was a long spit of black rubble, round which the lake curls, the cold waves fretted against it, lashed by a freezing gale from the north. In all the pools around, was green ice of great thickness—this too on the 27th June!

Our horses scrambled down the hill-side, and waded through a swamp of coal-black moss, crisp with ice splinters; a bit of rubble passed, and we drew rein by a hut, occupied in summer by three fishermen and a boy, the wildest fellows I ever saw. Hut is too good a name for the mud heap, four feet high and twelve feet long, in which these poor creatures lived. There was no opening for light, except the door, and through this, thin smoke curled from a fire, which the boy had lighted for roasting coffee. On one side of the cot was a pile of trout as high as the hovel itself, and the ground about it was strewn with their entrails.

The grey-bearded, fur-capped fishers kissed Grímr and the farmer; I took off my Glengarry to them, they bowed, and returned the salute,—so I escaped a kiss, and we galloped on.

Far, far away to the south stretched the enormous Lang jökull, forty-six miles long, ending in Lykla-fjall, faint and blue, with the quaint hunch Krákr projecting from its northern flank.

At the head of the lake, the Storisandr road branches off, and here is a little hut for the convenience of benighted travellers. My guide told me a long story about some post-Reformation bishop, who with a dozen theological students was snowed in there, and——what was the end of the story I forget; but I believe that the bishop ate the theological students, or *vice versâ*.

The Bútherá, which feeds the lake, shoots over a rock in a pretty cascade, and the road leads through the water just below it. A few marsh marigolds shone among the sedge on its bank.

Then we came out on a portion of the heithi, even more bald than any we had passed, for the grey willow roots were wanting now; and, far as the eye could reach, there was nothing but an apparently endless succession of slime and rock, snow patches and pools of water.

At six o'clock we saw turf again—brown and scanty, a

8

patch of about two acres, below a sand-hill, in a bend of the
Hólma-kvísl. Kvísl is the Icelandic for a feeder to a large
river. It was a satisfaction to see the water flowing north,
and to know that we had broken the neck of the heithi.

We rested our horses for half an hour. Grímr un-
strapped my fishing stockings from behind his saddle, and
shook out of one of them kaager, and from the other, a cold
leg of mutton.

After having satisfied myself, I was making for the sand-
hill with the intention of searching for fossil freshwater shells,
which are to be found in the sand formations between the
trap beds, when Grímr called me back, urging the necessity
of our not losing more time, as we should get no supper, if
we arrived late at Grímstúnga.

Near Hólma-kvísl the road for Víthidalr branches off to the
left. The mountains dividing that vale from Vatnsdalr rose in
greater majesty before us as we proceeded, but unfortunately,
their heads were shrouded in mist.

"In six hours," said Grímr, "we shall be at the head of
Vatnsdalr. It shall rain before we arrive."

CHAPTER VII.

THE VAMPIRE'S GRAVE.

Barren Wilderness—A Cairn—The Valley of Shadows: a Saga—A magnificent Gorge—Waterfalls—The Vatnsdalr—Arrival at Grímstúnga.

TOWARDS seven o'clock we reached perhaps the most repulsive portion of the heithi; scarce a blade of grass was visible. The land, for the most part, was a tract of mud and stone, with only here and there a patch of grey moss, covering though not disguising the hideous nakedness of the desert.

A pile of trachyte blocks indicated a road, which otherwise would have been undistinguishable, as no attempt had been made to clear the stones from the track; and it was only where these stones gave place to black mud, that by its kneaded filth we could ascertain that horses passed that way.

Mr. Briggs, Grímr, and I had ridden in silence for more than an hour, our spirits depressed by the revolting scene and by the dull quilt of cloud obscuring the sun; when suddenly Grímr drew rein, and pointing to a cairn distant about a quarter of a mile from the path, said in a solemn voice—
" There is Glámr's grave!"

Had there been any exhilarating object within sight, my guide would have been the last man to point it out. Much to his discontent I turned my horse's head and rode over the rocks to the spot.

The tumulus rises in a bend of the stream, and is composed entirely of stones gathered from the patch of ground

8—2

around it, which consequently is free from them, and able to produce a scanty crop of grass. The cairn may be fifteen feet high, and is bell-shaped: to the north, a coating of hoary moss has spread itself over it, and choked the interstices with its felty roots. A small tarn to the east, with cat-ice about the rim, is fed by patches of dirty snow, which seems hardly inclined to thaw this summer.

As I stand by the cairn, the wind soughs up from the north, lashing the viscous pool into ripples, rustling among the reeds, humming with a strange mournful note through the crevices of the dead man's home, then rolls onward, to furrow the snows on Eiriks jökull. A falcon wheeling overhead, with a harsh scream swerves in the blast, his wings flicker, and he soars aloft, to appear, but as a speck, against the whirling vapours. Not a plover nor curlew to be seen or heard. I draw my cloak closer about me, and pull my hood farther over my head. My companions shout, and, nothing loth to rejoin them, I spring upon my pony, and scramble back to the road.

" Pray, Padre, what have you to say about this Glámr, whose grave is in such an accursed spot ? "

" You shall hear, but if you get the blue devils by listening to my story, blame your own inquisitiveness."

" If they come on," replied Mr. Briggs, " I have a sovereign remedy; the blue devils shall be expelled by ardent spirits."

So I began the story of—

The Valley of Shadows.*

NEAR our halting place to-night, opens a glen, which, from its overhanging crags and generally sombre aspect, has, from time immemorial, been hight the Vale of Shadows. To-morrow we shall visit it.

---

* *Gretla*, chaps. 82—85. I give this story as a specimen of a very remarkable form of Icelandic superstition. It is so horrible, that I forewarn all those who have weak nerves, to skip it.

In the beginning of the eleventh century, there stood, a little way up this valley, a small farm, occupied by a worthy bonder, named Thorhall, and his wife. The farmer was not exactly a chieftain, but he was well enough connected to be considered respectable : to back up his gentility, he possessed numerous flocks of sheep, and a goodly drove of oxen. Thorhall would have been a happy man, but for one circumstance —his sheepwalks were haunted.

Not a herdsman would remain with him; he bribed, threatened, entreated, all to no purpose; one shepherd after another left his service; and things came to such a pass, that he determined on asking advice at the next annual council. Thorhall saddled his horses, adjusted his packs, provided himself with hobbles, cracked his long Icelandic whip, and cantered along this identical road; and in less time than we have taken over it, he reached Thingvellir.

Skapti Thorodd's son was lawgiver at that time, and, as every one considered him a man of the utmost prudence and able to give the best advice, our friend from the Vale of Shadows made straight for his booth.

" An awkward predicament certainly,—to have large droves of sheep, and no one to look after them," said Skapti, nibbling the nail of his thumb, and shaking his wise head,—a head as stuffed with law, as a ptarmigan's crop is stuffed with blaeberries. " Now, I'll tell you what—as you have asked my advice, I will help you to a shepherd; a character in his way, a man of dull intellect, to be sure, but strong as a bull."

" I do not care about his wits, so long as he can look after sheep," answered Thorhall.

" You may rely on his being able to do that," said Skapti. " He is a stout, plucky fellow ; a Swede from Sylgsdale, if you know where that is."

Towards the break-up of the council, " Thing " they call it in Iceland, two greyish-white horses belonging to Thorhall slipped their hobbles, and strayed ; so the good man had to hunt after them himself, which shows how short of servants he was. He crossed Sletha-ási—you remember the place, Mr.

Briggs; I made a sketch of Súlur from it, and close by is the Grettis-tak—well, thence he bent his way to Armanns-fell, and just by the Priest's-wood he met a strange-looking man driving before him a horse laden with faggots. The fellow was tall and stalwart: his face involuntarily attracted Thorhall's attention, for the eyes of an ashen grey were large and staring, the powerful jaw was furnished with very white protruding teeth, and around the low forehead hung bunches of coarse wolf-grey hair.

"Pray what is your name, my man ?" asked the farmer, pulling up.

"Glámr, an please you !" replied the wood-cutter.

Thorhall stared ; then, with a preliminary cough, he asked how Glámr liked faggot-picking.

"Not much," was the answer; "I prefer shepherd life."

"Will you come with me ?" asked Thorhall; "Skapti has handed you over to me, and I want a shepherd this winter uncommonly."

"If I serve you it is on the understanding that I come or go as pleases me.   I tell you I'm a bit truculent if things do not go just to my thinking."

"I shall not object to this," answered the bonder; "so I may count on your services !"

"Wait a moment ! You have not told me whether there be any drawback."

"I must acknowledge that there is one," said Thorhall; "in fact, the sheepwalks have got a bad name for bogies."

"Pshaw ! I'm not the man to be scared at shadows," laughed Glámr; "so here's my hand to it; I'll be with you at the beginning of the winter night."

Well ! after this, they parted, and presently the farmer found his ponies. Having thanked Skapti for his advice and assistance, he got his horses together and trotted home.

Summer, and then autumn, passed, but not a word about the new shepherd reached the Valley of Shadows. The winter storms began to bluster up the glen, driving the flying snow-flakes and massing them in white drifts at every winding of the vale. Ice formed in the shallows of the river, and the streams, which in summer trickled down the ribbed scarps, were now transmuted into icicles.

One gusty night, a violent blow at the door startled all in the farm; in another moment, Glámr, tall as a troll, stood in the hall glowering out of his wild eyes, his grey hair matted with frost, his teeth rattling and snapping with cold, his face blood-red in the glare of the fire which smouldered in the centre of the hall.

Thorhall jumped up and greeted him warmly, but the housewife was too frightened to be very cordial.

Weeks passed, and the new shepherd was daily on the moors with his flock; his loud and deep-toned voice was often borne down on the blast, as he shouted to the sheep, driving them into fold. His presence always produced gloom, and if he spoke, it sent a thrill through the women, who openly proclaimed their aversion for him.

There was a church near the byre, but Glámr never crossed the threshold; he hated psalmody, which shows what a bad man he was.

On the Vigil of the Nativity, Glámr rose early and shouted for meat. "Meat!" exclaimed the housewife; "no man calling himself a Christian touches flesh to-day. To-morrow is the Holy Christmas-day, and this is a fast."

"All superstition!" roared Glámr. "As far as I can see, men are no better now than they were in the bonny heathen time. Now bring me meat, and make no more ado about it."

"You may be quite certain," protested the good wife, "if church rule be not kept, ill-luck will follow."

Glámr ground his teeth, and clenched his hands: "Meat! I will have meat, or—— !" In fear and trembling the poor woman obeyed.

The day was raw and windy; masses of grey vapour rolled up from the Arctic Ocean, and hung in piles about the mountain tops. Now and then a scud of frozen fog, composed of minute spicula of ice, swept along the glen, covering bar and beam with feathery hoarfrost. As the day declined, snow

began to fall in large flakes, like the down of the eider-duck.
One moment there was a lull in the wind, and then the deep-
toned shout of Glámr, high up the moor slopes, was heard
distinctly by the congregation assembling for the first vespers
of Christmas-day. Darkness came on, deep as that in the
rayless abysses of Surtshellir, and still the snow fell thicker.
The lights from the church-windows sent a yellow haze far out
into the night, and every flake burned golden as it swept within
the ray. The bell in the lych-gate clanged for even-song, and
the wind puffed the sound far up the glen ; perhaps it reached
the herdsman's ear. Hark ! some one caught a distant shout
or shriek, which it was he could not tell, for the wind muttered
and mumbled about the church eaves, and then, with a fierce
whistle, scudded over the grave-yard fence.

Glámr had not returned when the service was over. Thorhall suggested a search, but no man would accompany him ;
and no wonder ! it was not a night for a dog to be out in ;
besides, the tracks were a foot deep in snow. The family sat
up all night, waiting, listening, trembling ; but no Glámr
came home.

Dawn broke at last, wan and blear in the south. The
clouds hung down like great sheets, full of snow, almost to
bursting.

A party was soon formed to search for the missing man.
A sharp scramble brought them to high land, and the ridge
between the two rivers which join in Vatnsdalr was thoroughly
examined. Here and there were found the scattered sheep,
shuddering under an icicled rock, or half-buried in a snow-
drift. No trace yet of the keeper. A dead ewe lay at the
bottom of a crag, it had staggered over it in the gloom, and
had been dashed to pieces.

Presently the whole party were called together about a
trampled spot in the heithi, where evidently a death-struggle
had taken place, for earth and stone were tossed about, and
the snow was blotched with large splashes of blood. A gory
track led up the mountain, and the farm-servants were following it, when a cry, almost of agony, from one of the lads

made them turn. In looking behind a rock, the boy had come upon the corpse of the shepherd—it was livid and swollen to the size of a bullock. It lay on its back with the arms extended. The snow had been scrabbled up by the puffed hands in the death agony, and the staring glassy eyes gazed out of the ashen-grey, upturned face, into the vaporous canopy overhead. From the purple lips lolled the tongue, which in the last throes had been bitten through by the horrid white fangs, and a discoloured stream which had flowed from it was now an icicle.

With trouble the dead man was raised on a litter, and carried to a gill-edge, but beyond this he could not be borne; his weight waxed more and more, the bearers toiled beneath their burden, their foreheads became beaded with sweat; though strong men, they were crushed to the ground. Consequently, the corpse was left at the ravine-head, and the men returned to the farm. Next day their efforts to lift Glámr's bloated carcase, and remove it to consecrated ground, were unavailing. On the third day a priest accompanied them, but the body was nowhere to be found. Another expedition without the priest was made, and on this occasion the corpse was found; so a cairn was raised over it on the spot.

"What! that which we have just passed?" asked Mr. Briggs.

"No," answered I; "Glámr was twice buried, as you shall hear."

Two nights after this, one of the thralls who had gone after the cows, burst into the stofa with a face blank and scared; he staggered to a seat and fainted. On recovering his senses, in a broken voice, he assured all who crowded about him, that he had seen Glámr walking past him, as he left the door of the stable. On the following evening a house-boy was found in a fit under the tún wall, and he remained an idiot to his dying day. Some of the women next saw a face, which, though blown out and discoloured, they recognized as that of Glámr, looking in upon them through a

window of the dairy. In the twilight, Thorhall himself met
the dead man, who stood and glowered at him, but made no
attempt to injure his master. The haunting did not end
there. Nightly a heavy tread was heard around the house,
and a hand feeling along the walls, sometimes thrust in at
the windows, at others clutching at the woodwork, and
breaking it to splinters. However, when the spring came
round the disturbances lessened, and, as the sun obtained full
power, ceased altogether.

That summer, a vessel from Norway dropped anchor in
Húnavatn. Thorhall visited it, and found on board a man
named Thorgaut, who was in search of work.

"What do you say to being my shepherd?" asked the
bonder.

"I should much like the office," answered Thorgaut; "I
am as strong as two ordinary men, and a handy fellow to
boot."

"I will not engage you without forewarning you of the
terrible things you may have to encounter during the winter
night."

"Pray what may they be?"

"Ghosts and hobgoblins," answered the farmer; "a fine
dance they lead me, I can promise you."

"I fear them not," answered Thorgaut; "I shall be with
you at cattle-slaughtering time."

At the appointed season the man came, and soon established himself as a favourite in the household; he romped
with the children, chucked the maidens under the chin, helped
his fellow-servants, did odd jobs for his master, gratified the
housewife by admiring her skyr, and was just as much liked
as his predecessor had been detested. He was a devil-may-
care fellow too, and made no bones of his contempt for the
ghost, expressing hopes of meeting him face to face, which
made his master look grave, and his mistress shudderingly
cross herself. As the winter came on, strange sights and
sounds began to alarm the folk, but these never frightened
Thorgaut; he slept too soundly at night to hear the tread

of feet about the door, and was too short-sighted to catch glimpses of a grizzly monster striding up and down, in the twilight, before its cairn.

At last Christmas-eve came round, and Thorgaut went out as usual with his sheep.

" Have a care, man ! " urged the bonder; " go not near to the gill-head, where Glámr lies."

" Tut, tut ! fear not for me.   I shall be back by Vespers."

" God grant it," sighed the housewife ; " but 'tis a wisht day to be sure."

" And pray, what does a wisht day mean ? " asked Mr. Briggs.

" It is a Devonshire expression ; wisht means anything ill-omened, desolate, dangerous."

" Go on then ; but don't put any more Devonshire expressions into Icelandic mouths," said Mr. Briggs.

Twilight came on ; a feeble light hung over the south, one white streak along this heithi we are crossing.   Far off in southern lands it was still day, but here the darkness gathered in apace, and men came from Vatnsdalr for evensong, to herald in the night when Christ was born.   Christmas-eve !   How different in Saxon England ! there the great ashen faggot is rolled along the hall with torch and taper ; the mummers dance with their merry jingling bells ; the boar's head with gilded tusks, " bedecked with holly and rosemary," is brought in by the steward to a flourish of trumpets.

How different, too, where the Varanger cluster round the Imperial throne in the mighty church of the Eternal Wisdom at this very hour !   Outside, the air is soft from breathing over the Bosphorus, which flashes tremulously beneath the stars.   The orange and laurel leaves in the Palace gardens are still exhaling fragrance in the hush of the Christmas night.

But it is different here !   The wind is piercing as a two-edged sword ; blocks of ice clash and grind along the coast of the Hunaflói, and the lake waters are congealed to stone. Aloft, the Aurora flames crimson, flinging long streamers to the zenith, and then suddenly dissolving into a sea of pale green light. The natives are waiting around the church-door, but no Thorgaut has returned.

They find him next morning, lying across Glámr's cairn, with his spine, his leg and arm bones shattered. He is conveyed to the churchyard, and a cross is set up at his head. He sleeps till the Resurrection, peacefully.

Not so Glámr—he becomes more furious than ever. No one will remain with Thorhall now, except an old cowherd who has always served the family, and who had long ago dandled his present master on his knee.

"All the cattle will be lost if I leave," said the carle; "it shall never be told of me that I deserted Thorhall from fear of a spectre."

Matters rapidly grew worse. Outbuildings were broken into of a night, and their woodwork was rent and shattered: the house-door was violently shaken, and great pieces of it were torn away; the gables of the house were also pulled furiously to and fro.

One morning, before dawn, the old man went to the stable; an hour later, his mistress rose, and, taking her milking cans, followed him. As she reached the door of the stable, a terrible sound from within—the bellowing of the cattle, mingled with the deep bell-notes of an unearthly voice, sent her back shrieking to the house. Thorhall leaped out of bed, caught up a weapon, and hastened to the cow-house. On opening the door, he found the cattle goring each other. Slung across the stone which separated the stalls, was something : Thorhall stepped up to it, felt it, looked close—it was the cowherd, perfectly dead, his feet on one side of the slab, his head on the other, and his spine snapped in twain.

The bonder now moved with his family to Tunga, the place where we sleep to-night; it was too venturesome living during the midwinter night at the haunted farm; and it was not till the sun had returned as a bridegroom out of his chamber, and had dispelled night with its phantoms, that he came back to the Vale of Shadows. In the meantime, his little girl's health had given way under the repeated alarms of the winter; she became paler every day; with the autumn flowers she faded, and was laid beneath the mould of the churchyard in time for the first snows to spread a virgin pall over her small grave.

At this time Grettir—of whom I have so often spoken—was in Iceland, and, as the hauntings of this vale were matter of gossip throughout the district, he heard of them, and resolved on visiting the scene. So Grettir busked himself for a cold ride, mounted his horse, and in due course of time, drew rein at the door of Thorhall's farm with the request that he might be accommodated there for the night.

"Ahem!" coughed the bonder; "perhaps you are not aware ——"

"I am perfectly aware of all. I want to catch sight of the troll."

"But your horse is sure to be killed."

"I will risk it. Glámr, I must meet, so there's an end of it."

"I am delighted to see you," spoke the bonder; "at the same time, should mischief befall you, don't lay the blame at my door."

"Never fear, man."

So they shook hands; the horse was put into the strongest stable, Thorhall made Grettir as good cheer as he was able, and then, as the visitor was sleepy, all retired to rest.

The night passed quietly enough, and no sounds indicated the presence of a restless spirit. The horse, moreover, was found next morning in good condition, enjoying his hay.

"This is unexpected!" exclaimed the bonder, gleefully. "Now where's the saddle, we'll clap it on, and then good-by, and a merry journey to you."

"Good-by!" echoed Grettir; "I am going to stay here another night."

"You had better be advised," urged Thorhall; "if misfortune should overtake you,—I know that all your kinsmen would visit it on my head."

"I have made up my mind to stop," said Grettir, and he looked so dogged that Thorhall opposed him no more.

All was quiet next night; not a sound roused Grettir from his slumber. Next morning, he went with the farmer to the stable. The strong wooden door was shivered and driven in. They stepped across it: Grettir called to his horse, but there was no responsive whinny.

"I am afraid——" began Thorhall. Grettir leaped in, and found the poor brute dead, and with its neck broken.

"Now," said Thorhall, quickly; "I've got a capital horse—a skewbald—down by Tunga, I shall not be many moments in fetching it; your saddle is here, I think, and then you will just have time to reach ——"

"I stay here another night," interrupted Grettir.

"I implore you to depart," said Thorhall.

"My horse is slain!"

"But I shall provide you with another."

"Friend," answered Grettir, turning so sharply round that the farmer jumped back, half frightened; "no man ever did me an injury without rueing it. Now, your demon herds-man has been the death of my horse. He must be taught a lesson."

"Would that he were!" groaned Thorhall; "but mortal must not face him. Go in peace and receive compensation from me for what has happened."

"I *must* revenge my horse."

"An obstinate man must have his own way! But if you will run your head against a stone wall, don't be angry because you get a broken pate."

Night came on: Grettir eat a hearty supper and was right jovial; not so Thorhall, who had his misgivings. At bedtime the latter crept into his crib, which, in the manner of old Icelandic beds, opened out of the hall, as berths do out of a cabin. Grettir, however, determined on remaining up; so he flung himself on a bench with his feet against the posts of the high seat, and his back against Thorhall's crib : then he wrapped one lappet of his fur coat round his feet, the other about his head, keeping the neck-opening in front of his face, so that he could look through into the hall.

There was a fire burning on the hearth, a smouldering heap of red embers ; every now and then a twig flared up and crackled, giving Grettir glimpses of the rafters, as he lay with his eyes wandering among the mysteries of the smoke-blackened roof. The wind whistled softly overhead. The clerestory windows, covered with the amnion of sheep, admitted now and then a sickly yellow glare from the full moon, which, however, shot a beam of pure silver through the smoke-hole in the roof. A dog without began to howl ; the cat, which had long been sitting demurely watching the fire, stood up with raised back and bristling tail, then darted behind some chests in a corner. The hall-door was in a sad plight. It had been so riven by the vampire, that it was made firm by wattles only, and the moon glinted athwart the crevices. Soothingly the river prattled over its shingly bed as it swept round the knoll on which stood the farm. Grettir heard the breathing of the sleeping women in the adjoining chamber, and the sigh of the housewife as she turned in her bed.

Click ! click !—It is only the frozen turf on the roof cracking with the intense cold. The wind lulls completely. The night is very still without.

Hark !—a heavy tread, beneath which the snow crackles. Every footfall goes straight to Grettir's heart. A crash on the turf overhead ! By all the saints in paradise ! the vampire is treading on the roof.

For one moment the chimney gap is completely darkened ; the monster is looking down it ; the flash of the red ash is reflected in the two lustreless eyes. Then the moon glances sweetly in once more, and the heavy tramp of Glámr is audibly moving towards the farther end of the hall. A thud—he has leaped down. Grettir feels the board at his back quivering ; for Thorhall is awake and is trembling in his bed. The steps

pass round to the back of the house, and then the snapping
of wood shows that the creature is destroying some of the
outhouse doors. He tires of this, apparently, for his footfall
comes clear towards the main entrance to the hall. The
moon is veiled behind a watery cloud, and by the uncertain
glimmer, Grettir fancies that he sees two dark hands thrust
in above the door. His apprehensions are verified, for with
a loud snap, a long strip of panel breaks, and light is
admitted. Snap—snap! another portion gives way, and the
gap becomes larger. Then the wattles flip out of their laces,
and a dark arm rips them out in bunches, and flings them
away. There is a cross-beam to the door, holding a bolt
which slides into a stone groove. Against the grey light,
Grettir sees a huge black figure heaving itself over the bar.
Crack! that has given way, and the rest of the door falls in
shivers to the earth.

"Oh, God!" exclaimed the bonder.

Stealthily the dead man creeps on, feeling at the beams
as he comes; then he stands in the hall, with the firelight
on him. A fearful sight; the tall figure distended with the
corruption of the grave, the nose fallen off, the wandering,
vacant eyes, with the glaze of death on them, the sallow flesh
patched with green masses of decay; the wolf-grey hair and
beard have grown in the tomb, and hang matted about the
shoulders and breast; the nails, too—they have grown. It
is a sickening sight—a thing to shudder at, not to see.

Motionless, with no nerve quivering now, Thorhall and
Grettir held their breath.

Glámr's lifeless glance strayed round the chamber: it
rested on the shaggy bundle by the high-seat. Cautiously he
stepped towards it. Grettir felt him groping about the lower
lappet and pulling at it. The cloak did not give way. Another
jerk; Grettir kept his feet firmly pressed against the posts,
so that the rug was not pulled off. The vampire seemed
puzzled, he plucked at the upper flap and tugged. Grettir
held to the bench and bed-board, so that he was not moved
himself, but the cloak was rent in twain, and the corpse

staggered back, holding half in its hands, and gazing wonder-
ingly at it. Before it had done examining the shred, Grettir
started to his feet, bowed his body, flung his arms about the
carcase, and, driving his head into the chest, strove to bend
it backward and snap the spine. A vain attempt! The cold
hands came down on Grettir's arms with diabolical force,
riving them from their hold. Grettir clasped them about the
body again; then the arms closed round him, and began
dragging him along. The brave man clung by his feet to
benches and posts, but the strength of the vampire was
greatest; posts gave way, benches were heaved from their
places, and the wrestlers at each moment neared the door.
Sharply writhing loose, Grettir flung his hands round a roof
beam. He was dragged from his feet; the numbing arms
clenched him about the waist, and tore at him; every tendon in
his breast was strained, the strain under his shoulders became
excruciating, the muscles stood out in knots. Still he held
on; his fingers were bloodless; the pulses of his temples
throbbed in jerks; the breath came in a whistle through his
rigid nostrils. All the while, too, the long nails of the dead
man cut into his side, and Grettir could feel them piercing
like knives between his ribs. Ah! his hands gave way, and
the monster bore him reeling towards the porch, crashing
over the broken fragments of the door. Hard as the battle
had gone with him indoors, Grettir knew that it would go
worse outside, so he gathered up all his remaining strength
for one final desperate struggle.

I told you that the door had shut with a swivel into a
groove, this groove was in a stone which formed the door-
jamb on one side, and there was a similar block on the other,
into which the hinges had been driven. As the wrestlers
neared the opening, Grettir planted both his feet against the
stone posts, holding Glámr by the middle. He had the
advantage now. The dead man writhed in his arms, drove
his talons into Grettir's back and tore up great ribands of
flesh, but the stone-jambs held firm.

"Now," thought Grettir, "I can break his back," and thrusting

9

his head under the chin, so that the grizzly beard covered his eyes, he forced the face from him, and the back was bent as a hazel-rod. "If I can but hold on," thought Grettir, and he tried to shout for Thorhall; but his voice was muffled in the hair of the corpse.

Crack! One or both of the door-posts gave way. Down crashed the gable trees, ripping beams and rafters from their beds; frozen clods of turf rattled from the roof and thumped into the snow. Glámr fell on his back, and Grettir staggered down on top of him. The moon was, as I said before, at her full; large white clouds chased each other across the sky, and as they swept before her disk, she looked through them like a pale saint in tribulation,—(I forget whose simile that is)— with a brown halo round her. The snow-cap of Jörundarfell, however, glowed like a planet, then her white mountain ridge was kindled, the light ran down the hill-side, the bright disk starred out of the veil and flashed at this moment full on the vampire's face. Grettir's strength was failing him, his hands quivered in the snow, and he knew that he could not support himself from dropping flat on the dead man's face, eye to eye, lip to lip, nose to where the nose *had* been. The eyes of the corpse were fixed on him, lit with the cold glare of the moon. His head swam, as his heart sent a hot stream through his brain. Then a voice from the grey lips said—

"Thou hast acted madly in seeking to match thyself with me. Now learn, that henceforth ill-luck shall constantly attend thee; that thy strength shall never exceed what it now is, and that by night these eyes of mine shall stare at thee through the darkness till thy dying day, so that for very horror thou shalt not endure to be alone."

Grettir at this moment noticed that his dirk had slipped from its sheath during the fall, and that it now lay conveniently near his hand. The giddiness which had oppressed him passed away: he clutched at the sword-haft and with a blow severed the vampire's throat. Then, kneeling on the breast, he hacked, till the head came off.

Thorhall came out now, his face blanched with terror, but, when he saw how the fray had terminated, he assisted Grettir, gleefully, to roll the corpse on top of a pile of faggots which had been collected for winter fuel. Fire was applied, and soon, far down Vatnsdalr, the flames of the pyre startled people, and made them wonder what new horror was being enacted in the Vale of Shadows.

Next day the charred bones were conveyed to the spot we have so lately passed, and there buried.

"Now then," said Mr. Briggs; "let us have a drop of whisky." Of course I assented.

"By the way," quoth my portly friend, "did the vampire's prophecy come true?"

"Yes, as you shall see when I send you my translation of the Grettis Saga."

"I shan't read it," said Mr. Briggs; "I like hearing a story, but I don't read much."

By this time our road lay among some small lakes which abound in trout and char. A couple of swans were sailing majestically on one of them. Crossing a river some dozen times at least, we descended from the heithi into a pretty glen down which the river rolled. The bottom reached, we changed horses, crossed the stream once more, scrambled up the opposite scarp, and keeping along the ravine edge, felt cheered by the reappearance of the pink lamba grass (*Silene acaulis*) and the white tassels of the cotton rush. The river too began to brawl over rocks in an excited way, making straight for a chasm of black crag, out of which I could hear the rumble of a considerable fall. After having climbed one of the cliffs we came on a view of the ravine which was particularly striking.

The mountains had been rent by an earthquake, and into the abyss, several hundred feet deep, the eye glanced from crag to crag, till it rested on a green pool into which the cascade plunged. Time was pressing and I was obliged to gallop on without making a sketch. I overtook the packhorses and Grímr, as they crossed two kvíslar or brooks parted from each

9—2

other by a narrow shred of rock. These streams rose far
apart, but here they flowed side by side and bounding down
the gorge met in foam at the bottom.

Clouds gathered now thicker around us, and a nasty
drizzle shut out all the prospect.

In an hour we came out above the broad green valley of
Vatnsdalr, and my guide could not refrain from a burst of
admiration at the verdure of the many túns scattered along it.
After the desolation of Kaldidalr and the bleakness and baldness of Arnarvatns heithi, it was a welcome sight to us. Close
at our feet lay the little church and farm of Grimstúnga.
Our horses brisked up wonderfully, the grey forgot that he
was bearing so fat a man as Mr. Briggs, the chestnut was
oblivious of his packs, and all at a swinging canter came up
to the farm door.

S. Harting Gould, del.

Edmund Evans, sc.

.

CHAPTER VIII.

THE VALE OF WATERS.

Vatnsdalr—Forsœludalr—A Dangerous Predicament—Thorolf wi' the Dead-face—An ancient Fort—Colonization of Vatnsdalr—Hóf—The Story of Hrolleifr: a Saga—MSS.—Old Nick—Icelandic Endearments—Music—Hankagil—Undirfell.

VATNSDALR (pronounced Vatzdalur) is considered one of the most fertile of northern valleys. It rightly deserves the name of the Water-vale. Through it winds a fine river formed by the confluence of two streams, that along which we had ridden yesterday, and another which sweeps through the Shadowy Dale, meeting below Grimstúnga. Cascades "spill light" down every mountain scarp, and their waters gurgle through impassable bogs to the main river. At the mouth of the vale are lakes of considerable extent, the resort of countless swans and wild ducks.

I rode with Grimr up the Vale of Shadows (Forsœludalr), and visited the site of Thorhall's farm; a few low foundations on a knoll, round which the river sweeps, mark the spot—they are, however, the ruins of a byre more modern than that which Glámr haunted. The hill on the left of the river is strangely barren, grass scantily covers it, like the skin on the ribs of a lean horse. As we ascended the dale, these terraces became more marked, and the cliffs being bare, showed the formation very distinctly. The mountains are formed of basalt beds overlying layers of sand; evidently the former have been erupted at intervals, allowing sand to accumulate on each successive couch of molten matter before a fresh overflow.

This process has been repeated again and again, and finally, the whole mass has been heaved up, forming the vast central heithies and terraced mountains at their skirts.

The vale narrows to a gorge, and the rocks, barred black and red, have a rich umbreous colour effect, which is very striking.

We lost our way, and in following a sheep-track, or what we took to be such, got into a sufficiently perplexing situation. The river boiled a hundred and fifty feet below us, and we were on a ledge canted over the gulf, the rock sheered up some hundred feet above our heads, and a fall of washed shale lay in a slant on the terrace before us. I despaired of getting the pony over this; it was perilous in the extreme scrambling across on hands and knees, and in venturing to do this I set the shingle in motion, so that though I drew quickly back, a shoot of rubble and sand whizzed into the chasm. "We must go on," said Grimr; "it is impossible to turn the horses round." It was so indeed, there was not room even for Grimr to pass to the head of my horse, which was foremost.

Knotting my bridle on the pony's neck lest it should slip and entangle his feet, I crept along the slant, supporting myself on my whip, which I drove at each step into the loose soil. "Bottle-brush," my piebald, put his nose to the ground, and advanced one foot, snuffed out a firm spot, and planted the other; then came a particularly critical slide of shale, which was wet with tricklings from the rock overhead. Bottle-brush pawed the earth away till he had scraped a hole through the rubble to the firm rock, and then fixed his hoof resolutely in it. Slowly and cautiously he advanced. Ah! the crumbling basalt gave way once, he floundered down, was up again, the dislodged rubbish puffed into the indigo abyss, and the little flat slatey fragments of clinkstone tinkled down the slope and leaped into the water.

The Icelandic horses are wonderfully sure-footed, they will climb wherever a goat can clamber, will trot over wastes of angular stone fragments, and tread fearlessly over bogs, supported only by a network of long grass.

When the dangerous slant was passed, my pony pushed his droll big head under my arm and rubbed it against my side, evidently expecting to have a word of praise.

We crept along the same ledge till a slope of turf allowed us to scramble up to the heithi, and as we were on the wrong side of the river, we had to descend the vale till we reached a convenient spot for crossing the stream.

We came out shortly on a bluff overhanging the junction of the Frithmundará and the Stránga kvísl. Both rivers pour through deep chasms; the former, leaping headlong into the rift over a lip of basalt, is broken by a ledge into a second fall halfway down, before it rains into the dark green well below. The left-hand river rolls through a wider gorge, up whose blue vistas the eye traces it from plunge to plunge of foam. At the point of confluence a sharp rock, gashed by the torrent, shoots to a point, and then widens to a triangular plateau, connected with the main hill only by a sharp neck of rubble, so narrow that one only can walk along it at a time. This is the spot chosen by Thorolf wi' the Deadface (he was so nicknamed from his livid complexion) for his castle; in caves burrowed in the sand strata he sacrificed men, and this pure river, once rolling bits of hacked flesh into the Water-vale, told of horrible atrocities committed here. Thorolf considered the spot impregnable, and for some time he lived here unmolested, preying on his neighbours' cattle. A thrall vanished from some near farm now and then, and for days the lone shepherd up the heithi heard shrieks which were echoed from crag to crag along these ravines. At last Thorstein and Jökull, brother chiefs of the Vatnsdalr, with a large body of retainers, assaulted the robber-fort. Thorstein stood where we are now, and fired arrows into the castle, attracting the attention of the besieged, whilst Jökull crossed the river-head, and tripping lightly over the neck of rubble, crept up the wall at the back of the fort, by thrusting the horn of his axe between the stones and dragging himself up after it. Thorolf was slain and his fort ruined.

I scrambled up to the plateau, picking heart's-ease at

every step, and found the traces of a circular mound—very
faint they were.   No wall remains now, though portions of it
existed at the end of the last century, till the undermined rock
fell, carrying them with it.   A huge node of crag, which is
now nearly severed from the cliff, seems to have been drilled
by caverns, the falling in of which has cut it off from the main
crag; one hollow, six feet deep, is all that remains of the
sacrificial vaults of the dead-faced Thorolf.

As rain began now to patter down pretty heavily, we made
the best of our way back to Grimstúnga.

Vatnsdalr was first colonized (A.D. 900) by Ingimund the
Old, who left Norway because some Finns had spaed that he
should settle in Iceland, and he knew that it was useless to
resist destiny.   The account of the arrival and establishment
of the settlers in Vatnsdalr is so simply and naturally told in
the Saga, that I am tempted to give it.

Ingimund had landed in the Borgar-fjord, had crossed the
heithi to Hrúta-fjord, and wintered in Vithidalr.   But when
spring came, and the snows began to thaw in the district,
Ingimund said, " It would be a satisfaction for me to know
that some fellows went up one of these high peaks, and took
a survey of the country, to see whether there be less snow
elsewhere ; for I do not much fancy erecting our home in this
dale.   A poor exchange, this, for the glens of old Norway ! "

Accordingly some men climbed a lofty mountain, and
looked about them far and wide ; then, returning to Ingimund,
they told him that the country was much less snow-covered
towards the north-east, and that it had a pleasanter look
altogether.   " As for this place," said they, " it seems to be
pretty much exposed to bad weather all the year round, and
the land yonder looks far more productive."

Quoth Ingimund, " Very well, we shall look up your green
lands."

So when spring was well set in, they busked themselves
for flitting, and crossed into Vatnsdalr.

" Ah ! " said Ingimund, " the Finns' spae is come true ;
I recognize the lie of the land from the description given me :

so here we shall settle. Certainly we have bettered our condition, for I see that the land hereabouts is well wooded,* and if it be fertile we shall build."

Now when they came to Vatnsdale river, Vigdisa, Ingimund's wife, said, " I must rest here a while, for I am in my pangs."

Ingimund said, " Very well!" And then she gave birth to a little girl, whom they called Thordisa, after Ingimund's mother.

" We must name this spot Thordisa's Holt," quoth Ingimund.

Then the party descended the vale, and seeing that the produce of the earth was good, that there was plenty of grass, and the hill slopes covered with well-grown woods, very beautiful to the eye, they were rejoiced, and their faces brightened up. Ingimund claimed all Vatnsdalr from above Helga lake and Urthar lake, &c. He pitched on a little dell for his home, and set about erecting a byre, with a great hall one hundred feet long. The mansion he called Hóf. Ingimund's men spread themselves all over the dale, and chose sites for habitations after his advice.

That winter much ice formed, and when men went on the ice they found a she-bear with two white cubs. Ingimund was on that trip, so he brought both whelps back with him, and said that the sheet of water should be called Húna-vatn (cub lake), and the frith which opened into it Vatna-fjord. Well! Ingimund built a fine house, and soon became chieftain of the dale and whole adjoining district. He had a great many head of cattle, cows, sheep, and small stock; during the autumn some of his sheep escaped, and they were found in spring in a wood now called Sheepsdale (Santha dalr), and it shows how productive the land must be that the sheep should stray thus. It fell out that some swine also broke loose, and were not found till next year, and they had increased to a hundred, and were quite wild. One old hog who

---

* There are at present no trees in Vatnsdalr.

followed them was called Beigathr.   Ingimund gathered his
men together, and they drove the swine towards the lake now
called Swinelake (Svina vatn), and wanted to bring them to a
standstill there, but the boar leaped into the water and swam
across, but was so tired that his hoofs dropped off; he
scrambled to a hill now hight Brigathr's hill, and there died.

Ingimund now took his ease in Vatnsdalr, several districts
were inhabited, and law and the rights of property were
established.*

Hóf is planted in a pretty spot, between shale mounds,
covered with heart's-ease of all varieties of colour—sulphur,
yellow and blue, deep purple.   The present farm is very poor,
its three gables sadly rickety, and its weathercocks too rusty
to turn in the wind.

A pretty little girl offered to lead me to the site of the
temple, as her sulky brother was too shy.   It is planted on a
knoll at the back of the farm, and consists of the foundations
of two chambers, some eight feet square.   Looking hence
across the river to the shingle hill of As (pronounced Ouse),
it was impossible to recall without emotion the stirring scenes
of the past, and especially one sad deed of treachery, here
committed.   Mr. Briggs saw that something was on my mind,
and asked me what it was?   " You shall have the story
nearly in the historian's words, as we ride hence," I answered.
We mounted our horses.   I did not reject the proffered
kiss of the little girl, and, as we rode to the river, I began—

The Story of Hrolleifr. †

INGIMUND THE OLD was the only man who would open his
doors to two such ill-conditioned people as Hrolleifr (pronounced Hrodlayver) and his hag of a mother, Ljót, when
they were driven from their own district on account of their
disreputable conduct.   Ingimund was a kind-hearted man,
and knowing that they were homeless, and were moreover
connected with a friend of his, he took them under his roof

---

* *Vatnsdæla Saga*, cap. 15.   † *Ibid*. caps. 22—26.

at Hóf, and nourished them as his own family. They were
disagreeable people to have in the house, and stirred up
bickerings and strife all day long. Ingimund's sons could not
tolerate them, and begged their father in vain to turn them
out of doors. At last the old chief made over some land and
the farm of As to the hag and her son. Hrolleifr had come
into the district without so much as a silver ring on his arm,
and now he had a farm with good buildings, sheep, horses,
cows, and a right of fishing in the river, all found him, out of
kindness, by Ingimund the Old. As for the fishing, there was
plenty of it in Vatnsdalr—salmon, trout, and char were abundant. Ingimund's four sons, Thorstein, Jökull, Högni, and
Thorir, had divided the farmwork between them, for in those
days the chief's sons were wont to do the drudgery as well as
others; and part of their work was fishing. The brothers
did not exchange many civilities with Hrolleifr, for they disliked the man and grudged his living at their expense, and
repaying every kindness with evil. Now Ingimund's arrangement with regard to the fishing was, that Hrolleifr might fish
whenever his own men were not throwing their nets in the
river, and then only. However, the fellow was too great a
scoundrel to pay attention to regulations of this kind.

One evening Ingimund's churls came down to the river
and found Hrolleifr drawing his net; so they warned him to be
off as their turn had come; but Hrolleifr replied that he cared
not a rap for what a set of thralls said. The men begged him
to be advised and not stir up a quarrel with Hóf-men, for that
was rather different from squabbling with the underlings of
petty farmers. Hrolleifr answered by bidding them—" Pack !
a set of rascally thralls that you are !" and he drove them
from the water's edge. They cried out, " You are acting most
wrongly; Ingimund is deserving of great courtesy from you,
for he received you when you were homeless, and everything
you have, land, house, live-stock, and fishery, you owe to
him; and a good-for-nothing fellow you are, so most folk
think !" Hrolleifr answered by picking up a boulder and
knocking one of the churls down, and bidding the whole crew
cease from arguing with him.

Away they scampered, and reached home, panting, just as supper was served. Ingimund asked why they were in such a flurry, and they replied by telling him how Hrolleifr had driven them from the river with blows and insults.

"Indeed!" exclaimed Jökull, the second son, a testy and hot-headed fellow. "The man wants to lord it over all Vatns-dalr; but it shall never be said that a troll like that has made us knuckle under."

Thorstein, the eldest, said, "This is too bad; however, I dare say we may make some compromise with Hrolleifr."

"Something of the kind is certainly necessary," said Ingimund; "do your best, my son, to allay the quarrel; if you come to blows you will find it hard to defend yourselves, for he is a cursed fellow, and ill befalls those who have to deal with him."

Jökull burst forth with, "I shall soon see whether he will leave the river or not!" and he started up from table.

Ingimund held his eldest son, and said, "Thorstein, my son! I rely on you; go with your brother."

Thorstein answered that he hardly knew how to restrain Jökull when his blood was up, and besides he doubted whether he could himself stand back if Jökull attacked Hrolleifr.

Well! they came to the river and found Hrolleifr fishing in it. Then Jökull shouted out, "Fiend! Away with you from the river; venture not to squabble with us!"

Hrolleifr answered contemptuously, "You are only two over there, and were you four I should not budge an inch; storm your utmost!"

"You trust through your mother's witchcraft to get all the fishing out of our hands!" cried Jökull, and then he rushed into the river and waded towards him, but Hrolleifr did not stir.

Then Thorstein called out, "Do lay this waywardness aside, Hrolleifr! you must see that you will get the worst of it if you meddle with our rights, and we are not the people to be borne down by violence, however well you may find that answer with others."

Jökull burst in with "Kill that devil there!"

Then Hrolleifr quietly stepped up on land, for there were stones there, and he pelted the brothers; they retaliated, but Hrolliefr remained unhurt.

Jökull wanted much to cross the river and punish the rascal, but Thorstein advised a retreat, for " It is all very well having to do with respectable people, but I do not fancy getting into the clutches of this fellow's mother, who is a downright witch, and her son is no better."

Jökull replied that he cared nothing for that, and begged his brother to keep up an incessant shower of stones whilst he was wading across.

In the meantime an eyewitness of this scene rushed off to Hóf, and told Ingimund what was going on. The old man called out, " Saddle my horse ! I must ride to the river." He was old and nearly blind, so that he had made over the management of his property into the hands of his sons. He had got a little country lad for his guide, and the boy led the horse. They came then to the river bank, and the white-haired chief was dressed in an ample blue cloak.

Thorstein exclaimed, " Here comes our father, Jökull ! let us desist, and then he will be pleased to think that we are falling in with his wishes. I am anxious about the old man, and do not like his being here. Jökull, do draw off !"

Ingimund rode down into the river, and called out, " Go away from the river, Hrolleifr; think of what becomes you ! "

Now directly Hrolleifr saw him, he flung his spear, and struck him in the middle. It was dusk, so that neither his sons nor the boy noticed the blow; but, directly he received it, Ingimund snapped the haft, folded his blue mantle round him, and rode up the bank.

" My little lad," said he, " lead me home."

So the boy guided him back, and the evening had quite closed in when they arrived. As Ingimund tried to get off his horse, he said, " I am getting rather stiff; you see we old folk are rather shaky."

Then it was, as the lad helped him down, that the boy saw the spear thrust through him.

Ingimund said softly, " You have long been true to me, child! Now go, as I bid you, to Hrolleifr, and tell him that my sons will certainly be on his track by morning, so he must fly before dawn. I do not wish my blood to be revenged on him ; it becomes me rather to screen the man whom I have taken under my protection as long as I am able." Then, leaning on the boy, he went into the hall, sat himself down in the seat of honour, and bade the servants not light the fire or kindle a lamp before the return of his sons.

The boy ran to the river, and saw Hrolleifr draw out a fine salmon, which he had just caught.

" Dog of a fellow!" shouted the exasperated boy. " You have done a deed for which no mulct will be asked. Now, reluctantly, I bring you a message from the dying lips of him whom you have slain. Fly at once, before the brothers are after you ; yet, on my honour, I long to see their axes rattling about your skull."

" Were it not for the news you bring me," answered Hrolleifr, " you should not slip away with a sound skin."

Now I must tell you that Ingimund's sons came home in the gloaming, and talked thus to each other on their way—

" The like of Hrolleifr is not to be found far and near ! " And Thorstein added, " My mind sadly misgives me about my father's ride to the river. There is no knowing what mischief that Hrolleifr might do ! "

They came in ; Thorstein walked to the end of the hall, and, whilst feeling along the ground with his hands, he asked, " How comes there any wet here ?"

The housewife answered that it ran in all probability from the old man's clothes; but Thorstein said, " It is gluey, like blood! Quick! kindle a light!"

When this was done, they found Ingimund enthroned in his high seat, dead, and the spear was thrust through him.

Jökull was like a madman. Thorstein could hardly restrain him from rushing off to revenge himself on the murderer. " You do not know the kind-heartedness of our father. The boy is not here. Undoubtedly our father sent

him to warn the miscreant to escape. It is then useless our
hastening to Ás; we must rather take counsel in the matter,
and not act precipitately. Truly! there is a great disparity
between our father and his murderer, and this will avail him
before Him who created the sun and all the heavens, and
who assuredly is Great, whoever He may be."

Jökull was quite frantic, and there was no pacifying him.

Just then the lad returned, and Jökull was wroth with him
for having warned the murderer. But Thorstein observed
that the boy was not to blame, as he had acted in accordance
with the father's wishes.

After the burial the brothers agreed not to occupy their
father's seat, nor to frequent public gatherings, till he was
avenged.

"And this," said Mr. Briggs, as we plunged into the
river, "is probably the very spot where the brave old man
received the blow. Look! yonder, on the opposite side, is
the bed of rolled stones from which the scoundrel—I can't
remember his name, much less pronounce it—picked up the
boulders wherewith to defend himself. Now go on with the story,
and don't add anything of your own, or you will spoil it."

"Instead of adding to the narrative, I am going to curtail
it; and I shall pass over the meeting of Hrolleifr with his
ugly mother, and the advice she gave him to flit at once, as
the first nights after a murder are the bloodiest, and I shall
land Hrolleifr in Skaga-fjord, whither he fled, at the house of
his kinsman Geirmund, of Sœmundarhlith."

"Well, what news?" was Geirmund's exclamation, as
Hrolleifr cantered up to his door.

"Ingimund, the bonder, has been slain!"

"Alas! there fell a right worthy man."

"He was rather shilly-shally though!" broke in the
murderer.

"But how was he killed?"

Hrolleifr told his kinsman the story. Geirmund waxed
wroth, and said, "I see clearly that of all ill-disposed fellows

you are the worst, a low, luckless blackguard. Be off, I'll not have you here."

"I am not going to budge from this spot," replied the other. "Here I remain, here I shall be taken, and here slain; then you shall bear the disgrace. Remember, my father fell in your service."

"Stay if you choose," quoth Geirmund; "but directly Ingimund's sons come here, I shall give you up."

There was a hay-barn close by, and Hrolleifr said that he should hide in it.

"As you please," from Geirmund.

Ingimund's sons remained quiet during the winter, and sat on the nether bench, and did not show at any pleasure parties, or at the Things; but as summer drew nigh, Thorstein collected his brothers and proposed that the conduct of the revenge should be entrusted to one of them. The brothers unanimously agreed that Thorstein, as being the coolest and most sensible of them all, should be their leader.

One morning he woke them early, and said: "Busk you for a ride!" The brothers rose and mounted their horses: no one accompanied them. Thorstein led them to Geirmund's house, and he received them cordially. They spent the night with him and were most hospitably treated. Next morning Thorstein told his brothers to have a game of draughts whilst he went out to take a quiet chat with their host. They did so: then Thorstein drew Geirmund aside, and said, "We have come hither in quest of Hrolleifr, whom we expect to find skulking hereabouts. You are the man to help us, as it was through your father that the scoundrel came to my father. Now I know that the deed done was far from your wish."

"It was so, truly," answered Geirmund; "you are quite right to hunt the murderer down. However, he is not in my house."

"No," said Thorstein; "but we are convinced that he is in your shed. Look here! I make over a hundred pieces of silver to you; dismiss Hrolleifr, and I shall take precautions that his blood be not shed on your lands, so that no stain may rest on you for suffering it. Track him down we *must*. Go to the barn and tell the fellow that we are here, and that you have no thoughts of keeping us from off him."

"I shall do so," replied Geirmund; "and advise him to decamp; and do you track him down when he has left my shelter."

"Be it so."

Geirmund then stole into the out-house and told Hrolleifr that Ingimund's sons were in full hue and cry after him.

"Then I must run for it," said the skulking rascal.

Geirmund returned to Thorstein and said, "Don't be too hasty, wait here another day." To this the young man agreed.

Next morning the brothers took themselves over the scaur, and lighted on the track of a man in the snow. "Let us sit down here," quoth Thorstein; "and I shall tell you my conversation with Geirmund. It is true that Hrolleifr was at his house whilst we were there."

"You are a fine fellow," burst forth Jökull, "to dangle about a farm composedly, whilst your father's murderer was within reach. Had I only known it, none of your wise reasons would have withheld me from my revenge."

"Be patient," said Thorstein; "it would not have done to let Geirmund be involved in the murder of his kinsman. We are in full scent now; here are the traces in the snow, pointing towards As. Now I guess that Ljót will be offering a sacrifice for the coming in of summer, according to her wont, and we must strike the blow before that is accomplished."

Jökull jumped up shouting, "Let us hasten on at once!" and as he was walking ahead, he looked scornfully over his shoulder, and said, "Bad luck to a man as frail of hand and heart as brother Thorstein. I am sure the chance of revenge will slip through our fingers through his procrastination, and we shall do nothing."

Thorstein answered calmly, "It is not clear yet that your indiscreet impetuosity would be more advantageous than my matured plans."

Towards undern they reached Hóf, and supper was ready.

10

Thorstein stepped outside, and calling a herdsman to him, said, "Run to As, tap at the door, and see how long they are before opening the door; sing in the meantime some verses, and give out as your errand, a search after strayed sheep. Should they ask whether we have returned, reply in the negative."

The shepherd went to As and knocked at the door; no one answered his rap, till he had trolled out twelve verses, then a house-churl came and asked what he wanted, and also whether the brothers had come home. He replied that they had not, and then asked after his sheep: he was told that they had not strayed thither.

So the shepherd returned, and told Thorstein how many verses he had chanted.

"Humph!" quoth Thorstein; "then they had ample time for arranging things within before they admitted you: but did you step inside?"

He said that he had, and that he had looked about him.

Thorstein inquired, "Was there a brisk fire on the hearth?"

The man replied that there was, "Just enough to be quickened up in a twinkling."

"Did you notice any peculiarity at all about the house?"

"Yes," answered the shepherd; "there was a great bundle in one corner, and a bit of red dress peeping out of it."

Thorstein exclaimed, "Then you saw the sacrificial robe! we must give chase at once, and do our best."

So he and his brothers hurried to As, and no one was outside the house; they saw a heap of logs on either side of the roof-ridge filling up the space between the gables, and a little house standing before the door of the byre.

"That is the temple," quoth Thorstein; "Hrolleifr will go thither when all his devilry is ready. Now, all of you hide behind the corner of the house, and I will sit up here over the door, holding a club in my hand. When Hrolleifr steps out, I will throw the log towards you, and then rush to my assistance."

"I see, brother!" exclaimed Jökull; "you want to get all the merit of the job yourself. I will sit up in the wood-stack with the stick."

"Have your own way," answered Thorstein; "but you are too rash to be well trusted; you may bring us into a hobble."

Jökull scrambled up into the pile of logs, and the others lurked behind the wall.

Presently out came a man, who looked round, and not seeing any one, gave a signal that all was safe, and out stepped a second man, then a third, and the last was Hrolleifr himself.

Jökull recognized him instantly, and, turning sharply round, the pile of logs gave way, yet he managed to fling the club towards his brothers; then he slipped down on Hrolleifr and caught him in his arms, but with the impetus both fell on the ground and rolled down the slope, so that one lay on top of the other.

Out rushed the brothers, and Högni exclaimed, "Look, only look, a troll is coming towards us, what can it be?"

What he saw was Ljót, who was running towards them with her head between her feet, and in a manner truly frightful; flashes of glamour shot from her eyeballs.

Just then Thorstein shouted to Jökull, "Kill Hrolleifr at once, now you have the chance!"

"That is what I am about," answered Jökull, smiting the murderer's head off, with a hope that evil might befall him elsewhere.

"Ah, ha!" yelled Ljót; "you sons of Ingimund have luck on your side, or I would have overmastered you!"

"Pray, how would you have done that?" asked Thorstein.

"If I had only seen you before you caught sight of me, I would have made your heads spin so that you would have grovelled like barrow-hogs on the ground."

"Fortune wills it otherwise," said Thorstein, cutting the hag down; and so she perished in her evil temper and her sorcery. So now these two are done for, and badly too.

At Grimstúnga I obtained two MSS. of interest; one was a 12mo volume of Sagas and Rimur, written in different hands, and at different times; of these, the Saga of Asmund the Viking is unpublished; the other MS. was the Harald's Hringsbana Saga, wanting the first leaf, also unpublished, and moreover, not mentioned by Müller in his *Saga Bibliothek.* At the end of the *Nitidar Rimur,* in the first book, are some curious broken lines, arranged like the well-known Latin,—

<pre>
          pit   em   pit   rem,
qui ca   uxor   ca   atque dolo
          ret   e   ret   re.
</pre>

The Icelandic *Vers brisés* are more ingenious, though less intelligible.

<pre>
Hni   ta   skun   skr   skr
     kars   li∂   dar   ef   eittum
nu   fa   þun   þr   þr
H   R
     alli   yma.
F   G
</pre>

Hnikars tali∂, or Odin's speech, signifies poetry. It is from this name of the Scandinavian god, that we get our vulgar appellation of "Old Nick," and the word is derived from some root signifying to rage with freakish violence, common to several of the Aryan tongues. Thus we find the Greek νίκη, victory; the old Norse, or Icelandic *hnika,* to agitate, strike; and its cognate verbs *hnikkja,* to thrust forward violently, and *hnekkja,* to repel: hence also the Anglo-Saxon *næcan,* to kill; the Latin, *necare;* the German, *knacken;* the Danish, *nykke,* whim, freak; and the English, *knack, knock.*

From having this meaning of violence accompanied with whim, it was early applied, in mythology, to the divinities connected with the elements; and, beneath the tempestuous skies and wild lashing seas of the north, Odin, as their ruler, was termed, Hnikarr, Nikarr, or Hniku∂r—the Mœso-Gothic form of which was *Nikuz.* The rainbow was, "regn bo∂i Hnikars."

Christianity overthrew the worship of Odin, but, like the polypus, though cut to pieces, he revived in each morsel an entire Nick, to frequent the waters throughout the north of Europe.

The Swedish *Neck* appears, generally, as a handsome youth with his lower extremities like those of a horse. In Norway, the *Nök* lives in lakes and rivers, and demands a human victim every year. There, any one approaching extensive sheets of water, must not forget to say, "Nyk! Nyk! needle in water! The Virgin Mary cast steel into water! Sink thou, I float!"

In Germany he is called *Nisc* or *Neck*—the river Necker is named after him. In the Netherlands he is to be heard of under the same appellation. In North Germany exists, under the surface of the water, the black *Nickle man*, or *Nick*, who is formed like a man as far as the middle, but terminates as a fish; he has very sharp teeth; his usual food consists of fishes, but he not unfrequently drags down human beings. In Thale, the country people were obliged, till lately, to throw a black cock into the Bode every year, for, if they omitted to do so, the Nick would catch and drown some one. *Nykr*, as a water-horse, frequents several Icelandic lakes, among others that in the Vatnsdalr. He is fond of getting human beings to mount his back, that he may plunge with them into his native element, and make them his prey. The Icelanders have a lay about a certain damsel named Ellen, who was thus carried off by *Nyk*. This ballad exists in other languages, such as Faroese, Norwegian, Swedish, Danish, German, English, Wend, Slovakian, Bohemian, and Breton.

Poor Nick! I pity him, for he is of honourable family, being, as you will observe, descended from the gods of Asgaard. Why should England have taken upon herself to bedaub him with layer upon layer of lamp-black?* I will tell one story about him, and have done.

Once upon a time, an old priest was ambling homewards on his nag, and as, towards even-fall, he neared a pool, to his astonishment he saw a lad, naked to the waist, sitting on

---

* Among the *grafiti*, or scrawls, on the walls of Pompeii, is a school-boy sketch of Pluto, in black chalk, armed (as heralds would say) with horns, hoofs, and tail, just the very appearance Nick has taken upon himself now.

the surface of the water, his long golden curls floating over
his delicate shoulders, from beneath a jaunty red cap. The
Neck held a shining harp in his hand, and from it rang the
sweetest harmony as he chaunted, " I know, I know that
my Redeemer liveth ! "

The old priest was indignant that a Neck should apply
these holy words to himself, and in his zeal, he cried to him :
" Why dost thou sound thy harp so gleefully, O Neck ?
Sooner shall this dried cane that I hold in my hand grow
green and blossom, than *thou* shalt obtain salvation." Thereupon the gentle minstrel flung aside his harp, and rocked
himself, bitterly weeping, on the water. The priest turned
his horse and continued his course. But, lo ! before he had
proceeded far, he noticed that tender shoots and leaves began
to bud forth from his old staff, soon bursting into most
glorious and fragrant flowers, so that, as the old man rode,
he seemed like some saint bearing a branch from Paradise.
This seemed to him a sign from Heaven, directing him to
preach redemption after another fashion. He therefore
hastened back, and found the sobbing Neck on his pool,
which was full of water, ready to trickle over, like an eye
full of tears just ere they fall.

He showed the Neck his green flowery staff, and said—" So
this old stock has grown green and blossomed as a young
branch in a rose-garden ; therefore, like it, may hope blossom
in the hearts of all created beings, for their Redeemer liveth."

Then the Neck caught up its harp, and long through the
night rang its gladsome song, and the little waves danced
around. The old priest was sorry to go, and, as he went, he repeated, " Praise the Lord ! His mercy is over *all* His works ! "

It will be observed from the foregoing popular tradition,
which is found in Iceland, Germany and Norway, that, in
other countries than our own, Nick is not considered synonymous with Satan ; but rather as a fantastic being with many
good points about him.

Sunday I spent quietly at Grimstúnga, hoping to see an
Icelandic service in the little barn-like church. But I was
disappointed. The priest of Underfell came, this being an

annexja, or chapel of ease, and was received by the widowed
housewife of Grimstúnga with the warmest kisses.   The
pastor was a small man, not taller than a boy of twelve, with
bandy legs, a large head and long arms, and swallow-tails
which swept the floor; he reminded me strongly of Quilp, but
his face was intelligent, though ugly.

Hardly had he been half-an-hour in the house when coffee
and cakes were brought him by the widow. The pastor's long
arms folded his sheep lovingly to his heart, and repeated
kisses testified his gratitude for the coffee.   The offer of
sugar-candy produced another outburst of affection and a
repetition of the same scene. When the coffee was drunk,
its excellence was acknowledged with more kisses; the delicacy of the cakes was also duly honoured.

"Another cup?" asked the housewife.

"You are too good!" kisses again. In came the coffee
once more, and all the kisses, before and after drinking it,
were repeated. Time for church arrived, some farmers dropped
in dressed in blue jackets, and hung about the room; four
sat down on my bed, and three on Grímr's, others sauntered
in and out of the door.

"I think we may do without service to-day," said Quilp;
so the parishioners returned to their homes.

"And pray why is there no service?" I asked of Grímr.

"There are not enough people here to form a congregation," he answered. Yet there must have been twenty!

On Sunday afternoon the members of the household assembled for a family service in the garret above my chamber.
This service consisted in reading a chapter of the Bible,
saying the Lord's Prayer and the collect for the Sunday, in
monotone, singing two hymns, and reading a portion of one
of Vidalin's sermons. The singing was execrable, it was like
the sound which might be produced by a chorus of Cochin-
china fowls, accompanied by a hurdygurdy, yet it was sufficient
to stir Grímr's enthusiasm.

"Is not this beautiful!" he exclaimed; "yet tourists
persist in saying that we Icelanders are not musical!"

The first tune was that of Luther's " Ein fester Burg ;"
considerably altered. The second struck me, notwithstanding
the intolerable way in which it was executed, as being peculiarly beautiful. It is not, I believe, a genuine Icelandic
melody, having been imported from Denmark, where it was
originally composed.

W. B.

Half-an-hour later the Yankee and Martin arrived with their guides and the baggage, and took up their quarters at the farm of Haukagil, a mile north of Grimstúnga.

They had not been remarkably successful in their sport, having caught no trout, but shot a few birds, among which were a red-breasted diver (*Colymbus septentrionalis*), and a scaup (*Fuligula marila*).

Haukagil is a spot of historic interest, as it was the scene of a struggle between the first Christian missionary and some Berserkirs; it was also the home of Olaf, one of his first converts to the true faith.

The farm is prettily situated on the scarp of a hill, facing the east, and has an extensive tún, very green, but more than half morass. The door of the house is curious, being of carved oak, with the Austrian eagle in medallions, and a border of grapes surrounding the panels. There are traces of vermilion and blue on the wood, which show that the door must have been originally painted. The farmer was a remarkably fine man, with long hair flowing over his shoulders; he was well built and muscular; he was dressed neatly in a short jacket and blue breeches, with his legs, from the knee downwards, encased in a wrap of sheep's hide bound round with leather thongs; and his feet were shod with the usual Icelandic shoes of undressed sheepskin.

On Monday morning I rode down the valley on my way to Hnausir, where we purposed spending the night, as we bore letters of introduction to the proprietor of the farm.

The Vatnsdalr is hemmed in between mountains, with their flanks like iron walls, on the right hand and on the left, and their tops enveloped in cloud, so that I had no opportunity of seeing them. Over the wall-like sides shot torrents in superb cascades, from snows wrapped in vapour, falling from one to two thousand feet without a break. We remarked especially one fall on the eastern side of the valley, where the stream leaped out of a grey cloud into a singular black groove, scooped out of the mountain side, and reached the base in a heavy shower.

Grimr and I paid a visit to the parsonage of Underfell, but found that Quilp was absent. We were, however, received by his seven sons and buxom wife, who showed us a volume of MSS., which her husband had borrowed from a farmer in Langadal. It contained the Saga of Erik red, the Atli Saga, several of the sagas relating to the bishops, and, finally, the Draplangar Sonar Saga.

The sketch of Vatnsdalr in Plate V. was taken from the door of the parsonage, and represents the church, which is a fair specimen of Icelandic ecclesiastical architecture. In the distance on the left is the smoke from Ingimund's farm Hóf, the smoke on the right proceeds from the byre of his murderer, which is now the richer farm of the two. After having finished my drawing, I bade farewell to Quilp's wife and seven sons, and rode to Helgavatn, celebrated for its opals. I had no time now to look for them, but had to press on to Hnausir. We crossed the river at its last ford, and rode up to the door of Hnausir farm.

CHAPTER IX.

FROM HNAUSIR TO EYJA FJORD.

A Tract of Slag Cones—An Icelandic Doctor—Part from my Friends—Giljá—Conversion of Iceland—Svínavatn—Icelandic Churches—A miserable Lodging—Slang—Harlequin Duck—Ford the Blandá—Vatnsakarth—Vithimyri — Purchase a Horse — Ford the Heradsvatn — Mikliboer — Oxnadals Heithi—Steinstathr—A Caravan—Strange Merchandise—The Princess Alexandra—A Death.

Nothing could be kinder than the reception I met with at Hnausir from Mr. Skaptason, surgeon and apothecary, a nephew of the excellent Reykjavik doctor, Hjaltalin.

His farm is perhaps the largest and richest in the north, and the house is certainly the best built in Iceland. Grímr's admiration of it was excessive, he evidently regarded it quite as a palace, and the doctor himself was deservedly proud of his house, which, comparatively speaking, was clean and comfortable. The tún is very large and productive; it lies on a flat between the river and a small lake, full of teal, wild-duck, and pintails, and a mile north of a large sheet of water thronged with swans. On the west of the river is a most singular district, a mile square, covered with countless sand and slag heaps devoid of vegetation, generally yellow and speckled with reddened stones. These heaps are perfectly symmetrical cones, and are alike regular in formation whether they are three or fifty feet high. On the level between their

bases, a little grass sprouts, but not a blade on the hills themselves. Some are capped with large stones, and, as a general
rule, the smaller stones are nearest the base and the larger
blocks crown the apex. These mounds have been thrown up
by an earthquake. Drs. Preyer and Zirkel remark that
similar heaps were raised in Chili during the earthquake
on the 20th November, 1822.

On Vatnsdals-fjall, immediately above Hnausir on the east,
are found large masses of petrified wood susceptible of a
bright polish.

The apothék of Mr. Skaptason little resembles an English
doctor's laboratory, as there were none of its neat phials and
carefully labelled bottles: instead of these the room was
blocked up with brown jars of purges, pills and confections,
barrels of gums, bundles of simples, old green wine-bottles
filled with mixtures, jam-pots containing ointments and salves,
kegs of boluses, besides shelves of German, Latin, and Danish
medical treatises of the last century and the beginning of the
present. The good doctor has a loft above the kitchen, in
which his patients are stowed; apparently the kitchen smoke
ascends to the infirmary and thence escapes through a hole in
the roof. The place was in a cloud of the acrid, pungent fumes
of sheep's-dung and peat, before each meal; so that I was
always made aware of the approach of a repast by the cough-
ing, sneezing, and grunting of the diseased population above-
stairs.

Mr. Briggs, Martin, and the Yankee were accommodated
with a den in a labyrinth of chambers downstairs; and as there
were more sick in the house than the loft would contain, the
compartments round my friends' beds were occupied by patients.
My companions' cabin was lighted by a window, which consisted of a single pane, hermetically sealed into the walls, and
was only ventilated through the roof of the sick loft and the
tunnel communicating through the kitchen with the yard. My
apartment was infinitely more commodious; it adjoined the
sitting-room, but, as it was also the passage between it and the
kitchen, the servant-girls were traversing it all the morning

whilst I was in bed, or undergoing the process of becoming presentable, and this, considering my innate modesty, was sufficiently harrowing to the feelings.

On the evening of July 1st, Grimr and I parted from our friends, as they proposed spending some time in Vatnsdalr, shooting and fishing. We left behind us every thing that we could possibly spare, and rode to Svinavatn, where we intended sleeping. We passed an inconsiderable hot spring at a farm called Reykr, below a singular conical mountain, and traversing the bogs which surrounded the beautiful Swine-lake, reached the little church of Svínavatn at ten o'clock.

The wind had been piercingly cold all day, rolling up from the Arctic Ocean without any break. Giljá, with its pretty brawling stream, dancing and foaming through a chasm, had stirred my heart with real emotion. It had been the home of that Thorwald, who was the first to introduce Christianity into his native land, and who, I believe, is the original of Fouque's hero in the exquisite romance, " Thiodolf, the Icelander." Many a time, doubtless, had the boy scrambled up that gill, leaping its rocks, and plunging into its vitriol-green pools, till the time came for him to travel. His tenderness of heart had amused, as well as gained the love of all who met him while he was still heathen; in viking expeditions he had freed his captives, and with his prize-money had ransomed prisoners. Whilst abroad he was converted to the true faith, and, full of zeal, he persuaded a certain Saxon bishop, Fredrick, to accompany him to his native isle. The first winter was spent at Giljá, in converting Thorwald's parents; after which, a mission tour was undertaken, but with poor results, the bishop preaching in his own tongue, and Thorwald interpreting what he said. The young man incurred the anger of his companion, by killing a person who had made some insulting verses on them, likening the bishop to an old woman, and Thorwald to her baby; and finally the two parted company. Fredrick and Thorwald, dissatisfied with the progress that was made, deserted Iceland, and the youth is believed to have visited Constantinople, and died in a monastery. This mission bore fruits, though they were not visible at first; it unsettled the minds of the heathen, it gave them ideas which were new to them, it inspired doubts in their minds as to the truth of their ancestral faith, and prepared the way for the missions of Thangbrand and Gissur the white, and the general conversion of the island.

We rode from Hnausir to Svínavatn at an amble, breaking into an occasional trot. This increase in speed was hailed by Grímr with an exultant shout of, " Now we are going like dee-vils ! " If this pace was diabolical, what must he have thought of the rate at which we scoured the country beyond Akureyri ! Svínavatn church is interesting, as it contains a curious diptych with mediæval figures in four compartments, painted on a gold ground, in the style of the fifteenth century; the subjects are—

First, in the top compartment on the right,—The Annunciation.

Second, below this,—The Nativity.

Third, on the left,—The Resurrection.

Fourth, on the left below,—The Last Judgment.

An Icelandic church is both externally and internally much like a barn; its plan is a parallelogram, with its eastern and western faces occupied by wooden gables surmounted by weathercocks; the sides are flanked with walls of turf, so thick as to resemble aisles. The roof is made of wood covered with turf, on which grass and buttercups grow in profusion, and are most attractive to the ponies. On one occasion little "Bottle-brush," my favourite riding pony, walked from me, whilst I was sketching, and proceeded to escalade the church; I had to bring it down by the bridle when the creature was half-way up to the roof-tree. These grass-grown roofs afford a valuable hint to the natives that the land would produce threefold if properly drained.

The church bells are usually suspended in the lychgate, which gives access to the graveyard. The yard is surrounded by a high turf wall, covered with a profusion of grass ; indeed, the soil is ready enough to produce herbage if relieved of the chilling influence of the water, which turns all grass-land into cold morass.

The church is the general receptacle of the farmers' clothes, saddles, and wool, which are stowed among the rafters. Bibles, sermon-books, and hymnals, are also stacked, out of the damp, along the cross-beams.

The nave is filled with open benches, but the Thing-man, or M.P., has a pew opposite the pulpit, in an Athal-kirkja, or Mother-church. The pulpit is modern, and is usually adorned with coarse paintings of apostles, evangelists, or Icelandic imaginings of tropical flowers. The screen extends to the rafters, and is painted; it is a lattice or palisade, more or less carved and coloured; within it, a seat runs round the four sides of the choir, interrupted only by the altar-rails and the rood-screen door. On this the men sit during service, whilst the women are accommodated in the nave. Hanging from a nail in the chancel is a large brass pan, something like an alms-dish, only deeper: this is the font. These basins are often exceedingly handsome, and are of German or Danish workmanship. That of Svinavatn represents Adam and Eve, on either side of the tree, in a garden of lilies and roses. Curled round the tree is the serpent, with a crowned female face, and long hair. Around the bowl is the motto, in old German,—" Ich bart geluk alzeit " (" I bear luck always! ") repeated five times.

A lantern without glass, frequently painted and gilt, is also hung in most chancels, or stands beside the richly-coloured box, serving as aumbry. When there is no such box, the altars are made to open, and disclose a shelf on which stand the Eucharistic vessels, together with the case of wafers for communion, each wafer stamped with a crucifix and SS. Mary and John. Below the shelf are heaped the vestments of priest and altar, the former consisting of alb and chasuble, the latter of frontal. These, with exception, of course, of the alb, are of various colours. The chasuble has a gold cross on the back, and is of a debased shape. The altars are of wood, not movable, nor at all resembling the tables which

are the disgrace of many English village churches. I never
saw an Icelandic place of worship in a neglected condition ; its
appropriate Christian symbols always give it a look of dignity,
notwithstanding its poverty.

The altars are furnished with crucifix or painting, and
with two or more candlesticks of brass or copper.

The churches are lighted by two windows at the east end,
and two at the west ; these are closed with shutters during
the week.

My lodging at the farm was none of the best. The
guest-room was miserable indeed. Three yellow smeared
panes, and a gap for the fourth, stuffed with an old peaked
cap, hardly lighted a chamber with a damp earthen floor,
deep in fish skins and bones, with nodes of rock sticking up
about two feet above the surface. A table under the window,
numerous trunks painted green with staring pink flowers on
them, or not painted at all, piles of clothes fit for rag-fair, a
locker in the wall, fashioned to hold a bed, but with the floor
overhead broken through, so that the eye looked up into
a garret full of refuse ; a host very old, and ingrained with
filth, his white hair luminous against his dingy skin ; one,
whose sedentary habits had utterly obliterated that screen
which society draws between man and the outer world,—and
you have a picture of my lodging and host.

The old gentleman brought me a MS. account of the
Holy Land, translated from the German by a bishop, in 1615,
and probably in his handwriting. I did not purchase it, as
the book was borrowed, and of no particular interest. Till
late at night I amused myself with filling in some water-
colour sketches, as the bed looked most uninviting, and it
required a struggle before one could resolve on plunging into
its densely populated recesses.

As I drew, the old man watched me, and assured me that
it afforded him " mikit gaman," or great pleasure. " Gaman,"
fun, is cognate with our slang expression, " gammon ! "
Other of our vulgar terms are closely allied to the Icelandic ;
thus, " gaby," a fool, is related to the verb " ath gaba," to

make a fool of; and the New Testament speaks of Herod, as seeing that, " Hann var *gabbathur* af Vitringunum," he was made a gaby of by the wise men.*

Next morning I found that a horse which had strained its foot in the bogs on the preceding day, was too lame to move, and I was obliged to leave him with a farmer from whom I purchased another horse. This worthy man, a connection of Grimr, took charge of my horse for a month, and, on my return, would take nothing for its keep.

My breakfast consisted of stock-fish, and cold mutton boiled a twelvemonth before, the fat of which was well nigh putrid. It was cut into junks and covered with hair and dirt.

---

* I add a few more examples to show what light is thrown on their derivation by this language.

Brag (to boast), Icel. *bragð*, rumour, renown.

Chap, Icel. *kappi*, a fighting man, a hero.

Dandy, Icel. *Dándi*, anything good; *dándis maðr*, a worthy fellow. The word has certainly changed its signification considerably.

Duffer (a stupid fellow), Icel. *dofi*, laziness, from the verb *dofna*, to be dull and stupid.

Fluke (a chance), Icel. *flux*, of a sudden, used similarly, derived from the verb *fljúga*, to fly.

Fellow, Icel. *félag*, a comrade, literally one who goes shares in money.

To go the whole hog. This I believe to be, to do all in one stroke—hog to be the Icel. *högg*. The Icelanders similarly speak of doing something " með höggi," all at once.

Lagged (outlawed), a contraction for *útlag*, outlaw.

Land-lubber. In the early part of last century the word was spelt *loper*; land-loper was a vagabond who begged in the attire of a sailor, and the sea phrase, land-lubber, was synonymous. Icel., *land-hlaupr*, one who runs on land.

Ninny-hammer (a silly fellow). The old Norse used *einn-hammar* to signify a man in his right senses; with the negative particle *nei* before it, it would have a contrary meaning, and may have originated our word. One who was not einn-hammar was possessed, and capable of becoming a weirwolf, or going into fits of madness on the smallest provocation.

Ransack, Icel. *ransacka*, has the same meaning.

Skulk, Icel. *skelk*, fear, from the verb *skelka*, to frighten, related to *skjálfa*, to tremble.

Skittles is derived from a verb *skjóta*, to shoot, whence the adjective *skjótt*, speedy; similarly, the word *brittle* is formed from a verb *brjóta*, to break; *fog* from a verb *fjúka*, to drive with the wind.

11

I was fain to eat what was given me; indeed, I became less and less particular every day whilst travelling. Some kaager or rye-cake, about the thickness and taste of the wood of which bandboxes are made, and dirt flavoured with butter, rather than butter itself, completed a meal which was washed down with glasses of corn-brandy, the taste of which very much resembles spirits of wine out of a castor-oil bottle.

The district about Svínavatn teems with wild-fowl. Kittiwakes and sand-pipers, teal, phalaropes, and snow-buntings abound. Ptarmigan poults, hardly fledged, started up under our horses' hoofs, and the mothers with a sad cry ran among the willow-tops for shelter. A dozen red-throated divers (*Colymbus septentrionalis*), in a batch, sailed away from the lakehead, but a magnificent harlequin garrot (*Anas histrionica*), as though conscious that we were unarmed, floated unmoved within stone's throw of where we were halting. This goodly bird is not uncommon in the Icelandic lakes and rivers, frequenting the latter during the day, and retiring for the night to still water, where it may rest from incessant swimming against stream. In its summer plumage it is a beautiful object. Patched with white and black, the latter of purplish metallic lustre, its colours are blended into the most beautiful harmony by cool greys and rich chestnut reds.

The farmer from whom I had bought my horse guided us across a ford in the Blandá. Grímr was glad to avail himself of his knowledge, as the ford was continually altering, and on a previous summer he had himself nearly lost his life in venturing across without a guide, reckoning on his remembrance of the spot where he had crossed the year before. The farmer's dog, when we reached the river, gave a jump and seated itself comfortably *en croupe*, a position which it retained during the passage.

We had a pleasant scamper to Blöndudalshlith, a new church gaudily painted, the doors and shutters vermilion, with diamonds of blue and yellow in the centres. The walls inside were red striped with blue, and the screen was one mass of yellow and blue bulls'-eyes on a scarlet ground. The only object of interest in the church is a brass chandelier in the nave, which has been ignominiously ejected from the chancel to make room for a frightful glass chandelier of ball-room type.

The Vatns-skarth, our next pass, began with a sharp scramble. The wind cut us to the bone, and blew a storm of snow in our faces. My stockings had been soaked in crossing the Blandá, and they nearly froze on my feet. We skirted a lake with a farm beside it, near the top of the pass: a most wintry spot for any poor souls to inhabit! and then in the teeth of the snow-storm descended along a wild mountain

KITCHEN AT VITHIMYRI.

torrent to the farm and church of Vithimyri, or "the extensive swamps." My feet and hands were so effectually numbed that I was obliged to beg permission to warm them at the miserable offal embers in the kitchen. As I thawed, the desire came

11—2

upon me to sketch; and I drew the interior of the apartment, to the amusement and surprise of some unkempt and unwashen urchins who crawled by dozens in and out of the cavities in the house, like so many maggots. The lopsided door represented in the woodcut, is so low that one has to bend double to pass through it; this opens out of the dark tunnel leading from the main entrance to the house. The rafters are not sufficiently elevated to allow of one's traversing the kitchen without ducking at every second step, and, as light is only admitted through the hole which serves as chimney, the kitchen is so gloomy that one stumbles repeatedly over pots and pans, or even over babies, wriggling and sprawling on the earthen floor.

The farmer was absent, but his man showed me a volume of MS. Rimur, founded on the Fostbroethra Saga, and composed by the great-grandfather of the present farmer.

Grimr and the young man were soon in an animated conversation on the subject of the absent farmer's merits.

"There is one thing for which I don't like him, and only one," quoth the man; "and that is, the way in which he uses my horse. I have a very nice chestnut, and master comes to me day after day, and says: 'Lend me your horse, will you?' He is short of horses himself, and I can't refuse; so I have to pay for the keep of the horse, whilst my master has the use of it."

"I'd spite him, if I were you," said Grimr; "I'd sell the horse."

"Ay! but horses are plentiful about here, and no one will buy it."

"What do you want for it? The gentleman whom I am guiding is not exactly in want of a horse" (I was so very much, though), "but I might persuade him to buy it, if it were cheap."

"Oh! I don't want much for it," answered the young man. "I shall only sell it, just for the sake of aggravating my master. Faith! I shall like to see his face when he returns and finds the horse gone."

"Come!" said Grimr; "suppose you say eleven specie-dollars (2l. 10s.)"

"Be it so," answered the fellow. And they kissed on the bargain.

I purchased the horse at this price, and it was my best— a beautifully made creature. I should have been tempted to bring him with me to England, had not his former proprietor chosen to mark him by slitting his ears into ribands, which danced and quivered in the most ludicrous manner, when the animal was in motion.

A curious old gentleman assisted at the purchase, who had no hair on head or chin, neither eyebrows nor lashes. He was the butt of numerous jokes, which he received with the greatest good-nature. I quite won his favour by remarking that he brought the great beardless Njal up before my mind's-eye, a compliment he repaid with a hug and kiss, which I would gladly have dispensed with.

Below the farm by the river's side are flats, covered with short grass, over which we had a good scamper, till we reached the ford. The river Heradsvatn must be as wide as the Thames at London Bridge, and it is swift as an arrow, so that the passage is dangerous. The horses could hardly keep their feet against the violence of the fierce cold water, which surged up to the saddle, and foamed over their backs. In keeping the eye on the reeling eddies, one is apt to become giddy and lose one's seat. As a remedy, Grimr called to me repeatedly, "Look to the shore!" A good maxim through life, surely, to keep the eye fixed on the shore of the true country, among the troublous waves of this mortal life. After crossing the river, we had bogs to toil through, till we reached Miklibœr (the great farm). A bog in Iceland is a formidable affair; it rolls and quakes underfoot as though one were riding over an air cushion. The surface is matted with long grass, and the ponies, with wondrous instinct, select the right places for planting their feet; they snuff the soil, with head to the ground, till they have ascertained where there is safe footing, and neither persuasion nor blows will make them tread where their instinct tells them there is danger.

In traversing these bogs, one must give the horses their heads. When they come to a red glistening patch or streak, they will leap, but should they consider the ground on the opposite side to be doubtful, they will track the seam up till they find a place where they can overstep it. This causes great delay on a journey, especially as considerable detours have to be made before the direct track can be regained. I have known my baggage-pony run half a mile up the hill-side to the source of a morass which spread as it descended, before rejoining my caravan, the horses of which, being less heavily weighted, had tripped over swamps which would have engulphed the sumpter-pony.

In the church of Mikliboer, is an old German wood engraving of the Dürer school, in bad condition. Near the church lives the probst, or archdeacon, a good, hospitable man, one who smokes, too, a rare accomplishment for Iceland. He had, Grimr told me, the satisfaction of having reclaimed a brother priest from drunkenness by the bribe of a cow, and emboldened by this success, he had just presented his daughter to another tippler, in hopes of restoring him to a right mind.

The archdeacon is one of the only ministers who wears any distinctive clerical attire: he had on a black suit and a white neckcloth.

The finest scenery that I had as yet passed through, was that of the Öxnadals heithi, which I crossed next day. After following the Heradsvatn for a short while, till we passed the prettily situated church and parsonage of Silfrastathir, we branched to the left through a noble valley, the Northrár-dalr, with strange terraced mountains on either side. We were detained for an hour in the midst of a swamp by the pack-saddle getting out of order. Our course then lay up a gorge of Alpine magnificence, out of which branched glens—mere rifts in the almost perpendicular mountain scarps, giving glimpses of pyramidal snow-covered mountains of most perfect symmetry, cushions of snow alternating with steps of basalt, to the summit. The Plate opposite represents one of these peeps, and the double pyramid in the Frontispiece stands

S. Baring Gould, delt.   Edmund Evans, sc.

IN ÖXNADALR.

London : Published by Smith, Elder, & Co., 65, Cornhill. 1863.

above the morasses at the summit of the pass. It was near
midnight when I sketched this scene, clouds eddied and whirled
around the mountains, condensing on the snow; and the sun,
shining through vapour, lighted the peak with a nebulous
glow. We were the first to surmount the Öxnadals heithi

MOUNTAIN IN OXNADALR.

this year, and the track had been completely obliterated by
the spring torrents. Long beds of shale rested loosely on the
heads of rock walls, which sunk to a vast depth, their bases
chafed by a roaring stream, which rolled past unseen in the
gloom of the chasm it had torn for itself. The horses passed

these slopes trembling, and with the greatest caution, dislodging quantities of rubble and sand as they planted each foot, sending them in avalanches down the gorge.

In the marshes at the head of the passes are the sources of the Northrá and Öxnará, which flow respectively west and east, the former flowing into the Heradsvatn, which pours into the Skaga-fjord, and the latter joining the Hörgá, which enters the Eyja-fjord some miles north of Akureyri. The Öxnadal, down which we rode, is girded in by bluffs of basalt reaching to the snows, overleaped by magnificent waterfalls; through the gaps they have torn one obtains peeps of snowy cones and pyramids. The porkpie-shaped mountain top I engrave, was hastily scribbled down whilst I was in my saddle, as the clouds momentarily parted around it.

It was four o'clock in the morning when we reached Steinstathr, drenched to the skin by the passage of the Thverá (cross-stream), a fretful torrent which lay athwart our path.

The farmer, an M.P. for his district, received us warmly, though we had to rout him out of bed on our arrival. A bed was made for me, but Grímr had, I believe, to share one with the farmer and his wife.

We were not down to breakfast till two o'clock in the afternoon, and then we had an excellent repast off roast mutton.

As the sun was burning brightly in the sky, I retired behind a wall which sheltered me from the bitter wind, to bask in the unwonted warmth. It was pleasantly hot, but too much for Grímr, who lay for a little while beside me, and was then so overcome by the sun's power as to be indisposed for the rest of the day, and obliged to go to bed.

Steinstathr is prettily situated immediately under a jökull, which rises up in one start to the snows, and without buttresses. Opposite the farm is a range of singular rocks, several thousand feet high, with a saw-like edge, apparently quite sharp, so that the snow can never lie on them.

During the night a caravan belonging to a Danish merchant arrived, consisting of thirty horses, laden with wares from the station of Hófsós (pronounced Hopsoase), at which

a ship had arrived with goods. These he was conveying to
Akureyri, the capital of the north. We all started together in
the morning, and I found the Dane very agreeable, as he
spoke a little English. I mentioned my regrets at having
deprived him of the guest bed at Steinstathr.

"Dank you!" he answered; "but I never slip in an
Island bed."

"Did you lie on the bench in the sitting-room?"

"Yes, I not like a bed in a native byre. It is so dirty,
and so full of insect." Then after a pause he said, "You
should come to Húsavik, and see of the fair; it will be dare
in dree veek."

"I wish that I could, but I shall not be in the neighbourhood then. What trade is carried on at the fair, may
I ask?"

"Oh, de merchants sells of crockery, of corn-brandy, and
of clodes."

"How do the natives pay? They have no money!"

"No," answered the merchant; "but dey have fleas."

"I am well aware of that, but how comes that to alter the
case?"

"Why!" replied the Dane; "dey takes of de vares, and
dey gives us de fleas. De people of Denmark likes of de
Island fleas very much."

"A singular taste!" I remarked. "Indeed, I may say,
very singular. What can there be in them so attractive?"

"Oh!" with enthusiasm; "dey is more big and more
long dan in any odder country."

"Woe's me!" I exclaimed; "your statement is corroborated by my experience."

"De colder de vinter, de bigger de fleas!" remarked my
companion.

"And they bring these abominations to the merchant
stations for barter!"

"When dey have vashed and dried de fleas. —— Dare!"
exclaimed the merchant pointing; "dare is a lot, lying in de
sun to be dried."

these slopes trembling, and ;with the greatest caution, ɖ
lodging quantities of rubble and sand as they planted e
foot, sending them in avalanches down the gorge.

In the marshes at the head of the passes are the source
the Northrá and Öxnará, which flow respectively west and
the former flowing into the Heradsvatn, which pours int
Skaga-fjord, and the latter joining the Hörgá, which ente
Eyja-fjord some miles north of Akureyri. The Öxnadal,
which we rode, is girded in by bluffs of basalt reaching
snows, overleaped by magnificent waterfalls ; throu
gaps they have torn one obtains peeps of snowy con
pyramids.   The porkpie-shaped mountain top I e
was hastily scribbled down whilst I was in my saddle,
clouds momentarily parted around it.

It was four o'clock in the morning when we
Steinstathr, drenched to the skin by the passage of the
(cross-stream), a fretful torrent which lay athwart our

The farmer, an M.P. for his district, received us
though we had to rout him out of bed on our arriva
was made for me, but Grímr had, I believe, to share
the farmer and his wife.

We were not down to breakfast till two o'clock in
noon, and then we had an excellent repast off roast

As the sun was burning brightly in the sky,
behind a wall which sheltered me from the bitt
bask in the unwonted warmth.   It was pleasantly l
much for Grímr, who lay for a little while beside ɪ
then so overcome by the sun's power as to be in
the rest of the day, and obliged to go to bed.

Steinstathr is prettily situated immediately u
which rises up in one start to the snows, and witho
Opposite the farm is a range of singular rocks, sev
feet high, with a saw-like edge, apparently quite
the snow can never lie on them.

During the night a caravan belonging to
chant·arrived, consisting of thirty horses, la
from the station of Hófsós (pronounced Hop·

a mill ... ... Then he was ...
Always ... ; ... well. We all came together.
the morning, and ... in Inuit ... again. A...
spoke a little English. I mentioned my regret at not...
deprived him of the great bed at Summoser.

"Dank you!" he answered; "but I never ... g. at
Island bed."

"Did you lie on the bench in the sitting-room?"

"Yes, I not like a bed in a native byre. It is so dirt;
and so full of insect." Then after ... he said, "You
should come to Hiumvik, and me ... ... be me
in dree veek."

"I wish that I could ... ...
hood then. What train ... ...
I ask?"

"Oh, de ... ...
of clodes."

"How do the ... ...

"No," answered he ... ...

"I am well ... ...
case?"

"Why!" replied he ... "I ... ...
dey gives us in ... ...
Island fleas very ...

"A singular ... ...
very singular. ... ...

"Oh!" vell ...
long dan it ...

"Were ...
rated or ...

"Be ...
...

... and I ...
clodes de ...

...
...
... it ...

My eye followed the direction of my companion's finger, and I saw a quantity of sheep's-wool—fleece—lying in the tún of a small farm.

At the time that I was in Iceland, the engagement of the Prince of Wales to the Princess Alexandra had not taken place, but the possibility of such an event had been mooted. I mentioned it to my friend, the Dane.

" Well! " said he; " ve should be very sorry to lose her for any odder nation dan England. Ve all love her in Denmark, very, very much; and, if ever she become Princess of Wales and Queen of England, you will learn to love her as we have, for her own merits."

Now," said I, "let us have the rights of the Holstein squabbles."

The discussion of these rights occupied the Dane some hours. I will spare the reader. I never did understand them, and fear that I do not comprehend them a bit better now.

Eyja-fjord broke on us in all the glory of a breezy noon, the bright quivering blue frith skirted by jökulls, and furrowed by white-sailed fishing smacks.

To our left we saw the stone mansion Fredriksgave, which is the residence of the Governor of the North—a sort of railway shed, undoubtedly much colder than the turf houses of the natives. This stone house occupies the site of the ancient monastery of Möthruvellir, founded in 1295 by Bishop Jörundr of Hólar, and burned down in 1316, but rebuilt by Bishop John of Hólar in 1328. Not a trace of the ancient buildings remains. Our horses snuffed the wind, and set off at a scamper for the beach, where they paused to drink the sea-water and nibble the weed.

We rattled along a shingly shore, and entered Akureyri, whilst the Danish flag was flying half-mast high from all the stores, and from some of the craft in the bay.

" Who is dead ?" I breathlessly asked, with forebodings of death in the Danish royal family. " Who is dead ?"

" The baby daughter of a priest, ten miles up the fjord-head, has died, poor little thing, of croup!" was the reply.

in the:

nt of is
at take
model

se her
ner is
inces
r her

tein

me
m.

t,
l

CHAPTER X.

AKUREYRI.

A Little Town—New Church—Site of the Town injudiciously chosen—The Post—Grimsey—An Island Cure—Danish Hospitality—The Newspaper Office—Supper—English Vessels—Jack Tar in Iceland—Trees—MSS. —Icelandic Poetry—The Dream-Ballad.

AKUREYRI on the Eyja-fjord, the second largest town in Iceland, contains eight hundred souls. It consists of a straggling line of tarred wooden shanties and hovels, extending along the beach, the doorsteps being just above high-water mark, and the backs of the houses abutting on a precipitous hill rising three hundred feet to an extensive plateau, out of which soar a chain of snowy mountains, called "The home of the wind Jökull," and the less elevated belt of Súlur, three thousand feet high, with its quaint peaks called the Bonder and the Old Woman.

To the north of Akureyri, the hill throws out a spur, which shuts off the sea-winds, and shelters the town. The fjord is thirty-eight miles long, and ten miles wide at its mouth, but contracts to one mile and a quarter opposite Akureyri. On the farther side of the fjord rises the precipitous Vathla-heithi, over which lead the roads to Húsavik and My-vatn. In the distant north-east, at the mouth of the estuary, like a pale blue cloud, is a jökull with its head almost invariably covered with mist. South of the fjord lies a mountain district through which flows the Eyja-fjord river, which enters the frith after a short course of thirty miles.

In the fjord is a holm named Hríms-ey or the Isle of frost, on which are two farms; and, twenty-six miles off the mouth of the frith lies Grimsey, a small island, two and a half miles long by one mile broad, on which live several fishermen and a pastor.

There is no church at Akureyri, but one is in course of erection, built of wood, with large round-headed sash-windows, and a turret over the sanctuary at the east end. The edifice is somewhat pretentious, but it is in no Christian style of architecture. It is much to be regretted that the Icelanders have no idea of the capabilities of wood for constructional beauty.*

There are six or seven merchant stores at Akureyri, long wooden buildings, fitted up with a counter and furnished with

---

* The natives were more skilful in olden times, apparently. The Páls Biskups Saga gives the following account of a steeple erected at Skálaholti in the year 1196:—" Paul the bishop soon saw, after his enthronization at Skálaholti, that it was necessary for him to strengthen, and improve, and finish the work which the holy Thorlak, the bishop, had planned. So he purchased what was requisite. Now he had to set up the bells which he had bought for the cathedral of Skálaholti, and they were the best in all Iceland. He had brought with him also four pine timbers twenty ells long. Paul the bishop sent for the most skilful artisan in wood in all Iceland, and he was Amund Arnason. He made him erect a steeple so cunningly that it surpassed all the woodwork in Iceland, and also the church itself. In the steeple he built a chapel, with a stair leading to it. He consecrated this chapel to S. Thorlak, the bishop, on the 10th day after Yule, and he adorned the chapel in the finest manner, providing it with everything that was necessary. He made Atli the priest paint all the roofing within the steeple and also the pediment of the gable; and he hung all the lower part with three sets of drapery, very beautiful; so, also, he had inscriptions set up over the tombs of those who slept in the steeple. He laid out as much money on the steeple as though it were for himself, spending four thousands of silver at the least. He bought three bells for the steeple—a treasure of bells they are!—from a Norwegian, hight Kolr. There were also several bells which he bought for the steeple; also two pitched on the same note for the church; and he decorated both church and steeple alike with all that could be fancied, with ornaments requisite for a church—crosses, scrolls, images, lamps, and glass windows—as also with all the episcopal vestments. He had also a stone coffin, made very skilfully, in which he might be laid after death; and he buried in the steeple, in the best style, all such men as he thought most deserving."—Biskupa Sögur, i. 182.

everything that an Icelander can want. At these stores can
be procured corn-brandy, rum, beer, fox and swan skins,
eiderdown, ready-made clothes, hats, shoes, saddles, crockery,
timber, ironmongery, and provisions.

Some of the merchants winter at Copenhagen and come
out to Iceland in spring, but two or three remain at Akureyri
throughout the year.

The average temperature of the whole year in Eyja-fjord is
Fahr. 32°, or freezing point; in winter the thermometer
sinks as low as—32°; but the houses are kept warm by
stoves, supplied with coal by English smacks from Leith.
The winter is spent in convivial parties, at which the principal
amusements are card-playing, and dancing to the sound of a
guitar. A library has been established, but, as yet, it is quite
in embryo, and contains only about a hundred volumes. The
houses of the merchants are comfortable within, well papered
and furnished; the walls are hung with prints, such as views
of Copenhagen, portraits of the king, or of Thorwaldsen or
Ohlenschlager. A few objects of vertu adorn the side-tables,
and pots of flowers, which do not blossom, stand in the
windows.

When Henderson visited Akureyri in 1815, he found it to
consist of three stores, and from fifteen to twenty fishing
hovels; since then it has much increased, though perhaps
not situated in the most advantageous position for a town, as
the routes to it lie over some of the worst passes in Iceland.
On the east is a thinly populated district cut off from the
settlement by a steep pass and two fierce and dangerous
rivers; on the west is a belt of snowy mountains pierced by
rugged passes at Hjaltadals, Öxnadals, Heljardals heithies.
On the south is one moderately populated dale of no great
extent.

The site of all others for a town would have been the
Skaga-fjord, within easy access of the rich Hjalta, Vatns, and
Langa dales, in the midst of Iceland's most populous and
fertile district, instead of being at its extreme limit, in Eyja-
fjord. The post arrives from Reykjavik once a month, except

during the winter, and takes about five days on the road.
The postman passes by Thingvellir, Kaldidalr, the Arnarvatn
heithi, then branches off from the road I had gone, and comes
direct to Akureyri, over the Storisandur. In autumn and
spring the postman has to do the greater part of the journey
on foot, and sleeps, when benighted, in snow-pits, which he
digs for himself. The privations which this poor man has to
undergo are often very great; his predecessors have perished
in the snows or have been lost in crossing half-frozen rivers.

There are two other ways to Reykjavik besides that taken
by the postman, these are the Kjal-vegr and Sprengisandur
vegr. The former follows the Eyja-fjord river to its source
near the Hofs jökull, then passes between that and Lang
jökull, skirts the large Hvitár lake, and passing the Geysir
and Thingvellir, enters Reykjavik. The Sprengisandur road
is longer and more arduous. Leaving Akureyri, it passes
the Vathla skarth, follows up the Fnijoská dale, then crosses
to that of the Skjalfanda-fljot, "or flood of quivering
waves;" this it follows to the little grass patch of Kithagil,
after which there is a gallop of twenty-two hours over a
lifeless desert of black sand to the roots of Tungnafells and
Arnarfells jökulls, where there is a grass patch called Eyvindarkofaver. Thence it passes along the Thjorsá, near Hekla, to
the Geysir, and so by Thingvellir to Reykjavik.

The great event of the year at Akureyri, is the arrival of
the first ship from Copenhagen. Many a lady expects by it
her spring and summer dresses, some article of ornament, or
a long wished for piece of furniture; the merchants await
some additions to their stock, and the natives are looking for
various articles which they have ordered through the Danish
traders.

Last autumn a farmer came to one of the merchants, with
the request that he would procure him a clock. The order
was transmitted to Copenhagen, and by the first vessel in the
spring there arrived a clock, but it proved to be such an
inferior article, that the merchant returned it and demanded
another. The new one had not arrived when I was at

Akureyri, and the farmer was told that he should have his clock by a vessel due in August, just nine months after the good man had given his order.

Once a year the men of Grimsey visit Akureyri to lay in stores for winter, and part with the oil, fish, and feathers they have collected in their lonely isle. In the autumn of 1861, these poor fishermen, after having bartered their wares, and laden their boat, started down the fjord on their return to Grimsey, sixty-five miles distant. As they rowed down the frith, some of the party remembered that they had friends at the little farm of Sauthanes (the Sheep-ness), and, considering that this was their only chance, for a year, of renewing old acquaintance, they persuaded the rest to put ashore. The whole party left the boat and adjourned to the farm, where they sat drinking and talking till midnight, when they thought fit to return to their boat. But, alas! the tide had risen and carried their boat away. Next day it was discovered stranded, keel uppermost, in a creek not far distant. All the stores were at the bottom of the fjord, but the boat was not much injured.

The poor fellows at once returned to Akureyri, where they related their piteous tale; the inhabitants raised a contribution for them, and furnished them again with all that was necessary to support life.

Grimsey is the smallest cure in Iceland; it has its church, that of North-garth, and priest. When Henderson was at Akureyri, he found the Grimsey priest and a mainland priest at loggerheads about a Bible, which the latter had lent to his island brother, and which had never been returned. The mainland pastor sent demands for the restitution of his book by the Grimsey boat when it visited Akureyri, but on its return the following summer, there was neither book nor message from the Grimsey parson. Henderson settled the dispute by presenting two copies to the island, one for the use of the church, the other for the minister himself.

The present priest is such an inveterate drunkard that the islanders kicked him out a few years back, and lived without a

pastor, till they found that they could get no other, so they
sent a boat to land and brought him back again.

The Grimsey fishermen are said to be a lawless, quarrel-
some set, and very different in temperament from the natives
of the mainland.

A considerable amount of driftwood is cast on their shores,
and their hearths are supplied with the mahogany of Honduras,
the palms of Haiti, and the costly woods from the venerable
forests of the Amazon and Orinoco.

The water drunk by these poor fishermen is that which is
left by the rain in bogs; this is neither pure nor wholesome,
and in order to be made at all palatable, it has to be given a
flavour, by an infusion of the juice of scurvy-grass, or the
squeezed berries of the samphire.

As there is but little herbage on the islet, only a few cows
can be kept, so that milk becomes a luxury. The people
suffer severely from scorbutic attacks and leprosy, which carry
them off very speedily, unless they are removed to the mainland and supplied with wholesome and nutritious food.

My guide, Grimr, was excitable on the subject of Grimsey,
and for the following reason :

After having passed his theological examination, a message
reached him from the bishop and governor, telling him that
the island parish was without a pastor, as the inhabitants had
expelled their ancient priest, and that he was to take the living.

On the receipt of this communication, Grimr caught up
his hat, and rushed to the residence of his Excellency.

" Mr. Governor ! am I requested or commanded to take
this post ? "

" Commanded, most certainly ! "

" But I decline the cure ! "

" You cannot help yourself; take it you must."

" This is quite unexpected. I—the best candidate of the
Theological College—to be bundled off at a moment's notice
to the smallest living in your gift—to an inhospitable island
cut off from the world; to a parish of lepers ! This is preposterous ! "

" Listen to me," said the Governor. " Grímr Arnason, this is only a first step to a better living."

" Ah ! but out of sight out of mind. When once I am banished to Grimsey, I am forgotten, as a dead man."

" The people of Grimsey are a savage, semi-Christian set, and we wish you to convert them."

" ' Evil communications corrupt good manners.' I should deteriorate wofully if I were among the Grimsey folk."

" We have always regarded you as the most religious young man in Reykjavík," said the Governor.

" More's the reason that I should not be sent to the place of torment before my time."

After a pause, Grímr observed—" Besides, I have a strong desire to be married ; and if it is once known that I am to be exiled to that hateful rock, not a woman could I get to join her lot with mine."

" Oh, Grímr !" said the Governor ; " any girl would marry you !"

" I think, Mr. Governor, that if you banish me to the island, you are bound to provide me with a wife."

" How is that possible ?" asked his Excellency in amazement.

" You have a very charming daughter, who——"

Grímr never finished the sentence, and ever after showed an invincible repugnance to setting his foot within the Governor's door.

The end of the matter was, that Grímr disputed the right of the Governor to send him, will he nil he, to Grimsey, and his appeal went before the King of Denmark, who decided that Grímr was in the right, as the law stood ; at the same time, he decreed that henceforth all theological candidates should be sent wherever the Governor chose.

The result of this decree was, that more than half the theological candidates withdrew from the college, and no fresh entries have been made. When I left Iceland there were eight livings vacant, and no pastors ready to fill them.

It is certainly only just that those trained free of expense

12

at the college, should enter the ministry or return to government the cost of their education.

I was most hospitably received by Mr. Havsteen, a Danish merchant, who volunteered to lodge me, as there is no inn in the town. My horses were driven up the hill, and turned adrift on the moor, with their feet hobbled. Grimr and I drew our boxes into the merchant's storehouse, and we then made ourselves as presentable as possible before entering the house. Mrs. Havsteen met us, and in the kindest manner welcomed us to Akureyri. Coffee and cakes were brought in, and we were introduced to the young ladies, whose cheerful faces and blooming complexions spoke well for the air of Eyja-fjord. After having shaken hands all round, I sallied forth on a visit to the printer, who lived on the way to the new church. I found his house to be a small wooden cottage, so close and stifling as to be quite insupportable, so that after having purchased a couple of books, I was glad to withdraw. A newspaper, the *Northri*, appears at intervals from this printer's establishment, edited by Svein Skúlason; but as this gentleman has left Akureyri, the periodical will in all likelihood cease to appear. It contains an epitome of the news of Europe, articles on the topics of the day, local intelligence, letters from correspondents, and advertisements, among which, for a wonder, I did not see a commendatory notice of Holloway's pills.

On my return to the Havsteen's house, I found that an ample supper had been provided, and that the table was covered with delicacies: these consisted of flakes of smoked salmon, slices of garlic-sausage, and ham; hot mutton flavoured with whortleberry jam and potatoes, cold smoked shark's flesh, steaks of whale and seal, good Bavarian beer, Rhenish wine, and corn-brandy clear as crystal. Hungry mortal that I was! I did ample justice to the meal; so too did Grímr, whose tongue was loosed under the influence of the good cheer, and he told the story of his Grimsey grievance *in extenso*.

There is no grace said before and after meals, but it is

the custom as soon as you rise from table to shake hands or kiss all round, saying, "Tak for mad."

I did not retire to my bed till I had taken another stroll on the shore, and looked upon the fjord in all the stillness of an Arctic midnight. The mountains were enveloped in mist; the sea, flowing in, soothingly lapped the shingle and played around a little one-masted English vessel, which had been wrecked and dismantled off the entrance to the frith, and had been towed into the bay to be broken up for building-timber. A considerable number of English smacks visit the north of Iceland for whale and seal oil: they are mere cockle-shells, manned by four or five sailors, without a chart, and calculating their way by the log.

Many have been lost. "Ah, sir," said a skipper; "the storms of these Arctic seas are enough to try the pluck of a man. I assure you I have stood at the wheel when I daren't have looked over my shoulder. If I'd seen the waves as was a rolling upon us, I'd have deserted the helm. Mortal man couldn't have stood the sight." The majority of these English boats come into harbour at Grafsarós or Hófsós, in the Skaga-fjord, where the mad pranks of the sailors have produced a panic among the native farmers. The jolly tars seize on their horses and ride them helter-skelter up hill, down dale, trampling down the tún, mount the backs of the cows and gallop them about, chase the sheep, worry the dogs, court the women, and play practical jokes on the men. The Icelanders can get no redress. Jack laughs at their remonstrances, which are couched in a tongue of which he does not understand a word; and as for their threats! phew! there is only one policeman in all Iceland, and he is at Reykjavík.

Akureyri is famous for possessing the largest tree in Iceland; this is a mountain ash, outside Mr. Havsteen's drawing-room window. It is twenty-six feet high, a straggling fellow without much foliage, overtopping the roof, to which, during the winter, its branches are secured by ropes. Garden-seats are placed at its roots, and, on a warm summer-day, the

Havsteens take supper around it, and imagine themselves in
the gardens of old Denmark. There is a second tree, not so
large, outside another house, and these two are considered to
be quite the most remarkable sights of the town. Their
roots are covered with straw during the winter, and the young
shoots are wrapped in wool. In my sketch of Akureyri,
the Havsteens' house appears: the view is taken from a
potato field on the hill above the town. The ridge on the
right is Vathla heithi; the gap between it and the distant
glacier heap is the opening of the vale through which the
Fnjoská enters the fjord; and the snowy mountain beyond is
the Kaldbak jökull, twenty-six miles distant, which stands up
like a sentinel to guard the entrance of the fjord.

I spent Sunday morning basking in the sun under the wall
of the unfinished church, and afterwards visited Svein Skúlason,
late editor of *Northri*. He showed me several volumes of
Sagas in manuscript. One of these, a thick folio bound in
vellum and beautifully written, contained the Sturlunga Saga;
he produced also three octavo volumes of "Kvœthi," a more
perfect collection than that published by the Nordiske Litera-
tur-Samfund.

Icelandic poetry has gone through four stages; the first
or Edda period, when the wording was plain and vigorous,
and the metre simple. The second is the age of verse-
smiths, who hammered out stanzas full of epithet, simile, and
periphrasis, so obscure that none but the initiated could
extricate the meaning.

The third period is that of the Kvœthi, or ballads. These
are mostly reproductions of well-known and widely-spread
popular songs. That, for instance, of Olaf liljurós is the same
as our "Clerk Colvill and the Mermaid," and the German
"Peter von Stauffenberg und die Merfeie." It exists also
in Faroese, Norwegian, Swedish, Danish, Wendish, Bohemian, and Breton.

The fourth period is that of the Rímur. These are simply
the Sagas set to jingling rhyme. This fashion came into vogue
during the last century, and is popular now. The Rímur are

chanted to a tune varied according to the taste of the singer, but always strongly resembling a Gregorian melody.

The following ballad belongs to the third period : I give it as a specimen of the style which, to my taste, is peculiarly musical, and suited to the character of the language. I have preserved the characteristics of the original as nearly as possible.

The Dream Ballad.

" Fagurt sýngur svanrinn !
Um sumar lánga tíð,
þá mun lyst að leika sèr,
Mín liljan fríð !
Fagurt sýngur svanrinn ! "

Sweetly swans are singing
In the summer time.
Let us lightly laugh and play,
Lily maiden !
Sweetly swans are singing !

I.

" Rede my dream right, mother mine !
In the summer time.
I will give thee golden shrine !
Lily maiden,
Sweetly swans are singing !

II.

First, methought the moon did smile,
In the summer time,
Softly over Skáney isle ;
Lily maiden,
Sweetly swans are singing !

III.

Then methought a rowan-tree,
In the summer time,
Louted lowly unto me ;
Lily maiden,
Sweetly swans are singing !

IV.

Then a swan as silver white,
  In the summer time,
Lay upon my bosom light;
  Lily maiden,
Sweetly songs are singing!

V.

And I planets twain did see,
  In the summer time,
Lie a-rocking on my knee;
  Lily maiden,
Sweetly swans are singing!

VI.

Next I saw the tide rise fleet,
  In the summer time,
Sweeping o'er my little feet;
  Lily maiden,
Sweetly swans are singing!"

VII.

" As thou saw'st the moon arise,
  In the summer time,
Royal husband be thy prize,
  Lily maiden!
Sweetly swans are singing!

VIII.

As the rowan bent, I trow,
  In the summer time,
Many folk to thee shall bow,
  Lily maiden!
Sweetly swans are singing!

IX.

As thou claspedst cygnet fair,
  In the summer time,
Thou a princely son shalt bear,
  Lily maiden!
Sweetly swans are singing!

X.

As thou saw'st two planets shine,
  In the summer time,
Lovely daughters shall be thine,
  Lily maiden!
Sweetly swans are singing!

XI.

As around thee stole the flood,
  In the summer time,
Shall thy lot be ever good,
  Lily maiden!
Sweetly swans are singing!

XII.

This thy dreaming, daughter mine,
  In the summer time;
Keep thyself, thy golden shrine,
  Lily maiden!
Sweetly swans are singing!"

CHAPTER XI.

UXAHVER.

A Second Guide—Icelandic Horse-calls—A theological Candidate—Vathla-skarth—Wood in the Fnjoská dale Háls—Ljásavatn—The Raven—Myth regarding it—Extent to which the Myth has spread — Góthafoss— Flowers—Cross the Flood of quivering Waves—Acquaintances—Lava Cracks — Grenjatharstathr — Uxahver — Boiling Springs — The Wild Huntsman—Origin of the Myth—Myvatn.

I HAD now two guides, for Grímr was helpless, never having been beyond Akureyri. Jón, my new acquisition, was an honest, cheerful fisherman; very reasonable in his terms, as he came with a horse of his own for one dollar per diem. The fellow afforded me much amusement; his legs and arms were in continual vibration, like the wings of a bird, and his body lurched from side to side as he inflicted stripes with his long whip, first on the horse to his right, then on that to his left; yet Jón knew how to keep his seat as well as any man. His calls to the horses were quite original, and differed widely from those in ordinary use. Instead of "Afram, yho !" or "Ahr-r-r !" to urge the horses forward, the former signifying, "Go ahead !" and the latter, "The dogs are after you !" Jón compressed his lips and trumpeted forth, "Prrmp, prrmp !"

" Jón !" said I; "a man should never deviate from the customs of his forefathers, nor oppose traditional uses, without mature consideration, and careful balancing of the matter. For old uses and customs have generally been founded on rational grounds, and, like proverbs, contain a kernel of sound truth."

"Well, sir! I was Mr. Metcalfe's guide in '60, and that reverend father objected to my repeated 'Yhko!' so I was obliged to change my note, and adopt 'Prrmp!' which has become a second nature to me; I could not shake it off now!"

Many of our English country horse-calls may be recognized in this Ultima Thule. The "Bok-a-ooff!" of Nottingham, or "Bok'n-waay!" of Yorkshire, East Riding, are forms of the Scandinavian Bug-af, bend aside; the same verb, *ath buga*, to curve, is preserved in our word *bucket*, literally a curved receptacle. So, also, the common shout "Gee! Gee uup! Tzch!" are forms of Gá—walk; and the Northumbrian Heck! is the Icelandic Hoegr! (pronounced Haikir).

Jón carried a baby's weaning-bottle in his pocket, and constantly replenished it with water: he could not get on without a drop of something, and when fire-water was unattainable, he contented himself with imbibing ordinary water from this artificial mother.

I soon experienced the advantage of having an active fellow like Jón with me; he drove the horses with spirit, and kept them all day at a trot or canter. Grímr's perverse, "Now we shall go slow," said invariably, when I urged the horses out of a walk, was now unregarded; we trotted in spite of him. Grímr had picked up a dingy theological acquaintance at Akureyri, a spare young man, who looked as though he had recently emerged from a dust-bin. His hat was wondrously tall and very shabby; his long coat-tails flapped against his calves; his waistcoat—black, too—was buttoned to the chin; his sombre inexpressibles, with leather seat, were very old, and seemed to have been worn by successive generations. He rode his own pony, and joined Grímr for the purpose of accompanying us over the next mountain ridge, so as to hear all the Grimsey story, and the last news from the capital. As the river at the head of the fjord was low, we crossed its seven mouths with ease; all but the theological candidate, whose horse, falling into a quicksand, flung its rider on his tall hat, which was thereby effectually flattened.

With unruffled gravity, the budding divine picked himself up, and shook the great hat into shape; then, solemnly presenting his coat-tails to Grímr, asked him, with his head over his shoulder, to wring the water out of them. This Grímr did with equal gravity.

Jón was a long way ahead, scrambling up the scaur (skarth is the Icelandic name for a mountain pass: it is cognate with scaur), and we had to trot after him, in order to catch him up.

After a good pull up the steep mountain side, we got among the clouds, which gathered over all the high lands, as they rolled up the fjord. They parted once at the top, and we obtained a glorious vision of river, dale, and ridge on ridge of snow and rock; then the clouds closed over the scene, and we had to make the best of our way through them, until we descended into the Fnjoská dale, through which whirls a swift deep river.

My Indian-rubber stockings were worn through, so that in crossing rivers I was obliged to brave the cold, and pull off shoes and stockings; but on reaching the further bank, my feet became so numbed, that I was often unable to stand. Travellers in Iceland should be provided with fisherman's boots.

We rode through a forest, the finest in Iceland, some of the trees being quite twenty feet high. The fresh green of the birch, the fragrance and rustle of leaves, were most exhilarating, and we cantered, singing, over the light sandy soil, without drawing rein, till we reached Háls, where a new church was in course of erection.

The priest received us kindly, and gave us coffee and thin pancakes of native wild corn, powdered with cinnamon, and eaten cold. His pretty daughter was greeted affectionately by my guide, as an old acquaintance. The fellow has friends everywhere! and the Grimsey grievance was gone through in detail, notwithstanding all my entreaties that it might be cut short, as we had a long journey before us.

At last the story is done, and we gallop through the Ljósa-vatn skarth, till we reach the "Light water" lake, whose

pale flood is full of undissolved snows, brought down from the
white-crested mountains on either side. Seven Northern
divers on it ! Ducks, grebes, mergansers, in scores ; a white
gerfalcon watches us from yon pile of stone, a bowshot off.
The Icelandic raven flits around us, and runs among the
stones in a bold contemptuous manner, flinging us a disdainful
croak when we pelt it. No bird is more common in Iceland
than the raven (both *Corvus corax* and *Corvus leucophæus*);
it throngs all wild and desolate spots, and lays its five or six
greenish speckled eggs among the mountain gorges and clefts,
early in March, a month earlier than other birds. It feeds
on anything which it can digest, worms, which abound in the
morasses, whortle-berries, eggs, fish, insects, carrion—even
dung. It is a source of terror to the young of the sandpiper
and plover ; it perches on the backs of the sheep, and fills
its crop with the ticks abounding in the long wool; it hovers
round the breeding-places of the eider-duck, waiting to tap
the eggs, and in winter it flutters about the byres, to seize
on any refuse which may be flung from the doors. The raven
is regarded with much the same superstitious feeling in Iceland
as elsewhere.

> Hrafn sitr á hárri staung
> Hildar mark á taki ;
> Ei þess verðr œfin laung,
> Sem undir byr því þaki.

Which signifies :—

> Raven croaks on gable tree ;
> Watch ! Death is onward creeping :
> Short the life of him will be
> Who 'neath this roof lies sleeping.*

The ravens are said to hold formal wardmotes in autumn,
and to appoint captains and watchers in their respective
districts.

It is curious to find in Iceland a version of a world-wide
receipt. The natives tell one, that there is a stone of such

---

* Cf. also *Tháttr Hrómundar halta,* chap. 5.

wondrous power that the possessor can walk invisible, can, at
a wish, provide himself with as much stockfish and corn-
brandy as he may desire, can raise the dead, cure disease,
and break bolts and bars. In order to obtain this prize, one
must hardboil one of the green eggs in a raven's nest, then
secrete one's self till the mother bird, finding one of her eggs
without prospect of being hatched, flies off and brings a black
pebble in her beak, with which she touches the boiled egg,
and restores it to its former condition. At this moment, she
must be shot, and the stone be secured. Albertus Magnus
gives a similar receipt in his *De Mirab. Mundi* (ed. Argent.
1601, page 225). "If you wish to burst chains, go into the
wood, and look out for a magpie's nest, where there are
young; climb the tree and choke the mouth of the nest with
anything you like. As soon as she sees you do this, she flies
off for a plant which she lays on the stoppage; this bursts,
and the plant falls to the ground under the tree, where you
must have a cloth spread for receiving it." The same story
is told in the Talmud, but there, the moorhen takes the place
of raven or magpie. According to the Rabbinical story,
Schamir is a worm which preserves the stone of Wisdom:
Benaiah son of Jehoiada having found a moorhen's nest, laid
a plate of glass over the poults. The mother-bird fetched
Schamir and snapped the glass.

According to another version—"Solomon went to his
fountain, where he found the dæmon Sackar, whom he had
captured by a ruse, and chained down. Solomon pressed his
ring to the chains, and Sackar uttered a cry so shrill that the
earth quaked.

"Quoth Solomon, 'Fear not, I shall restore you to liberty,
if you will tell me how to burrow noiselessly after minerals
and metals.'

"'I know not how to do so,' answered the Jin; 'but the
raven can tell you; place over her eggs a sheet of crystal, and
you shall see how the mother will break it.'

"Solomon did so, and the mother brought a stone, and
shattered the crystal.

" ' Whence got you that stone ?' asked Solomon.

" ' It is the stone Samur,' answered the raven. ' It comes from a desert in the uttermost east.' So the monarch sent some giants to follow the raven, and bring him a suitable number of stones."

The same stone reappears in several household tales, and is mentioned in the thirty-ninth story of the *Gesta Romanorum.*[*]

Up the mountains are low birch bushes, and a line of blue smoke, from among them, showed that charcoal-burners were at work,—a farce surely, burning these twigs for charcoal.

" Now we shall go slow !" called Grímr from behind ; but Jón regardlessly cracked his whip, and we spun along. A pitch-black rock at the end of the lake, scooped into by the stream, and continually crumbling away, was surmounted, and we rode through moss and fen to the Skjalfanda fljót, or Flood of quivering waves. A mile up the river is Gótha-foss, a noble waterfall, bearing a striking resemblance to Niagara, in miniature. It is a horse-shoe, and has its Goat island, to which it is possible to wade ; and then a quaint peep of the landscape is obtained through a watery arch, spouted from a hollow, into which one arm of the river pours. Below the falls, the grotesqueness of the rocks, and their ironblack colour, add wildness to a scene, in itself, very impressive.

---

* I wish that I had space to give the origin of some of our popular superstitions and nursery tales, and to show what assistance is afforded by Icelandic literature in the clearing up the difficulties with which they are surrounded. I can only instance one :—

" Jack and Jill went up a hill
To fetch a pail of water ;
Jack fell down, and broke his crown,
And Jill came tumbling after."

These two children are mentioned in the *Younger Edda*, under the names of Hjúki and Bil (which have become, in course of time, Jack and Jill), as fetching water from the well Byrgir in the bucket Sœg, on the pole Simul. These children were taken up into heaven to follow the moon. Hjúki signifies " the quickening," Bil " the failing ;" and their attendance on the moon simply means that the moon becomes full and wanes. By the bucket of water I presume is signified the effect of the orb on the weather.

Gotha-foss is generally considered to be the finest fall in
Iceland, but the priest at Háls assured me that there is another
in the desert east of Myvatn, incomparably its superior.

The river below the falls is too deep to be forded, it is also
of considerable breadth. We called to a ferry-man on the
farther shore, and whilst he was rowing across, I amused
myself with gathering flowers.

In swampy spots clustered the white heads of the mountain
asphodel (*Tofieldia palustris*), and the drier ground was starred
with white and pink Alpine flea-bane (*Erigeron Alp.*), looking
like large daisies; butter-grass is the meaning of its Icelandic
name. Clumps of rich purple flowers, against a bed of low
whortleberry, attracted me, and I gathered my hand full of
Alpine bartsia (Icel. *Lokasjóths-bróthir*). Iceland may almost
be called a land of flowers, for they cover the soil, where grass
will not grow. The brave little moss campion (*Silene acaulis*),
flourishes everywhere, pushing its bonny pink face close to the
snow, dappling sand tracts otherwise barren, clinging to rock
crannies, where no other plant can get a footing, or planting
itself on mud slopes which are torn and swept into gullies by the
melting snows in spring. The traveller falls quite in love with
the pretty pink flower, the last to take leave of him on his entering a lifeless wilderness, the first, fresh and smiling, to greet
him as he reaches the desert's limits. The blossoms grow in
dense clusters, of all shades, from carmine to white, growing
close to the root, so that no gale can injure them; some of
the flowers are male, others female, and others seem to be
hermaphrodite. This catchfly blossoms about the end of May,
and remains in flower till August. The pale mountain avens
(*Dryas octopetala*), with its sunny heart, opens rather later,
and withers earlier; it is the traveller's second love. In the
ground where it blossoms, are the tremulous dancing flowers
of the Alpine meadow-rue (*Thalictrum Alp.*); also the loveliest
of Icelandic beauties, the Alpine speed-well (*Veronica Alp.*),
the loveliest and frailest, for its intensely blue petals fade in
the hand as one gazes admiringly at them. Where the soil is
lighter appear the common blaeberry and bog-whortle (*Vac-*

cinium myrtillus and *Vacc. uliginosum*), whose white flowers, pink-tipped, stuff the ptarmigan's crop, and in this half-digested condition are used by the Icelanders for tea. The berries are employed for flavouring *skyr* or curd, and are eaten as a preserve with mutton. Dwarf willows grow with the whortle, and of these there are seventeen kinds in the island.

The dandelion and devil's-bit gild depressions in the *skog* or thicket, and a quiet strawberry-flower, yellow centred, lies in the birch shadow on its leaves, so dear to Gothic architect. Horsetails are abundant in Iceland; I found six or seven varieties; of these, the commonest (*Equisetum Arvense*), or *ellting*, as the natives call it, grows on the roofs of houses and cowsheds to a considerable size. Its roots are very useful, as the long red filaments, thrown from its root, bind the loose soil together, and prevent it from being swept away when the snows thaw. The root-fibres have tubers like cherries hanging to them, covered with a dirty black skin, but are white within, and have a sweet taste. Grimr told me that horses go mad if they eat *ellting*, but I had no opportunity of verifying this statement.

Below the birchwood, in spots too swampy for the willow to grow in, and where buttercups and marsh-marigolds can hardly take root, the broad leaves of the buck-bean (*Menyanthes trifoliata*, Icel. *Alftarkólavir*, or swan's clapper) float on the red water, and the slim stem shoots high, holding its whorl of ruffed white flowers beyond the reach of soil or stain from the mire. Close by, tufts of cotton grass (*Eriophorum capitatum*) flicker in the wind, their pods bursting with silver hair; this the native plaits into wicks for candle or lamp, and stuffs into pillows when he cannot afford eider-down.

Hereabouts the leafless horsetail (*Equisetum limosum*) thrusts its spears, the rods like those of an ordinary equisetum, with the whorl of leaves stripped off. The bladder campion (*Silene inflata*) grows in profusion on the sand by the water-edge; the golden Ranunculus glacialis studs the

bank; and among the pebbles grows the red alpine catchfly
(*Lychnis alpina*). I picked one of the latter, and taking it
to Grímr, asked its Icelandic name.

"Lambagrass," he replied.

"Nonsense, Grímr!" I exclaimed; "you told me that
the moss campion was so called."

"Both are pink!" said my guide, pouting at having his
opinion disputed.

"Yes; but compare the flowers, the shape is different."

"My nose is not like your nose, yet they are both noses,"
answered Grímr, in a sulky tone.

Gótha-foss is the scene of one of Grettir's exploits. He is
said to have plunged under the falls, and to have discovered a
cave behind the water. I asked the boatman who ferried
across to us, whether any one had seen this cave since the
times of Grettir.

"A few winters back," he answered, "the cascade was
frozen, and then we went below it and looked for the cavern,
but it was not visible. This does not disprove anything! it
only shows that the undermined rock has fallen in from the
press of water rolling over it."

We now removed packs and saddles from our horses, and
stowed them in the boat, then drove the horses into the river;
they waded on till the stream lifted them off their legs, but
then turned with one consent, and came back towards the
shore. With stones and shouts, and cracking of whips, we
sent them back, and then they swam boldly for the farther
bank, their heads showing like dark specks above the water.

The ferryman rowed us across, and we supped, as it was
eight o'clock, on German sausage and hard biscuit. Then,
having saddled the horses, we ascended the heithi, and got a
view of long rolling hills stretching north, without a mountain peak. From the side of one a column of steam rose into
the air, and was blown in a southerly direction, as it reached
the top of the hill.

"That is Uxahver," said Jón.

As we passed farms at a gallop, the tinkle of the bell on

our leader must have startled the sleepers; it certainly roused the dogs, which barked furiously behind the closed doors. A bell on a horse was an innovation which produced great searchings of heart among the Icelanders, and it was only after two days' deliberation that Jón came to the opinion that it was a success. At eleven o'clock we met two natives on horseback, who stopped and asked our names. On our compliance with this request, they gave theirs, and then flung their arms round Grímr, with every demonstration of affection.

Jón dashed ahead with the packs, and I followed, leaving the model student locked in the embraces of his friends.

"Pray who were those loving people?" I asked when Grímr rejoined me.

"I know the elder of the two very intimately," he answered; "he held me in his arms twenty-four years ago, when he visited Reykjavík."

"How old were you then?"

"Six months, and I have not seen him since."

"And the other man?"

"He is likewise an old acquaintance; his father was apprentice to my grandfather."

"Have you met before?" I inquired.

"Never: it was quite delightful to meet him now."

At twelve o'clock we skirted strange parallel lines of lava blocks, running with such regularity that at first sight I thought they must be the fallen stones of an avenue like that of Karnac or Avebury. On riding up to them, and on examining the blocks, I found that they overlay rifts, now all but choked up, miniature gjás, and that the force which had snapped the lava bed had tilted the fragments of the broken edges, so that some blocks lay wedged in the crack, whilst others had dropped across it, and others again had fallen against each other. Having ridden along one of these lines for half a mile, I came upon a point where they diverged; and on climbing over the rocks I found a circular patch of turf girt with stones, another point of resemblance to the Druidical stone avenues.

13

The lava hereabouts is very old, and is covered with moss
and grass, except along these lines.   They run parallel to the
direction of the flow of lava, and were formed, I believe, by
the edges of the molten stream cooling and resisting the
tension of the still viscous centre.

It was twelve o'clock when we reached Grenjatharstathr,
and rapped loudly at the door.

A thermometer over the window stood at 2° cent., but as
neither my breath nor the application of my hand raised it
half a degree, the reading was worthless.

Tired after a ride of twelve hours, I soon fell asleep in the
clean and comfortable bed of the guest-room.

The church of Grenjatharstathr—pronounce that who can !
is particularly interesting.

There are Runes in the graveyard which I copied, rubbing
them with a German sausage, as I was unprovided with heel-
ball.

RUNIC TOMBSTONE GRENJATHARSTATHR

The fashion of having stone staves laid on the graves is
very prevalent here. From one I copied, "Hjer hvilir Jdrottni
Thurur . . . ."—"Here sleeps Idrottni Th——'s daughter."
Another probably bore Runes, but the inscription is effaced.
The iron-work of the church door is most exquisite, I have
seldom seen it surpassed in boldness and beauty of design and
execution.

"The sea must have washed that up," quoth the priest;
"never was anything so fine made in Iceland."

On one of the bells, hung in the lych-gate, were the
lines—

> " Ans dem Feuer ben Ich gegosen,
> Hans Meyer in Kopenhagen hat mich geflossen.
> Anno 1668."

The date on the other bell is 1740. Within the church are three altar-pieces of different dates, the newest one over the table, a very paltry affair.

Above the screen, viewed from the east, is a curious Last Supper, the figures in low relief and painted, date circ. 1620. At the back of this, facing the west, is a fine old triptych, rich with gold, but sadly mutilated. In the centre is the Sacred Trinity; God the Father supporting the crucifix, upon which the holy Dove descends. On the sides are medallions containing the annunciation, the nativity, the coronation of the Virgin, and subjects from the life of a bishop, probably S. Thorlac. The font is a brass bowl, the chalice of silver-gilt is most exquisite, and the paten to match has the Agnus Dei encircled in the centre.

Nothing could exceed the kindness of the priest; an intelligent, polished gentleman. He offered to guide us to Uxahver, as Jón had never been up to the boiling springs, and would be unable to thread his way to them among the bogs by which they were surrounded. On reaching the morasses, the priest himself was at fault; he tried several places, but his horse floundered in to the girths, and we were obliged to ride to the head of the valley, where there is a little byre. The priest asked the farmer to lead us up to the springs, which he willingly consented to do;—" but," said he, after we had got a little way nearer to the volumes of steam, which burst from the ground, and rolled away before the wind; the central jet shooting into a column, which veiled the whole hill-side, and then dying into an insignificant fume. " But," quoth the farmer, " I have not been up to the springs myself this year, and the bogs change repeatedly. My son can guide us best. I will call him."

So the young man guided the farmer, the farmer guided the priest, the priest guided Jón, Jón guided Grímr, and Grímr guided me, after the fashion of the old nursery jingles.

In this vale there are four or five springs of boiling water. Three only are of any size; the small ones lie in and near the river, one a pool of simmering red clay, another of boiling

18—2

water, which thumps and throbs underground, before each ebullition. The first of the great springs, or Northrhver, has two bores, really unconnected, but with their encrusted mouths close together, and discharging their waters down the same channel. The volumes of steam were so dense, that I found it difficult to get at the basins of these fountains, and I had to tread cautiously among rills of scalding water, and over pools of steaming red bolus ; then, through the hot vapour, I could see a surface of still water filling a basin measuring thirty-five feet by thirty-three feet.

Henderson, who saw this in 1813, describes it as—
" Simmering, and emitting large columns of steam for about the space of four minutes, when a few gentle concussions ensuing, a violent ebullition took place, and the water was raised in the middle of the basin to a height of a foot above the brim, which it immediately overflowed. In less than half a minute the ebullition began to subside, and the contents of the basin were almost instantly diminished to the same quantity that it displayed while in a more quiet state ; but in stormy weather this fountain is said to send up lofty and frequent jets." At the time when I saw the spring, it overflowed quietly, and exhibited none of the phenomena described by Henderson, and the farmer assured me that now it never erupted.

Henderson measured the pipe, and found it to be ten feet in diameter. I made it seventeen feet by six feet. This alteration in the bore may have something to do with the change in the character of the spring.

The other pipe opens in the same mound of incrustation ; it is about eight feet in diameter, and is very irregular in shape ; it has no basin, and one can stand close to the lip and watch the water boiling furiously. Spouts of steam break through, and fling scalding drops in all directions ; the surface is never at rest, now throbbing up and down and letting steam bubbles escape, then dashing upwards, or flying with a rumble and thump against the sides its own water has built up, so that one has to leap aside as the boiling jets squirt suddenly

at one over the red beslubbered rim. Tired with its violence,
the water sinks in the pipe a few feet, only to rage to its brim
with redoubled vehemence, and shoot its superfluous water
through an arched passage it has bored in its own encrusted
lips.

We leave this fountain with reluctance, and go to the
second, the real Uxahver, situated one hundred and fifty
yards to the south. It acquires its name from a tradition of
an ox having fallen into the northern spring, and having been
shot up out of this one boiled to rags.

THE SECOND SPRING.

This spring is particularly curious, being intermittent. At
the moment that we reached it, its crater was empty, and I
stepped into it to examine the pipe. About six feet below
the margin, the water was gently agitated. It slowly rose
during the next minute, bubbles of steam burst upwards, and
the water poured into the basin, filling it, and driving me
before it over the edge. The water next began to boil immediately over the bore, and then, with a premonitory concussion,
which made me retreat to the bank, a column of water, and
then a succession of jets broke from the orifice, accompanied

by dense pillars of steam nearly obscuring the exploded water.
Drops, however, were spirted violently into the sunlight from
the white opaque whorls of vapour, and now and then a higher
jet was tossed many feet above the cloud. The eruption lasted
for a minute and a half, and then subsided. The fountain
shoots up from twenty to twenty-five feet, and Gaimard's picture, representing a fountain some forty feet high, is a great
exaggeration. It intermitted with great regularity, at intervals
of four minutes forty seconds to a nicety, whilst I watched
it during five or six explosions. A stream of cold water flows
down the hill immediately behind the spring, and on reaching
the mound of incrustation, bends to the south, and rattles
down to join the river Helgá, which winds among the swamps
in the bottom. The work of an hour would divert this stream
into the mouth of the boiling spring, an experiment worth
making, as it might change the fountain into a great geysir,
and there could be no danger of injuring the mechanism, as the
cold water could always be turned off again. Two hundred
yards beyond this spring is the Sythastrhver. Henderson's
account of it is as follows :—" It consists of three apertures,
one of which is always perfectly quiet, though at the boiling
point, and is that used for the bending of hoops, and the other
two, situate at the distance of fifteen feet from one another,
regularly alternate, which circumstance compensates for the
diminutive size, and renders them scarcely less interesting
than the Uxahver. The largest can only be measured to the
depth of five feet. It is about half as much in diameter, and
jets for about two minutes to the height of six feet, when all
remains quiet nearly five minutes; after which the smaller
one throws up three curious oblique jets through three holes
in the thin crust with which the pipe is arched. Having
acted its part, the water instantly subsides, and, in the course
of two or three minutes the larger one again commences.
This was the only instance of alternation I observed about
these springs, though I have since found that Horrobow
remarked a regular rotation in all the three. I am sorry I
did not then know of the circumstance alleged by the same

author, otherwise I might have made the experiment, namely, that when the water of the largest is put in a bottle, it continues to jet twice or thrice with the fountain, and if the bottle be corked immediately, it bursts in pieces on the commencement of the following eruption of the spring ! ! ! "

The character of this spring is completely altered ; it now consists of a steaming pool, through whose clear blue water the three jagged openings are visible. No eruption takes place there, but bubbles wriggle to the top, and hot vapour rolls from off the surface. All explosions must long since have ceased, for moss grows now to the edge of the pool, which it could never do if the level of the water were subject to fluctuations.

Having satisfied myself with the wonders of Uxahver, I remounted my horse, and, bidding farewell to farmer and son, we galloped over moor and sand to Langa-vatn, a pretty lake, at the head of which we baited, and where the worthy pastor left us. After a rest of an hour, we crossed at a hand-gallop the Hóla sandur, a desert tract of black sand, where no green leaf shows.

Grímr called to Jón and me repeatedly, urging us to go slow, but we paid little attention to his remonstrances. I asked him whether we were going like dee-vils now ? as he had called a jog-trot a fiend-like pace, on a former occasion. He did not answer, but Jón shouted that our troop was sweeping over the country like the Yule host.

" Pray what is that ? " I asked. So Jón entered into an explanation, which showed me that the Icelanders have a superstition about a wild rout of phantom horsemen careering over the country at certain times of the year ; especially at the winter solstice and at Yule. This is none other than the German Wüthendes Heer. Jón could not give me any very particular details ; still, the certainty of the existence of this myth in Iceland was satisfactory. Perhaps I may give a short account of it without being tiresome to the reader.

Odin, or Wodin, is the Wild Huntsman, who nightly tears

on his white horse over the German and Norwegian forests and
moor-sweeps, with his legion of hell-hounds.   Some luckless
woodcutter on a still night, is returning through the pine-
woods; the air is sweet-scented with matchless pine fragrance.
Overhead the sky is covered with grey vapour, but a hush is
on all the land; not a sound among the fir-tops; and the
man starts at the click of a falling cone.   Suddenly his ear
catches a distant wail.   A moan rolls through the interlacing
branches : nearer and nearer comes the sound.   There is the
winding of a long horn waxing louder and louder; the baying
of hounds, the rattle of hoofs and paws on the pine-tree tops.
A blast of wind rolls along, the firs bend as withes, and the
wood-cutter sees the Wild Huntsman and his rout reeling by
in frantic haste.

The Wild Huntsman chases the wood spirits, and he is to
be seen at cock-crow, returning with the little Dryads hanging
to his saddle-bow by their yellow locks.   This chase goes by
different names.   The huntsman in parts of Germany is still
called Wôde, and the chase after him, Wüthendes Heer.*
In Danzig, the huntsman is Dyterbjernat, *i. e.* Diedrick of
Bern, the same as Theodoric the Great.   In Schleswig, he
is Duke Abel, who slew his brother in 1250.   In Normandy,
in the Pyrenees, and in Scotland, King Arthur rides nightly
through the land.   In the Franche-Comté, he is Herod
in pursuit of the Holy Innocents.   In Norway the hunt is
called the Aaskarreya, the chase of the inhabitants of Asgarth.
(Hence, perhaps, our word skurry).   In Sweden, it is Odin's
hunt.   This is the Netherland account of it :—In the neighbourhood of the castle of Wynedal, there dwelt, a long
time ago, an aged peasant, who had a son that was entirely
devoted to the chase.   When the old peasant lay on his
death-bed, he had his son called to him, for the purpose of
giving him a last Christian exhortation.   He came not, but
whistling to his dogs, went out into the thicket.   At this the

---

* The German word *wuth* is cognate with the name Odin. Our old English
word wood, equivalent to mad, is similarly related.

old man was struck with despair, and he cursed his son with
the appalling words : " Hunt, then, for ever ! ay, for ever ! "
He then turned his head, and fell asleep in Christ. From
that time the unhappy son has wandered restless about the
woods, and the whole neighbourhood re-echoes with the noise
of the huntsman, and the baying of dogs.

In Thuringia and elsewhere, it is Hakelnberg, or Hakeln-
bärend, who thus rides, and this is the reason—

Hakelnberg was a knight passionately fond of the chase.
On his death-bed he would not listen to the priest, nor hearken
to his mention of Heaven. " I care not for Heaven," growled
he. " I care only for the hunt ! " " Then hunt until the
Last Day ! " exclaimed the priest. And now through storm and
rain, the Wild Huntsman fleets. A faint barking or yelping
in the air announces his approach, a screech-owl flies before
him, called by the people *Tutösel.* Wanderers who fall in his
way throw themselves on their faces, and let him ride over
them.

Near Fontainebleau, Hugh Capet is believed to ride : at
Blois, the hunt is called the Chasse Macabée.

Children who die unbaptized often join the rout. Once
two children in the Bern Oberland were on a moor together.
One slept—the other was awake. Suddenly the wild hunt
swept by. A voice called—" Shall we wake the child ? "
" No ! " answered a second voice; " it will be with us soon."
The sleeping child died that night. Gervaise of Tilbury says,
that in the thirteenth century, by full moon towards evening,
the wild hunt was frequently seen in England, traversing
forest and down. In the twelfth century it was called in
England the Herlething. It appeared in the reign of Henry II.,
and was witnessed by many. The banks of the Wye was the
scene of the most frequent chases. At the head of the troop
rode the ancient British Herla.

King Herla had once been to the marriage feast of a
dwarf who lived in a mountain. As he left the bridal hall,
the host presented him with horses, dogs, and hunting gear ;
also with a bloodhound, which was set on the saddle-bow

before the king, and the troop was bidden not to get off the horses till the dog leaped down.

On returning to his palace, the king learned that he had been absent for two hundred years, which had passed as one night, whilst he was in the mountain with the dwarf. Some of the retainers jumped off their horses, and fell to dust, but the king and the rest ride on till the bloodhound bounds from the saddle, which will be at the Last Day.

In many parts of France the huntsman is called Herlequin, or Henequin; and I cannot but think that the Italian harlequin on the stage, which has become a necessary personage in our Christmas pantomime, is the Wild Huntsman. It is worth observing that Yule, or Christmas, the season of Pantomimes, is the time when the wild hunter rides, and his host is often called the Yule troop.

I have said that the Wild Huntsman rides in the woods of Fontainebleau. He is known to have blown his horn loudly, and rushed over the palace with all his hounds, before the assassination of King Henry IV.

On Dartmoor in Devonshire, the same chase continues: it is called the Wisht hunt, and there are people now living who have witnessed it.

Now for the names Wôd, Herod, Hackelnbärend, &c.; perhaps Icelandic will help us to explain the myth. Wôd is evidently Woden; the name is derived from the preterite of a verb, signifying to rage.

	Infinitive.	Perfect.	Hence the Names
Icelandic . . .	Vatha	Oth	Othr, Othinn.
Old High German .	Watan	Wuot	Wuotan, Wodin.
Old Saxon . . .	Wadan	Wôd	Wôd, Wôdan.

Hackelnbärend is the Icelandic Hekluberandi, the mantle-bearer. Herod is derived from Her-rauthi, the red lord. This name is known in the North (Heranth's Saga, *Kormak Saga* and *Fornmanna Sögur*, ii. 259). But Dr. Mannhardt derives the name from Hrôths—rumour, fame. The name of Chasse Macabée is given from the allusion to it in the Bible (II. Macca-bees, v. 2—4) :—" Then it happened that through all the city, for the space almost of forty days, there were seen horsemen running in the air, in cloth of gold, and armed with lances, like a band of soldiers. And troops of horsemen in array, encountering and running one against another, with shaking of shields and multitudes of pikes, and drawing of swords, and casting of darts, and glittering of golden ornaments, and harness of all sorts. Wherefore every man prayed that that apparition might turn to good."

When men began to name the different operations of nature, they called the storm, from its vehemence, its *rage*— " The Raging" Wuothan, Wôden; or from its coming at regular *times*—tempestas; or from its *outpourings*—λαῖλαψ (cogn. λαπάξω, λαπάσσω, λαπτω); or again from its *breathing* —storm (styrma, Icel., to puff; Sturmen, Teut., to make a noise; thus: Gisah trumbaro inti meniga sturmenta; *Schilt. Thesaur, sub voce*—Christ saw the musicians and the multitude making a noise). Our word gale comes from its whistling and *singing;* the root is also preserved in nightingale, the night-singer (gala, Icel. cogn. yell), and from this Odin (the storm) got his name of Galdnir or Göldner, and Christmas tide was hight Yule; or from its *gushing* forth like a flood, we get the word gust (Icel. geysa and gjôsa), or, once more: from the storm *cloaking* the sky, covering the fair blue with a mantle of cloud, it got its name of Procella (cogn. celo, προκαλύπτω—I screen with a cloak); and so we find the Wild Huntsman, who, you see, is the storm, called Hackelnbärend, from Hekluberandi, the cloak-bearer.

Now, in the first ages, there was no intention whatever of making the raging storm into a god, nor expressing a divine act in saying that the storm chased the sere leaves; yet, by degrees, the epithet Wôden was given form and figure, and became personified as a deity; then, too, the idea of the storm chasing the leaves became perverted into a myth representing Wôden as pursuing the yellow-haired wood nymphs.

The same thing has taken place in Greek mythology. For instance, Nyx (night) is represented as the sister of Chaos, and the mother of Hemera (day): in plain words, that is—In the beginning the earth was without form and *void*, and *darkness* was over all; then light was created, and *day* sprang out of darkness.

Now, after this long digression, I must return to my narrative.

After a scamper of some hours we saw mountains rising before us; the red charred top of Hlitharfjall became our way mark; presently the beautiful lake Myvatn, or Midge lake, opened before us, studded with countless lava islets; beyond was the sulphur range, yellow as though the sun ever shone on it; near us was Vindbelgr, a mountain perfectly resembling a Tyrolean cap.

After a troublesome scramble over an arm of fresh lava, over which the horses floundered and stumbled, past a pretty tarn cut off from the lake by an arm of lava (Plate X.), over a hill covered with the grass of Parnassus, we came down at a canter to the farm and church of Reykjahlith.

Pl. 9.

R. Baring Gould, del.

Edmund Evans, sc.

MYBAIN.

CHAPTER XII.

A REGION OF FIRE.

Myvatn—Lava Streams—Reykjahlith—Sand Columns—A Plain of Boiling
Mud—Chaldrons—Krafla—Obsidian Mountain—A Ride over a Desert
—Eylifr—A wretched Farm—Dettifoss—A magnificent Fall—Volcanic
Cones—Return to Reykjahlith.

STAND with me one moment on the slope above Reykjahlith,
and scan Lake Myvatn.*

The horizon to the south-west is indistinct, for the lake
winds, and is so studded with islands that its low, swampy
shore is indefinable from this point.

The sheet of water is seven or eight miles long. Yon
black speckles on its surface are lava points, glorious breeding
places for ducks.

If we were in a boat we should see that the bed of the lake
is full of rifts and splinters, among which glide char and
trout. The water is not so cold as that of other lakes in the

---

* My, a midge. Norse dialect, Smikka; Lithuanian, Musa; German,
Mücke; Danish, Myg; Russian, Múkà; Slovakian, Muka; Sanskrit,
Makskd; Bengal, Makjeka; Afghan, Mac; Hindustan, Makki, Magas;
Latin, Musca; Greek, μυια. From the Latin musca (a fly) came the term
muscatus (speckled), and the French moucheté. From its spotted plumage
the sparrowhawk was called mousquet in French, moschetto in Italian, and
musket in English.

"How now, my eyas-musket!"

Merry Wives of Windsor.

When fire-arms took the place of these birds in the chase, the name was transferred to them.

island and does not freeze in winter, from the existence of hot
springs in its depths, and from the fact of the lava having
never thoroughly cooled.

To the right is a hill like a dust-heap heaving itself out of
the morasses which surround it, with a thread of vapour
creeping along its base. This is Vindbelgr, or "The Bellows."

Now turn to the left, and you see the indigo chain of
Bláfell (Plate IX.), beyond which is a field of sulphur and
boiling mud, called Fremri Námur, not visited by travellers,
as it is difficult of access, and inferior in interest to the Námarfjall springs. Nearer at hand is Hverfjall, which was thrown
up in 1748-52; it is a crater, dipping conveniently on one
side, so that we can see into the bowl and admire its symmetry.
Perhaps you can distinguish a black line along its base; that
is a fissure in the lava, similar to the Almannagjá, only on a
smaller scale. More distant is Vilingafjall, a crater much like
Hverfjall. Both are built up of shale and dust, and have
never erupted lava.

Now turn your eyes to the strip of land between us and
the water. Below us is the farm, with its emerald patch of
tún, and the church, the latter encircled by lava which has
flowed towards it in an undivided stream, parted into two
arms, and met beyond. This took place during the last
eruption of Krafla between the years 1724—1730.

The mountains then vomited flames and matter in a state
of fusion, which rolled down in torrents, and inundated the
neighbouring fields, overlapping older beds of lava. In the
lake, where the matter burned like oil for several days, it
killed all the fish and dried up the greater portion of the
water.

The largest branch of this river of fire ran nine miles
from the mountain, and was three miles in breadth; whilst
another torrent overwhelmed the parsonage of Reykjahlith,
which was swallowed up without leaving the slightest trace
behind. The volcanic matter advanced slowly, destroying
everything in its progress, without undergoing the least
change. During the day it emitted a blue flame, like that

of burning sulphur, though the smoke that rose from every
part prevented it from being often seen.   During the night
the horizon was marked by a line of flame, and the clouds
rolled fiery overhead.   Globes of flame rose from the mass
amidst deafening explosions.   Whenever the torrent stopped,
its surface was soon covered with a crust similar to the skin
formed on hot milk; this cake, which might be from one to
two feet thick, soon hardened into stone; but when new
waves of fire swept over it, they broke, melted, and carried
off the crust as a thawing stream dislodges, and bears off the
ice which has formed on its surface.   In cooling, the lava
assumed the most fantastic shapes, such as flowers or
sculpture.

After the volcano had ceased belching fire, the core of
the lava remained a long time in fusion, and continued to
run under the crust in such parts as were sloping; in forcing
its passage, the fiery substance generally broke the crust, and
thus occasioned many crevices and caverns, internally vitrified
with stalactites suspended from their roofs.

Two streams of lava have descended the hill on which we
stand, one on either side of us.   Of these, one is older than
the other, and is coated with thick gray moss, which vainly
struggles to veil its deformity, whilst lichen paints the stone
with blood-streaks and orange stains, as if to relieve its
gloom.

The other is in all its nakedness.   Blocks as big as houses
are propped among vitreous snags; slabs, whose upper surface is scored with corrugated, concentric wrinkles, like coils
of rope, and whose nether side is spiked with stalactites of
olivine-coated stone, are canted up with their teeth ready to
rend you as you scramble past.   Caverns gape among the
ruins; into these snow has slipped and become discoloured,
but their intricacies you shrink from exploring without a light.
Everywhere one sees hummocks of angular fragments clashed
together; blisters which have burst, their cankered lips gashed,
and their throats blocked with the cakes from which they have
been unable to free themselves; cracked domes with holes in

them, just large enough for the foot to slip through, and the
jagged edges of which would mangle it if you drew it back,
down which holes you look into the utter darkness of a cavern
without an adit; jaws which have gnashed together till they
have ground their teeth to powder; horns, spikes, shavings,
polygons of inky rock, shivered, ripped, spurned aside, welted
and crushed, as the fiery mass, restrained from doing further
injury, has mangled itself in its writhings.

See! the work of regeneration and restoration has begun.
On yon tilted block, one tremulous saxifrage has taken root,
and lifts its white face to God and man; the forerunner of
other plants, which are to subdue and reduce to powder this
iron rock, to fill its grizzly hollows, and make the rough places
plain. The rains will honeycomb its shoulders, the frosts
chip off its angles, the winds fret its sides; the birds will bring
seeds to it, plants will spring up and dissolve its tissues,
willow will take root in its crannies, birch plant itself and shed
leaves into its crevices, till a good mould is formed, and the
wilderness becomes a fruitful field.

Now look behind, and you will see the red gable of
Hlitharfjall standing up soft and rosy in the evening air,
with a fleck of white on its apex. Several of the neighbouring
mountains have the same charred hue, like slag from a furnace.
If we climb a little higher and get a glimpse of Námarfjall, we
shall see a chain of bright red and yellow mountains with
steam curling from the gullies on their sides.

But I think that it is time for us to descend to the farm,
whence Grímr is signalling that supper is ready.

The guest-room at Reykjahlith is a curious sample of Icelandic taste. It is well boarded, and there are curtains to the
recess which contains the bed; the walls are painted gam-
boge to the height of three feet, above which they are ultra-
marine; the cornice is composed of ideal green and pink
flowers, and the ceiling is flesh-colour.

The farmer called me out just as I was going to bed, to
observe a curious phenomenon. At the farther end of the
lake was a moving russet pillar, the head of which reached

HLITHAR_FJALL.

London, Published by Smith, Elder & Co 65, Cornhill. 1863

the clouds. The column advanced slowly against the wind, till it reached the marshes around the lake, when it split in the middle, the upper portion being absorbed into the clouds, and the lower sinking to the earth. This was a sand column on the Sprengi-sandur, and was distant about fifteen miles. To the best of my judgment, the diameter must have been over two miles.

Grímr assured me that he had only once seen a sand pillar on such a gigantic scale, and that was during the eruption of Kötlugjá.

"It is not impossible that there may be an explosion somewhere," said the farmer; "indeed, I have heard a report that Skapta is in eruption."

Grímr reminded me of a peculiar noise resembling thunder which we had heard on the Eagle-tarn heithi, and told me that the same sound had been noticed at Akureyri. The Skapta had thrown up sand in 1861, and it was by no means improbable that it should be active now.

The next day I spent in visiting and sketching the immediate neighbourhood of Reykjahlíth. There is a hot spring called Thurásbath, in the lava at no great distance from the farm, but it is unimportant.

Hverfjall is worth a visit; a sharp scramble of a quarter of an hour takes one to the top, and the great bowl lies before one, slightly raised in the middle. The hill is composed of Palagonite tuff. The church of Reykjahlíth is interesting. The font basin is the same as that of Svínavatn; there is a coloured lanthorn in the chancel; the altar is poor; the chasubles are three in number, one rather ancient, of white silk, figured with pomegranates; the other two are modern, one of red damask, the other of crimson velvet. The two latter have gold crosses on the backs, but the former has simply a coloured bar in front and behind. Hanging against the screen is a curious candlerack of wrought iron foliage of considerable beauty. The workmanship is delicate, but the iron has suffered sadly from rust. At the door is a bronze ring, said to have been the sacrificial ring from

14

the temple, on which, when dipped in blood, vows were taken.

The ride to Námarfjall is over old lava, and must be traversed cautiously, lest the horses should break their legs, as the track passes across bubbles of stone, cracked at the top, and with holes just large enough for the hoof to enter. Heaps of burned and erupted stone on either side of the way look like exaggerated cinders from a smelting furnace, and are coloured red, black, and brown.

In half-an-hour we reach the sulphur mountains, a chain of red hills perfectly destitute of vegetation. We dip into a glen, and find it full of fumaroles, from which steam is puffing, and sulphur is being deposited. These run along the dale in a zig-zag. There are no considerable jets. Some way up the mountain, a gush of steam escapes with a whistle from a wild crag much rent by heat.

By the roadside I noticed a block of pure sulphur, from which every traveller breaks a piece, so that in time it will disappear altogether. The soil is composed of soft bolus full of splinters of trachyte, all of a brick-red hue. Wherever there is a hot spring in Iceland, the same clay reappears; it is, I am convinced, formed of trachyte, or trap, disintegrated by the action of steam and water at a high temperature. This theory is borne out most clearly, by the phenomenon of the Námarfjall, which is being gradually resolved into bolus; so that the stages of the process may be distinctly traced. On examining the steam jets which issue from splits in the rock, I found a soft deposit of mud in each fissure. Passing through the Námar-skarth, a winding cleft in the mountain, I came upon a most appalling scene.

Picture to yourself a plain of mud, the wash from the hills, bounded by a lava field; the mountains steaming to their very tops, and depositing sulphur, the primrose hue of which gives extraordinary brightness to the landscape. From the plain vast clouds of steam rise into the air and roll in heavy whorls before the wind, whilst a low drumming sound proceeding from them tells of the fearful agencies at work.

We left our horses to cross a patch of dwarf willow at some distance from the steam clouds, as they became very frightened and restive, and we then advanced on foot. My guide pointed to my feet and laughed. I was wearing india-rubber goloshes.

"You must take those off," said he; "or they will melt on your feet."

As we walked we sank in the clay, which became warmer as we approached the jets of steam; and, at last was so hot, that I could not bear to hold my finger two inches below the surface.

MUD CHALDRON. NÁMAR-HLITH.

Let me recommend you, reader, to get an iron pan or tray, and fill it three inches deep with clay and water, worked to the consistency of paste; place the tray on a stove, and watch the result. After a few minutes you will notice air-

14—2

holes opening in the clay, like craters, out of which steam will escape. These will become enlarged, and the fluid will be seen agitated in them, whilst crevices form in other portions of the surface from which steam issues. On a small scale you have the operations taking place on this plain.

It is not pleasant walking over the mud; you feel that only a thin crust separates you from the scalding matter below, which is relieving itself at the steaming vents. These vents are in great numbers, but there are, especially, twelve large chaldrons, in which the slime is boiling. In some, the mud is thick as treacle, in others it is simply ink-black water. The thundering and throbbing of these boilers, the thud, thud of the hot waves chafing their barriers, the hissing and spluttering of the smaller fumaroles, the plop-plop of the little mudpools, and above all, the scream of a steam-whistle at the edge of a blue slime-pond, produce an effect truly horrible.

In some of the chaldrons the mud is boiling furiously, sending sundry squirts into the air; in others, bells of black filth rise and explode into scalding sprinklings; in one, a foaming curd forms on the fluid, and the whole mass palpitates gently for a minute, then throbs violently, surges up the well, and bursts into a frenzied roaring pool of slush, squirting, reeling, whirling, in paroxysms, against the crumbling sides, which melt like butter before its fury. One or two of the springs have heaped themselves up mounds around their orifices; others, however, gape in the surface without warning, and the steam is so dense, and the sulphurous fumes so suffocating, that one becomes bewildered and can hardly pick one's way among them.

But for the readiness of my guide, the son of the Reykjahlith farmer, it would have been all up with me, as I recoiled before a scalding splash in my face, whilst making the accompanying sketch, and nearly slipped into another seething mud-pit behind me. The young fellow caught me by the shoulders, with his strong hands, just in time, but I sent an avalanche of sulphur and caked bolus into the abyss. I can assure you, the sensation of my foot slipping through the greasy marl, the thrumming of the hot flood behind me, into which I felt that I was dropping, together with the consciousness that there was no firm rock to which I could cling, disturbed my sleep for many a night, and come over me still in dreams, again and again, so that I awake with a start.

Around and among these chaldrons are small slobbering holes of all sizes, out of which issue steam and slime; some widen in time into large boilers, and the old ones fall in. These changes are continually occurring; Mr. Shepherd, who visited this scene on two consecutive years, assured me that the position of the chaldrons was altered on his second visit; and the farmer of Reykjahlith added confirmation to this statement.

I broke up some of the soil with my whip, and found that it was composed of blue, saffron-yellow, bright red, and white layers; it was impossible to preserve any of it without a tin case, as it fell to the minutest powder on drying.

Krafla (pronounced Krabla), to the north, is an insignificant mountain, far inferior in dignity to Hjörendr or Hlitharfjall. Krafla spoiled his beauty in his outbreak last century, when he tore himself in half.

There are some steam-jets on the side, one of which rushes forth with a whistle, but the Geysir in the crater, seen by Henderson, is extinct; and the pool in which it played is now still and green.

Hrafntinnufjall, close by, is an interesting spectacle to the mineralogist, as it is an obsidian mountain,—it looks like a mountain of broken wine-bottles. A magnificent block, clear as glass, which stood at the bottom, was split by Mr. Shepherd, who brought off half of it to England.

The scene from either of these mountains is very fine. The stately cone of Hjörendr rises to the left, the steaming heaps of Námarfjall to the right, and away, in the "Lava of evil deed,"—a tract like a troubled sea turned to stone, as big as Devonshire, and never crossed by mortal man—rises, massed on rocky flanks, the magnificent snow pile of Herthu-

breith, 5,290 feet high, bathed in the tenderest purple and
gold. To the east and south-east, is the barren desert of
Myvatns Orœfi, over which lies the track to Vápna-fjord, a
small trading station on the coast, distant three long days'
journey. The uniformity of the waste is only broken by
Möthrudalsfjall in the south-east.

Enough for one day! To-morrow I am bound for Dettifoss, the mightiest waterfall in Iceland; one, too, which no
European has ever visited, and which has been seen by very
few natives, and was unknown to the compilers of the great
map of Iceland. I heard of it only by accident from the
priest at Háls, and, on inquiring at Myvatn, found that the
farmer and his son were the only individuals who had been to
it, and could guide me : they confirmed the parson's opinion,
that it was unequalled in Iceland.

Next morning I started in a storm of wind and rain, with
the prospect of a long ride; the farmer said it could not be
done in less than seventeen hours, but his son, who was to
guide me, and who loved a rattling pace as much as myself,
promised that we should take less time about it.

We crossed the sulphur range, and then turned north,
skirting a lava stream which has flowed in serpentine windings
from Leirhnukr, below a hill purpled with the wood-crane's
bill (*Geranium sylvaticum*), growing in the greatest profusion,
on account of the warmth of the soil.

We passed a sandy ridge above Hjörendr, in a dense fall
of snow, which prevented me from seeing much of the surrounding mountains; but I observed that the wash from their
sides was yellow with sulphur. We then descended to a plain
covered with volcanic sand, extending, apparently, to the
horizon. I noticed that the cinders erupted from Krafla
diminished in size, with the greatest regularity, as we got
farther from the mountain. Our course was north-east, I
believe, but I found that my compass would not act over the
igneous rock; the needle was violently agitated, quivering on
its pivot, and vibrating from side to side; it finally settled with
its austral pole resting on the dial, pointing nearly west.

After a tedious ride of some hours, through snow, sleet, and drizzle, we drew rein in an oasis, beside a lake, beneath the mountain called Eylífr. This is the region of reindeer, and we saw traces of them in several places. A fine pair of horns lay on the roof of a farm just established in this doleful spot ; and I was told that there was a herd ranging the mountains behind the lake.

The farm where we rested had no tún ; it was nothing but a heap of turf, grubbed out into holes for rooms. These had not even windows, and the light was admitted in a somewhat novel manner. A pole had been driven between wall and roof at their junction, then forced downwards, and made fast with a stone. This lever tilted the portion of the roof against which it bore, so that light and air were admitted ; but, as soon as the stone was removed, the pole flew up, and the turfs slid into their places again, leaving the room in obscurity.

I purchased some stockings and gloves of the farmer's wife, a woman with bright eyes and dimpled cheeks. After allowing the horses to graze for half an hour, we remounted, and struck between north and north-east, over soil as barren as that we had passed before reaching Eylífr. After a while we came to a tract of lava, not marked in Olsen's map. It seemed to have issued from some red cones, rising about forty feet above the lava, complete miniature volcanoes.

Presently a white column on the horizon, reaching to the low trailing clouds, came in sight as we topped a swell in the waste. My guide pointed to it, and said, " Dettifoss ! " We had three or four miles to ride before reaching a spot near the falls where our horses were to be left ; a patch of dead sodden wild corn, our common sea-reed, or marram (*Psamma arenaria*), of the size of a drawing-room table. No grass anywhere within sight !

The young man led me to the upper cascade, which is disappointing, an inferior Góthafoss. The river Jökulsá breaks over a ledge, in a horse-shoe fall of no great height. The walls on either side are of columnar basalt. It is difficult to obtain a good view, without leaping over some of the rills, into which the stream breaks; and, as the stone is very slippery, one is pretty sure of soaking one's feet in the milky water.

The Jökulsá is the longest, and probably the largest river in Iceland; it rises in the extensive Vatna Jökull, and, after receiving tributaries from Trölladyngja and Herthubreith, empties itself into the Axarfjord after a course of about a hundred and twenty-five miles, having passed scarcely a human habitation on its way to the sea.

The lad conducted me next to the second fall. The sight was so overwhelming as I came out above it, through a natural door in the dislocated trap-wall on the side of the river, that I could only stand lost in amazement. I have never felt so thoroughly the helplessness of man, when nature puts forth her strength, as at that moment when standing amidst the wreck of creation, in a waste and howling wilderness, where no grass can find root, nor flower blossom, above an awful chasm, into which the mighty stream plunges, with a roar like a discharge of artillery.

In some of old earth's convulsions, the crust of rock has been rent, and a frightful fissure formed in the basalt, about 200 feet deep, with the sides columnar and perpendicular.

The gash terminates abruptly at an acute angle, and at this spot the great river rolls in. The bottom of the abyss is invisible from the point at which I am standing, and I have to move a couple of hundred yards down the edge, before I can see to the bottom of the gulf, and make a sketch.

The wreaths of water sweeping down, the frenzy of the confined streams, where they meet, shooting into each other from either side at the apex of an angle; the wild rebound when they strike a head of rock, lurching out half-way down; the fitful gleam of battling torrents obtained through a veil of eddying vapour; the Geysir spouts which blow up about seventy feet from holes whence basaltic columns have been shot by the force of the descending water; the blasts of spray which rush upwards and burst into fierce showers on the brink, feeding rills which plunge over the edge as soon as they are born ; the white writhing vortex below, with now and then an ice-green wave tearing through the foam, to lash against the walls ; the thunder and bellowing of the water, which make the rock shudder underfoot, are all stamped on my mind with a vividness which it will take years to efface.

DETTIFOSS.

The Almanna-gjá is nothing to this chasm, and Schaff-hausen, after all Turner's efforts to give it dignity, is dwarfed by Dettifoss.

My sketch gives but a poor idea of the falls, the majesty of which is beyond human skill to portray. One man only could have given a true version of its magnificence, and he is dead—that man was Turner. I have no hesitation in saying that Dettifoss is not only the finest sight in Iceland, but is

quite unequalled in Europe: it amply repays the toil of a
journey to it in its fastnesses; and I am sure that any future
visitor will be of opinion that I have underrated its wonders.

Our ride back was very monotonous; we traced the Jökulsá
up, till we came in a line with the Sulphur range, and then
rode straight for it, passing low craters in the desert, the
bowls, more or less complete, composed of sand and cinder;
and cracks in the ground formed by earthquakes, over which
the horses stepped cautiously.

We were delayed a short while by my guide catching a
sheep and ripping off its wool—in Iceland, shears are never
used; this he tied into a bundle, and inserted between himself
and his saddle.

The country brightens up a little on nearing the lava flood
from Krafla and Leirhnukr, and we passed a *sel*, or cot for
summer pasture, belonging to the Reykjahlith farmer.

Some juniper bushes, a little whortle, and some scanty
grass, grew about it, and plenty of moss coated the steep
sides of a stream, which had furrowed itself a way through
the sandy soil.

A thin fog came on as we sighted the lava torrent, and it
was curious to notice what shapes the blocks assumed through
the film of mist: it seemed to me as I rode along the brow of
a low hill above the flood, thoroughly exhausted with my
day's toil, that I was looking down on a Devonshire landscape,
from the top of a Dartmoor Tor. I seemed to distinguish
rolling wooded hills, towns and churches; but, as I approached,
hills and buildings became contorted, and appeared to rear
themselves up in new forms, monstrous and fearful, such as
perchance Schiller's diver might have seen at the bottom of
Charybdis, when he says,—

> " The purple darkness of the deep
> Lay under my feet like a precipice ;
> And though the ear must in deafness sleep,
> The eye could look down the sheer abyss,
> And see how the depths of those waters dark
> Are alive with the dragon, the snake, and the shark.

There, there, they clustered in grisly swarms,
Curled up into many a hideous ball;
The sepia stretching its horrible arms,
And the shapeless hammer, I saw them all;
And the loathsome dog-fish, with threatening teeth,
Hyæna so fierce of the seas beneath."

After a ride of fourteen hours and a half, we drew rein at the door of Reykjahlith, cold, hungry, and tired.

CHAPTER XIII.

A SNOW PASS.

I hold a Levee—My Plans upset—The Church of Thverá—Taking French Leave—Swimming across a River—MSS. Sagas—Jón's bills—Runaway Ponies—Rocky Spires—Perverseness of Grímr—A Mountain Sel—A fearful Pass—Snowbridge—A desperate Scramble—Stone Bog—Midnight on the Snow—A crumbling Snowbridge—Arrival at Hólar—An alarming Proposal.

I WAS awakened on the following morning by the entrance of a young man with some goose-eggs for sale. He was not long solitary, for his father pushed in after him, to see that his son was fairly remunerated, and his younger brothers followed, that they might have a look at the Enskrmathr (Englishman) who was going to buy goose-eggs. Grímr crawled out of bed to inspect the articles, my host came in to argue about them, followed by his better half and the red-haired servant-girl, both filled with feminine inquisitiveness, the latter pursued by her admirer, and the admirer in turn followed by his brothers. Then my host's little snub-nosed daughters, who were carrying on a flirtation with the brothers of the maid's admirer, poked their snub-noses in at the door, and presently, becoming emboldened, entered the room, with four dogs which they had been feeding. Finally, some folk from Möthrudalr, who had arrived on the preceding day with a train of horses, finding that my room was the general rendezvous, edged themselves in as well. As every one had something of his own to say on the subject of the eggs, and was perfectly indifferent to the opinion of the others, the room became a perfect Babel.

My comb and brush were in the window, and, as I knew that
these would be tried upon the different heads in rotation, if
once any of the throng caught sight of them, I had to stifle
all my sense of propriety, jump out of bed, dash through the
crowd, and capture my goods. I went through my dressing
operations without drawing attention from the eggs, which
were still the engrossing topic of conversation.

Later in the morning, the men from Möthrudalr gave me
information which completely upset my plans. That these
plans may be understood, I must enter into some geographical
details, which I hope will not exhaust the patience of the reader.

The south-east of Iceland is occupied by a vast heap of
mountains, shrouded in eternal snows, which discharge themselves by glaciers into the sea, on the south. The extreme
length of the mass is 115 miles, the width 60 miles. This
huge district of volcano and snow is named Vatna, or Klofa
Jökull, but the points in its fringe, some of which are of considerable height, go by separate names; of these the most
important are, Hofs, Heinabergs, Breithamerks, Orœfa,
Skeithar and Skaptár-Jökulls. Some of these have been the
foci of most appalling eruptions.

The northern fringe is hardly known, as it meets the
Odatha Hraun, or "lava of evil deed," extending, as may be
seen on reference to the map, from the Jökulsá in the east to
the Skjalfanda-fljót in the west, and northwards as far as
Myvatn. Out of this fearful tract rise mountains, standing up
almost like islands above a wild black sea; they can be seen,
though not reached, and these have caused the devastation of
this enormous district. Their names are Herthubreith, and
the two Trölladyngjas.

To the west of this lava district lie the sand-deserts of
Sprengi and Stori-sandur, quite destitute of vegetation. To
the east, again, beyond the Jökulsá river, are dreary wastes,
relieved only at long intervals by grass patches.

The northern scarp of Vatna Jökull has once been skirted
by some Danes, but in the attempt they lost a number of their
horses, through cold and starvation.

My object in coming so far east was mainly to explore this
tract, to ascend the higher points of the northern edge of the
Vatna Jökull, and endeavour to get a glimpse of the unknown
fastnesses of ice and fire beyond.

My plan was to push up the Jökulsá to its source near
Kverk-fjall, then, if I found grass, to remain there a day and
rest the horses, to allow for the ascent of the mountain. After
this I intended crossing the spurs of the glacier to the other
source of the Jökulsá under Kistufell and Trölladyngja, where
I hoped to find grass again. If, however, there were none
sprouting, I should be obliged to give up all prospect of
advancing farther; but if I found sufficient for the support of
my ponies, I intended making an effort to reach Eyvindarkofaver, a small patch of grass land between Hofs Jökull and
Tungnafells Jökull, where I should rest a day to recruit the
horses before proceeding south. This would make it a matter
of eight or nine days from Möthrudalr to the first inhabited
spot.

The news which the men from Mothrudalr brought spoiled
my plan completely. They said that on account of the coldness of the spring the grass had hardly begun to sprout anywhere along the Jökulsá vale, and that it was quite out of the
question expecting to find the least herbage under the roots of
the Vatna. Upon this, the farmer of Reykjahlith chimed in
with his story of the eruption of Skapta, which volcano, being
close to Eyvindarkofaver would effectually destroy the grass
there; adding, that he expected that something of the kind
must have happened, as no caravans had crossed the Sprengi-
sandur this summer from the south. As Eyvindarkofaver is
the last spot of grass which is met with in a journey north
over the Sprengi-sandur till Isholl is reached, which is distant
twenty-two hours' hard riding, it was evident that caravans
coming north, if they found no herbage at Eyvindarkofaver,
would turn back without venturing on the desert. Thus, the
non-arrival of the caravans showed that something was wrong
at the other extremity of the sand desert.

It was a bitter disappointment for me to have to postpone

the execution of my scheme for another and more propitious summer, but it would have been insanity to have persisted in it ; no guide would have accompanied me, and I should in all probability have lost my life with that of my ponies.

I now changed my plans completely, and determined on returning at once to Akureyri, and after visiting Hólar and Mithfjord, to hasten on to the scene of eruption, supposing it to be the Skapta.

Without more ado, determined on losing no valuable time, I ordered the horses to be brought round and saddled ; but it was soon discovered that they had run away, having found next to nothing to eat around Reykjahlith, and it was not till evening that they were recovered.

We started on our return to Akureyri in rain, and had showers till we reached the Laxá or Salmon river, which flows from Myvatn through a bed of lava, and empties itself into the sea in the Skjalfanda frith. At the farm of Thverá we remained the night, as I was too knocked up to proceed ; the long ride of the preceding day and night, and the short rest, broken by the clamour of the egg-fanciers, had given me a racking headache.

The church of Thverá possesses only one object of interest ; this is an exquisite font-bowl, representing a crowned hart in high relief on a low raised background of oak-leaves, symbolizing the panting of the unregenerate soul for the water-brooks of baptism. The rim of the bowl is stamped with grape bunches and intertwining tendrils. A more spirited and vigorous design can hardly be imagined.

I bought of the farmer a copy of the Sturlunga Saga, published in 1817–20, which I had been unable to procure in England or Denmark, as it is out of print. This is one of the longest of the historical Sagas of Iceland, and is, in fact, a compilation of Thatir or "Historiettes." It fills about 950 pages of close printing, square 8vo.

As Grímr was intent on letting every one know that I had seen Dettifoss, my sketch was in constant requisition, for Dettifoss is a mysterious scene of wonder, which has not been visited by a dozen living natives, and of which strange tales are told.

On the following morning, as I came to the door, I was surprised to find one of my horses mounted by an elderly Icelander with a long beard, of whom I knew nothing; but supposing him to be an extra guide engaged by Grimr, for a bit of ground over which neither he nor Jón knew the way, I made no remarks, till we came to the Skjalfanda ferry, where he dismounted. I then inquired who he was, and learned that he had arrived at Thverá on the preceding evening, from a farm near the ferry, and that his horses had run away during the night, and had probably returned home. He had consequently taken one of mine, without asking leave, partly to save himself the trouble of asking, and partly to save me the trouble of granting or refusing permission.

At Háls we missed seeing the priest, so that I could not thank him for having advised me to visit Dettifoss, as he was engaged in hearing confessions. People in their holiday attire were awaiting their turn at the door and in the passage.

In the church, which is quite new, is an inferior wooden Madonna of the fifteenth century, and undoubtedly of German workmanship. The altar-piece is modern and very poor. A good deal of gilding has been expended on the screen and altar-rails, but without judgment and with no effect.

On reaching the Eyja-fjord river, we found that the tide was up, and the estuary so deep, that we should be compelled to make the horses swim. As I had no change of apparel, and had no wish to be drenched, I stripped, and rode through with my clothes under my arm, under the shelter of an india-rubber poncho, to satisfy Grimr's scruples, as he was dreadfully afraid lest the ladies of Akureyri should be promenading on the opposite bank. The poncho was, however, inconvenient, as it floated out, and nearly blinded my swimming horse.

The river is a mile and a half wide at this point, branching into seven mouths, with grassy holms between them. We passed without danger, as the current was slight, but the sensation of cold water gurgling between one's self and the saddle and then rising to the waist was anything but agreeable. I fortunately kept my clothes perfectly dry, but my collection of dried flowers, which I had forgotten to remove from my saddle-bag, was soaked, and the flowers were completely spoiled, as the colour was washed out of them, and the paper was so sodden that they mildewed between the sheets.

I met with the same kind reception at Akureyri as before, every one vying with the other in showing me courtesy.

I bought here some MSS. of considerable interest from a native who was reduced to great poverty, and only parted reluctantly with the volumes.

"These Sagas," said he, " are our joy; without them our long winters would be blanks. You may have these books, but, believe me, it is *prava necessitas* alone which forces me to part with them." As he spoke, the tears came into his eyes, poor fellow! The volumes I purchased were, 1st, a copy of the Sigurgarthar Saga Sigurgartharsonar, in 8vo, bound in vellum, and written about the end of last century; 2nd, a folio in calf, written in 1713–15, containing :—

1. Völsunga Saga.

2. Ragnar Loðbrókar Saga ; both together occupying 60 pages.

3. Ænea ok Trojumanna Saga, with an account of S. Oswald, and with the history brought down to the end of the reign of K. Athelstan ; 32 pages.

4. Raisubok Bollings, with appendix ; 62 pages.

5. Bärings Saga fagra ; this Saga relates the history of a certain Bäring, son of Walter, Duke of Holstein, and grandson of the Grand Duke of Saxony. It exists in MS. on vellum at Copenhagen, one of the copies being supposed to be of the fourteenth century. The Saga is a translation from a lost German romance, half history perhaps, but mostly fable, probably made by order of King Hakon Hakon's son. It has never been printed and published. 16 pages.

6. Saga af Ambrosio ok Rosamunda ; fabulous ; 10 pages.

7. Saga af Remund ok Melucinœ ; fabulous ; 5 pages.

8. Skjolldunga Saga ; a restoration of the Saga from which Saxo Grammaticus compiled his history ; 4 pages.

9. Saga af Drauma Jóni ; unpublished.

10. Saga af Hakoni Harekssyni ; 4 pages ; historical. This is followed by a list of Icelandic Sagas.

11. Saga af Ulfari Sterka ; fabulous ; 10 pages.

12. Saga af Illuga Gryðarfostra ; mythical ; 5 pages.

15

13. Saga af Elys hinnum frœkna. If this be the same as the Saga of Elis and Rosamund, it was translated from the French, in 1226, by Robert the Monk, by order of King Hakon Hakon's son ; unpublished ; 21 pages.

14. Saga af Bárði Dumbssyni ; a wild, half-mythical tale of one of the first settlers in Iceland ; 19 pages.

15. Saga af Hjálmtyr ok Olver ; fabulous ; 20 pages.

16. Saga af Halfdani Eysteinssyni ; fabulous ; 22 pages.

17. Tyma Rymar.

18. Saga af Alexandro Magno ; translated by Bishop Brand Jónsson, by order of King Hakon Hakon's son ; 80 pages.

19. Saga af þiðrik af Bern ok Köppum ; containing 21 or 22 þatir or chapters.

20. Brjef Petri Pauli Abota til Consenta, til Printzins af Bisignono.

21. Saga af Fortunato ok sonum, translation of the favourite mediæval story of Fortunatus ; 49 pages.

22. Thornessinga or Eyrbyggja Saga ; 56 pages.

23. Saga af nafnafrœga sterka Hercule.

24. Saga af Asmundi Víking ; a very beautiful, romantic story ; unpublished ; 28 pages.

25. Saga af Asmundi ok Eigli einhenda ; 11 pages.

26. Egils Saga Skalagrímssonar ; 54 pages.

3rd. I obtained a square 8vo in calf written during this century, containing the Víglundar Saga, Flóamanna Saga, Jökuls Saga Búasonar, Vatnsdœla Saga, Laxdœla Saga, Hrafnkels Saga, and others. These volumes are now in the British Museum.

On Monday we started for a long ride, over what my host at Akureyri told me was the most accursed pass in all Iceland, and which I was particularly anxious to see, as Henderson had given such a thrilling account of his perils in traversing it, during the summer of 1814. He then took two days about what I purposed accomplishing in one.

Grímr had never been that way himself, so I was obliged to hire Jón to come on with us as far as Hólar. He shook his head at the prospect of having to cross Hjaltadals-heithi at the close of a long day's work, but he laconically observed that he could do his part, and if the horses were injured—that was not his affair but mine.

Grímr overawed him too ; and Jón was terribly afraid of Grímr, for a good reason, as, on the previous summer when they were together bound for Hólar, across the Heljardals-skarth, and poor Jón had refused to go any farther, the snows being in a dangerous condition, Grímr had caught him by the collar, and laid into him with his leathern whip, till the poor fellow, with shrieks and tears, had promised to go wherever his master chose. Ever since that event, whenever he was in company with Grímr, Jón had been most docile, and now, when any difference of opinion arose between them, he judiciously kept on the farther side of a horse, beyond the reach of Grímr's whip.

Before starting, Jón shuffled up to me with the request that he might be given at once part of his wages for the time he had already been with me.

" By all means," said I ; " I will hand you over all that I owe you, if you wish it." I was going to give him the sum, but he pushed my hand aside, saying with vehemence,

" Do not pay it to me, I should spend it all in snuff and brandy. Will you kindly go round to the different merchants, and pay them my little bills ?"

" Surely, Jón, you can do that best."

" Ah ! sir, you don't know me ; I should pay my debts at one store, perhaps, but I should then become so horribly drunk, that I should never reach the second."

" Well, but the shops are a stone's throw apart."

" Temptation is strong, sir."

I gave the money to Grímr, who settled Jón's little accounts for him, and these I found had been run up by his wife, for necessaries of life. I shall be glad if any one travelling in the north will employ Jón, for he is a most trustworthy and obedient fellow, and very moderate in his charges.

The system of guides in Iceland wants putting on an entirely new footing; at present, the Reykjavik *fylgi-mathr* charges exorbitantly, and is incompetent to fulfil his duties, being idle, and ignorant of a great part of the country, so that an additional guide becomes necessary in many portions of the island. The theological student is a fine gentleman, and the traveller had better not be encumbered with him.

We could not start early on Monday, as there was no getting Grimr off. Seeing that I was bent on starting betimes, he dawdled about without putting his hand to anything, so that it was noon before we were under way.

Scarcely, moreover, were we out of the town, than one wretched pony bolted at right angles to the direct road, and Grímr and I had to gallop in pursuit. To make matters worse, the girths of Grímr's saddle burst, yet he kept his seat over rough ground with the greatest indifference.

The runaway trotted through a gang of turf-cutters, but not one of the men attempted to stay him. We finally caught the truant in a bog, into which he had floundered, and could not extricate himself without assistance. Grímr shifted his saddle, and immediately the horse he had been previously riding started off at a canter, on his way back to Akureyri, so that we had to gallop back to the town before we could catch him. Grímr then tied the thong of his whip around his lower jaw, and led the brute along as a prisoner.

We retraced the road by which we had entered Akureyri, till we came to the point where Horgár-dalr branches off from Öxnadalr. Here, upon the river sand, I found growing large patches of the pink Sidum villosum, and on the grassy banks great quantities of the common spotted orchis (*Orchis maculata*), and two greenish white orchises (*Perystilus viridis* and *Perystilus albida*); the moonwort (*Botrychium lunaria*) was growing in great profusion, and the sand knolls were covered with heartsease (*Viola tricolor*) and wild thyme (*Thymus serpyllum*).

The scenery in the Horgár valley is magnificent; to the right open gorges leading to the Myrkár Jökull, or the "murky river glacier mountain;" to the left, is a strange serrated ridge with snow resting on its ledges. One peak of this range is represented in the accompanying illustration (Plate XI.), pointing to heaven like a church spire, and supported by precipitous walls of basalt, whose columnar formation gives them the appearance of being built up of reeds, and pierced here and there with gaps like pigeon-holes.

Pl. 11.

S. Baring Gould, delt.   Edmund Evans, sc.

IN HÖRGADALR.

London: Published by Smith, Elder and Co., 65, Cornhill. 1863.

According to popular tradition, Grettir ascended to the highest point of this ridge, and placed his belt and knife on the rock needle as a prize for the next who could reach the summit. Grímr had a rooted objection to my making sketches or gathering flowers, and when he saw me engaged in either of these pursuits, it was his wont to urge the horses on, and not draw rein till he was out of sight, thereby causing me considerable annoyance, as I repeatedly lost my way in following him, whilst he, with eagle eye, watched me from a rock, behind which he had secreted the baggage-horses.

During the first part of my journey I used to give way to my irritation, but eventually I hit upon a better plan. I rode up to the model student, singing or whistling, as though nothing had happened to disturb my equanimity, and he immediately sulked because *I* was not put out. Whilst making the sketch of the rock needle, Grímr attempted his usual trick, but was restrained by Jón, who feared losing me altogether. As it was, I missed the guides and passed through a couple of farms, whilst they kept down by the river. The good farm-people, who were then hay-making, stared with astonishment at me, much puzzled to know who the stranger in knicker-bockers could possibly be, who was riding towards the worst pass in Iceland without guide or baggage. I found my train of horses at last, halting on the shingly bank of the river, Jón having positively refused to advance till I rejoined them. We were detained for half an hour on this spot by one of the ponies deliberately wading across the river and scrambling up the mountain on the opposite side, whilst another horse set off as hard as it could gallop on its way back to Akureyri. Grímr and I pursued one, and Jón went after the other. When we had brought them back, the rest of the ponies were found to have strayed, though fortunately to no great distances.

On reaching the head of the vale, we stopped to bait at a little *sel* or mountain cottage, used only during the summer, when the cattle are driven up the heights for pasture. Here we supped, cooking our food and eating it in the *Bath-stófa*, seated on the beds and using horn spoons neatly carved, provided by the peasants. The place was unfortunately so swarming with vermin that all my peace and comfort for many a day were destroyed by what I carried away with me from that sel.

We were told here that Hjaltadals-heithi was quite impassable; one man only had crossed it this summer, and he had been on foot. Jón began to show signs of recusance, but I protested that go on we must and should, and Grímr brandished his whip ominously. So Jón submitted, and we saddled and bridled our horses, put on the packs, and started up the ravine which opens out above the sel.

After we had toiled for some time over all but precipitous rocks, which would have strained the backs of any but Icelandic horses, I asked hesitatingly whether the worst were over.

"Over! bless you!" exclaimed Jón; "why, we have not reached the pass yet! only wait till we come out upon the snow!" At the same moment a white ridge soared up before us, crowned by terraced sugar-loaves of basalt. "There!" said Jón, pointing to the snowy range; "we have to cross that."

My heart almost failed me at the sight, for the horses were tired, and there was before us the work of several hours.

We came upon the basin of a dried-up lake, traversed by a river wending towards a portal of black rock which it had cut for itself to the level of the silt which filled the bed of the lake.

Suddenly, we found ourselves in the angle between two gorges, down which roared torrents of milky water, floating off masses of dislodged snow, and sharp fragments of ice. That on the right cut us off from a mountain cone shooting up several thousand feet, and jauntily capped with snow, though too precipitous to allow of any resting on its sides. The ravine immediately before us was arched over with a snow bridge, in a very insecure condition. In winter the whole gorge had been choked with snow, through which the torrent had worked itself a tunnel. During the spring and

summer, portion after portion of this snow-bed had fallen
through, and been swept off by the stream, leaving at intervals, white bridges about twenty feet wide, spanning the gulf.
Jón boldly rode over the bridge, and reached the farther side
in safety; seeing this, the rest of the horses were driven over
by Grímr and me. This was succeeded by a desperate clamber up a crag, slippery with ice. The baggage horses could
scarcely get on, and we were obliged to stand by the poor
brutes, as they rested on the ledges in the ascent, and support
the box which was towards the precipice, clinging on to the
rock with one hand, and lifting the box with the other. We
allowed the unencumbered ponies, and our own riding horses,
to find their own way, being very careful first to tie up the
bridles and halters, lest the animals should get them entangled
in their legs. On reaching a long steep slant of loose rock-
fragments, which had to be surmounted, the most heavily
laden of the pack-horses seemed inclined to give it up in
despair, so Jón dragged at his halter, ascending just before
him, whilst Grímr and I put our shoulders to the boxes and
relieved him, to a certain extent, of their weight.

On reaching the top, we found that we had a tract of
stone-bog, through which to trudge. The stone-bog, as
mentioned in my Introduction, is formed by the thawed snow
percolating through the crevices of the rock, then freezing
and splitting it up, till complete morasses of stone and mud
are formed. These are not dangerous, but are very trying,
as one sinks in them to the knees, and the angular fragments
of stone cut the horses' legs and rend one's own shoes and
stockings. After labouring through the bog, we came upon
unbroken snow, and paused to take breath.

The view was striking. Looking back, we saw that the
vale from which we had come was now filled with white fog,
and looked like an extensive winding lake. To our left rose
the cone already mentioned, and to our right lay swell and
sweep of undinted snow in soft grey shadow, rising in smooth
slopes to some curious horns or pyramids, barred black and
white, behind which the slanting crescent moon shone golden

in the sky. It was midnight, and the sun was down, but the
heavens were still lighted with his rays, which turned their
blue to the tenderest ice-green. An eagle (*Falco albicilla*),
perched on a crag, watched our cavalcade, and then plunged
down through the mist, disappearing in it like a stone
dropped into water. The air was intensely cold, and the
line of demarcation between mist and mountain was extraor-
dinarily sharp. The snow was soft, so that we sank to our
knees at every step. We were obliged to walk, as the horses
had as much to do as they could well manage to pull themselves along. Sometimes the poor brutes refused to advance,
and stood up to their bellies with their desponding heads bent
to the white surface. Ponies will, at times, make up their
minds to go no farther, and then there is no stirring them;
they will stand in the same position till they are frozen,
and then fall over on their sides to die. Happily none of
my train became quite in this condition, though they were
very near it, and we had the greatest difficulty in getting the
baggage horses to move on. My pony floundered into a hole,
and sank over the saddle, so that he could not get out without
assistance. The snow is very cavernous in the neighbourhood of protruding rocks, and, as the horses invariably make
for the least black speck in the white waste, as a drowning
man would strike out for a skerry, the consequence is, that
they are continually sinking into holes which are perfectly
concealed beneath a smooth surface. We had now a steep
incline to overcome; up this we were obliged to crawl in a
zigzag course, dislodging masses of snow, which slid down
and vanished in the mist below, forming miniature avalanches.
As we reached the top, the sun broke over a marquee-shaped
mountain opposite, and cut off from us by another lake of fog.

We rested for half an hour on the summit of the pass,
lying thoroughly exhausted on the snow, beside our fagged
horses, which stood before us in a line rapt in a brown study,
without moving a muscle.

The descent was through dense fog, and was so precipitous that we were obliged to leave a considerable space

between each horse, lest one should slip down upon the other.

When we reached the bottom of the mountain we had to pass a torrent, which descended in a noble waterfall, and then rolled angrily away under an arch of snow. This bridge was so rotten, that Jón hesitated for some time whether we could venture upon it, and at last sent over a horse as an experiment. The snow bore, and we drove the baggage pony across; cakes of snow fell off the bridge into the gorge, but the arch remained unbroken, so we all passed without accident. We had taken between five and six hours in crossing this heithi, and had been on our legs for a great part of the time. It was seventeen hours since leaving Akureyri, when we drew up at the door of the archdeacon's farm at Hólar.

The old man received us very kindly. It was morning, and he with his wife and servant were already up, brisk as bees, and ready for a long talk. I was tired out and longed for bed, but hardly liked to get in before them, yet Grimr had just begun the Grimsey story, and that with all his grievances would last an hour at least. I fairly fell asleep with my head on the table, and was roused by Grimr, who recommended me to go to bed.

"But," said I; "these good people are in the room."

"Oh, don't mind us!" said the priest.

"Pray go to bed!" said his wife.

"Do let me pull your breeches off!" volunteered the maid.

I started up at the proposal, fully roused, and, with a flying leap, buried myself under the feather bed, then pulled off coat and waistcoat and curled myself up.

"Don't English people undress more than that when they go to rest?" asked the priest, who had been watching me gravely.

"He has got his breeches on," said the wife.

"I'll pull them off, if he likes," chimed in the maid with alacrity.

"Never, *never!*" I cried in desperation; "Grimr, save me!"

Poor Ebenezer Henderson, the Bible Society delegate! the
Icelanders still have a good laugh over his dismay, when first
the ladies of the house insisted on dismantling his legs.
This was according to etiquette in his time, though now
happily falling into disuse. In his book he tells the story
of his wild struggle to preserve his nether garments, but he
neglects to mention the compromise which was effected, he
coiling himself up in the coverlet, and letting the ladies pull
at the strap-buttons. Henderson was a very good fellow, but
he had no notion of a joke, and he only mentions the incident to found on it moral and pious reflections. Among
themselves it is still a common practice for the women to
peel the men after their day's work, but the Icelanders have
learned that strangers do not particularly relish this sort of
attention, and they now seldom offer it.

After my first nap, Grímr came to bed; he was to share
mine, so a pillow was put at the bottom of the bed for his
head to rest upon, whilst his feet lay on the pillow by my
head.

"Oh, Grímr!" said I; "this is dreadfully cramped!"

"Bless you!" he answered; "we sometimes sleep five in
a bed of this size, head to foot, lying on one side and not
stirring all night long."

It is not pleasant to have a calm dream of home interrupted in the middle of the night by the descent of a cold
foot on one's face. Reader! may you never experience it!

235

CHAPTER XIV.

HÓLAR.

The Cathedral—Altar Vestments—Triptych—Portraits and Tombs—MSS.—
The Ancient Possessions of the Church—Repairs—Flowers—The Skaga-
fjord—Drángey—Birds—Gannet—Puffin—Skua.

Hólar is situated in a noble valley between mountains covered
with snow. The soil is peculiarly fertile, and Hjaltadal is
regarded as the garden of the north. There are two houses
near the church, one the residence of the archdeacon, the
other of a farmer. At the time that I was at Hólar, all hands
were engaged in the hay harvest, reaping down the grass with
sickles, and raking it into little heaps. Although the tún is
considered to produce some of the finest hay in Iceland, I
believe that few blades of grass were longer than my fingers.
The church, dedicated to S. Mary the Virgin, is a stone
building sixty-four feet eight inches long, by twenty-nine feet
four inches broad, and twenty-seven feet high. Its plan is a
parallelogram, without either constructional tower or chancel.
A bell chamber and porch are formed by partitioning off the
west end of the church, and the ritual choir is separated from
the nave by a screen. On either side of the building are
seven windows: there is no opening at the east end, and the
western gable is pierced by two windows, in which are hung
the bells. Within the porch lies the largest bell of Hólar—
cracked. It is said to have tolled of itself when Jón Arnason,
the last real bishop of Hólar, suffered martyrdom.

The building was raised in the last century, and is devoid

of all architectural merit; it is wonderfully like the railway
station at Grangemouth, but the fittings within are full of
interest.

To begin with the altar. This is the old stone altar
belonging to the ancient cathedral, and measures five feet
nine inches by three feet one inch high. It is enclosed within
faded curtains of chequered blue silk and lace. The altar is
vested: first, in a green leather frontal stamped with gold
flowers, and a super-frontal to match, both falling to pieces,
but very handsome; secondly, in an admirably preserved
embroidered cloth, with five full-length figures on it, worked
in colours on a buff ground. These represent—First, an
angel with censer; second, Bishop Gúthmundr, in white alb
with red apparel, red stole, dalmatic striped blue and yellow,
red chasuble flowered with gold, and blue orphrey, blue mitre
and crozier particoloured red and blue; third, S. John of
Hólar, vested in white alb with blue apparel, red and yellow
striped dalmatic, blue stole with brown fringe, blue chasuble
with red orphrey, and violet mitre; fourth, S. Thorlak,
vested like Bishop Gúthmundr; fifth, an angel with censer
and book. All the bishops have episcopal rings, pink gloves,
red boots, puce fillets to their mitres, and brown maniples.
They are represented as closely shaven; their hair, as well
as that of the angels, is red. The angels are vested in blue
and red, with hoods or tippets, and have bare feet; the
chasuble is very full, almost circular. The altar is also
covered with white linen, embroidered in red and blue thread,
with a representation of the animals entering the ark.

On the holy table stand two brass candlesticks, one branch
candlestick, also of brass, for three lights, dated 1679, and
another similar stand, somewhat smaller. The priests' vestments hang over the curtain rods, and consist of an alb plain,
a gold-coloured chasuble, and two of velvet, one crimson, the
other green.

Above the altar is an immense triptych. The doors are
painted on the outside, with Christ appearing as the gardener
to the Magdalen, and with the martyrdom of S. Sebastian.

When these doors are flung open, the appearance of the altar-piece is most striking. It is carved with the greatest delicacy in full relief, in the style of German art in the fifteenth and sixteenth centuries, and is profusely coloured and gilt.

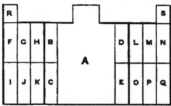

DIAGRAM OF TRIPTYCH.

In the centre, A, is a noble representation of the Crucifixion; Jerusalem is visible in the background, the sun and moon are being obscured, crowds are thronging the foreground, the centurion pierces the sacred side with his lance, the Marys and S. John are at the foot of the cross, and the Blessed Virgin has fainted into the arms of the beloved disciple. Angels with chalices receive the blood from the five wounds.

On either side of the central subject are tabernacles, or niches, containing single figures; these are—B, S. Katherine; C, S. Margaret; D, a female saint with panniers, and a child leaping up to them; E, a female saint holding a tower, containing the Host; F—Q, the twelve apostles. R and S contain groups of figures, but what they represent I was unable to distinguish.

This triptych is quite a masterpiece of carving, the figures are full of spirit, the faces are expressive, the drapery is carefully executed, the details of foliage delicately wrought; and the whole is as fresh and uninjured as it was when first erected in the cathedral. The weakest point is the colouring, as it is certainly overdone with gilding, and there is a deficiency of pure bright colour.

A second altar-piece of alabaster, picked out with gold

and colour, of a much older date, stands above the rood
screen. It consists of a parallelogram, divided into seven
compartments, the first of which contains a figure of S. John
Baptist; the second, a representation of the Betrayal; the
third, of the Flagellation; the fourth, of the Sacred Trinity;
the fifth, of the Entombment; the sixth, of the Resurrection;
and the seventh contains S. Katherine.

The screen is of wood coloured; it is formed of pilasters
much resembling the initial I's in old English MSS. The font
stands north of the screen in the nave, and is of stone, with a
circular bowl, carved with subjects from our Lord's life, such
as the circumcision and the baptism. At first sight I took it
to be of great age, as the style was much like our transition
work from Norman to Early English, but on closer examination it proved to be of a date as modern as 1674. In one of
the compartments is the inscription—

TYPUS BAPTISMI, XXX;

the drift of which I do not understand.

The pulpit is an ugly modern erection of dark pine, standing
on the south side. Near it I noticed a singular old lanthorn
for three candles, with gabled sides. Above the font, hanging
against the wall, is a life-size crucifix; over the door at the
west end is a crucifix with SS. Mary and John; another
hangs in the chancel. The choir is adorned with portraits of
the Protestant bishops. Gissur and his wife are represented
kneeling before a crucifix, with the legend around the picture—

" Præsulis externum Gissuri prospice vultum,
Aurea sed claræ mentis imago deest;
Ad latus effigies speciosæ conjugis astat,
Interior cujus promicat axe poli."

There are two portraits of Guthbrand, the second Protestant bishop, who translated the Bible into Icelandic. One
of these is an embroidered portrait in wools, worked by his
illegitimate daughter; the other is in oils. Certainly the old
gentleman showed his judgment in not suffering posterity to
judge of his personal appearance' only by the needlework pro-

duction, between which and the painting there are only two points of resemblance, the ruff and the cap. This individual is buried in front of the altar beneath a large slab, on which is inscribed—

EXPECTO RESURRECTIONEM CARNIS
ET VITAM ETERNAM,
GUDBRANDUS THORLACIUS, JESU CHRISTI PECCATOR.
ANNO CHRISTI 1627
20 JULII.

It is certainly satisfactory to notice that this tombstone bears testimony to the doctrine—that man's hope lies in the resurrection of the *body*, not in any translation of the soul to heaven after death, according to the popular modern opinion, which is at variance with the teaching of the creeds, Scripture, and Catholic antiquity, and which is probably a revival of long latent paganism.

There is also a portrait of the coarse and fleshy Thorlak, third post-Reformation ruler of the see of Hólar, with his three ugly wives.

In the nave are also some oil paintings in a bad condition; these are the portraits of King Christian V. and his gylden-löve, also of Paul Gaimard the French traveller.

The church once possessed a magnificent golden chalice, which was removed at the spoliation of the Icelandic church to Copenhagen, and is now in the Fruenkirke of the Danish capital. There remains, however, a very beautiful silver-gilt chalice, 7⅜ inches high, and the bowl six inches in diameter. The knob is wrought in spirals of flower-work and beads. The design is late, probably of the sixteenth century, but the execution is inimitable.

The paten is very plain. There remains also an ancient crimson velvet burse, embroidered with pearls; on one side is the Crucifixion, on the other the Annunciation.

Between the ceiling and roof of the church is a lumber garret, in which the priest keeps his books, his husbandry tools, and his coffin. Among his volumes I noticed a MS. written at the end of last century, containing the Laxdœla,

Eyrbyggja, Flóamanna, Vatnsdœla, Egils Skallagrímssonar,
Bjarnar, Fostbrœthra Sagas ; and a second volume in folio
containing Islendingabók, and the Kristni, Thorfins karlsefnis,
Hœnsa-Thoris, Gunnlaugs, Valla Ljóts, Kórmaks, Finnboga,
Ljósvetninga, Vemundar, Thorsteins stangarhöggs, Thorsteins
suthufari, Thorsteins frótha, Vápnfirthinga, Hrafnkels, Brandkrossa, Droplaugarsona, Egils Sithuhallssonar, and Gunnar
Keldugnupsfyfls Sagas. These transcripts, even though of
modern date, are valuable, as they are exact copies of older
texts.　An Icelander reads his sagas aloud winter after
winter, till the book is ready to fall to pieces, when he carefully transcribes it, and then casts the well-worn volume aside.
I one day saw an old MS. of the Hrafnkels Saga in a byre, and
offered to purchase it, but the farmer would not part with it at
any price, because he had not yet copied it.　In the 12mo.
volume which I obtained at Grimstúnga, is the last page of
the Ajax Saga ; the rest had been gradually thumbed away, but
the loose pages had not been lost till the farmer's daughter
had carefully recopied them word for word.　Arnœus Magnœus collected all the MSS. he could procure in the island, and
pretty thoroughly ransacked it of all early vellum or parchment
MSS.　I am sure that the majority of these were transcribed
before their owners parted with them.　The library of Arnœus
Magnœus was burned down, and a vast number of these precious MSS. perished.　Their contents may, however, in a
great measure be restored by collecting the copies still existing
in Iceland, and comparing them with those in the Copenhagen
libraries.

The see of Hólar was founded in 1104, by Gissur White,
Bishop of Skálaholti, and to fill it, Jón Ögmundson was unanimously elected. He was canonized by law in 1200. A story
told of him is, that he once, as a layman, entered a church in
Denmark, and heard the priest reading the Gospel so badly,
that it roused his indignation, and snatching the book from
him, he read it in the most beautiful manner to the edification
of the congregation.

He was succeeded by a line of prelates distinguished for

their piety and learning, till the series was closed in the zealous and faithful Jón Arnason, who fell beneath the sword of the executioner, and with him fell the Icelandic Church.

In præ-Reformation times, the Bishop of Hólar possessed 300 farms, pasturage for 15,000 cows, and the flotsam and jetsam along a considerable line of coast; besides these, he owned the island of Drángey, and two *tolfæringir*, the largest sized Icelandic vessels. But whenever the Church begins to lay up for herself treasures on earth, the State will watch her till the time comes to despoil her of her ill-amassed wealth; and so it fared with the Church of Iceland. Everything belonging to the see of Hólar was confiscated, with the exception, by an oversight, of the rights of seizing on the drift; and now the pastor of Hólar receives only sixty-five dollars (7*l.* 6*s.* 8*d.*) per annum. As there are no church-rates, the sacred edifice would have fallen into complete ruin if the present archdeacon had not expended 800 dollars from his own pocket for several years in necessary repairs. The venerable man, now in his seventieth year, laments for the old times, and bewails the wholesale spoliation of ecclesiastical property, which took place in the unhappy period of the sixteenth century. He took me into the church, and, looking sadly round on the damp walls, said—

"When I am gone, who will care for God's house, or have the means of keeping it in repair! I have spent all that I could afford on the building to keep it water-tight, yet the pictures moulder, the vestments decay, the rot eats into the wood, and the walls begin to crumble."

The old man presented me with a book which I had long wished to possess, namely, Markusson's Nockrir Marg-Fróthir Sögu-Thættir, in quarto.

Behind Hólar I gathered two varieties of gentian (*Gent. campestris*, and *G. bavarica*), and as much moonwort (*Botrychium lunaria*) as I wanted. On going to the foot of the precipitous mountain at the back of the farm, I came upon a number of hillocks of rubble brought down from the mountain; these were covered with flowers, and I spent a pleasant

16

hour in clambering over the mounds and re-stocking my book, as my former specimens of plants had been spoiled in the passage of the Eyjafjord river. The thrift (*Armeria maritima*) with its rose-coloured flower heads was very abundant, so also was the pretty orange alpine cinquefoil (*Potentilla aurea*). I picked several varieties of the speedwell, viz. the alpine, flesh-coloured and thyme-leaved (*Veronica alpina, V. fruticulosa, V. serpyllifolia*), also the vivarous alpine buckwheat (*Polygonum viviparum*), whose pale pink spire of flowers has gone on forming, and ripening its seeds between my leaves of blotting paper, so that I have been able to plant them since my return to England, though with little expectation of producing a crop, as the plant increases generally by the bulbs. In the grass below the rubble heaps, grew the common hemp nettle (*Galeopsis tetrahit*) in great profusion, along with the dry greenish white flowers of the spurless coral-root (*Corallorhiza innata*). As I picked my way among the marshes, I found the alpine cat's-tail-grass (*Phleum alpinum*), the kidney-shaped mountain sorrel (*Oxyria reniformis*), and both the common and sheep's sorrel (*Rumex acetosa* and *R. acetosella*). I wish that the daisy were more common in Iceland, the grass land sadly wants that friendly little face to brighten it up; I wonder, too, whether the primrose would be out of place against this black gloomy soil!

Next day, Wednesday 16th, Grimr and I left Hólar, bidding farewell to the patriarchal archdeacon and his wife. The old man would receive nothing, save thanks, for his most hospitable treatment of me and my guides, so that I was obliged to content myself with feeing the servant.

We followed the river nearly to its mouth, and then, crossing it, ascended a low hill which commanded a view of the magnificent Skagafjord, still and blue beneath a pure deep sky. The mountains on the left rise perpendicularly almost from the water's edge and are powdered with snow; one of them, Tindarstöll, is famous for its minerals; zeolite, onyx, chalcedony and opal being found in abundance among its ravines. The eastern horn of the bay is formed by the headland Thor-

tharhöfthi, to the right of which the pale blue of a lake is just
visible, and beyond that again, the faint line of Málmey isle.

The frith is ten miles from side to side, and, about half
way across is the islet Drángey, eighteen miles from the point
whence my sketch (p. 246) is taken. This rock starts up from
the water-line to the height of 600 feet. North and south of it
stand rock-needles, called the Old Man and Old Woman; the
former was once a very lofty spire, but it has fallen in some
of the volcanic throes of the country. In most parts, the
islet springs directly out of the water, but towards the west
there is a line of beach along the base of the crags. It is
impossible to ascend this isle without a ladder. On reaching
the summit, it is found to be covered with grass, and to be
as extensive as the tún of Hólar; sheep are brought to the
foot of the cliffs and drawn up by ropes, that they may eat
off the herbage in autumn and winter.

Drángey belonged originally to several proprietors, but
these made over their rights to one man, when Grettir was
on the island, and eventually it became the property of the
bishopric of Hólar, along with all the rights of fishing and
fowling around it. Bishop Gúthmundr visited the island, and
celebrated mass on a rock near the landing-place, called
Gvendar-altari; and it is the custom of those who ascend the
crag to stand beside it for a few moments, cap in hand, and
offer up a prayer, before commencing the perilous ascent.

In spring, Drángey is visited by a great number of men
who descend the cliffs, slung by hair or leathern ropes, and
rifle the nests of the numerous sea-birds which build upon
the ledges. They also catch the birds themselves, by floating
snares formed of boards and provided with nooses, at the
bottom of the rocks. The unsuspecting creatures fly down
and perch on the planks, when they are caught by the threads
and held till the fowler visits his fleke in the evening.

The birds which frequent this island are, the common
Guillemot (*Uria troile*), the Puffin (*Mormon fratercula*), and
gulls and skuas. The Gannet (*Sula bassana*) is said to breed
on Drángey. This fine bird when young is black, sprinkled

16—2

with white spots, as though snow had fallen on it; but after
two years it becomes perfectly white, with the exception of
the wing primaries, which remain black. It flies very evenly,
and in a direct line, for its rock, when returning after a day's
fishing; but it can stop instantly in its course, if it observes
a fish in the water below it; then it wheels once or twice,
folds its wings, and drops like a stone into the waves, which
close over it. Up it comes, a minute after, with its prey
firmly held in its strong beak; it flings back its head, and with
one gulp, the fish is swallowed.

Another bird, which lays upon the shelves of Drángey, is
the Razor-bill (*Alca torda*), called in Icelandic, *Alka*: the
mother bird is said to take her young on her back, and bear
them down to the water, then she dives and leaves them to
take their first lesson in swimming by themselves.

The Puffin (*Mormon fratercula*) is a pretty, gay little fellow,
with his brightly-coloured beak. He is called *Lundi* in Icelandic, and probably our island of Lundy in the Bristol
Channel receives its name from this bird.* He has a glossy
black back and cap, black wings, and a ring round the neck,
the breast and face are white; his legs and feet are yellowish
red, and his beak is painted red, yellow, and blue. The Puffin
makes a capital dish, stuffed with raisin pudding, and baked.
I tasted it in the Faroe isles when I dined with the Catholic
missionary on a Friday. "It is not fish, you know," said he;
" but it feeds on fish!"

On this island breed also the Tern, that most graceful
of all sea-birds (*Sterna arctica*), called by the Icelanders, after
its note, *Kria*, and its sworn enemy the Skua, of which there
are four or five varieties in the island : (*Lestris Catharractes,
L. pomarina, L. parasitica*, and the *L. thuliaca*, concerning
which see p. 107.) The Skua is a piratical fellow; he does
not fish for himself—not he! but he lets the gulls and gannets
do that; and then, when he sees them full fed, he rushes

---

* The rocky islet on the left of my view of Skagafjord is called Lundey,
or the Puffin Isle. The word *ey* remains as a termination in many English
names, *e. g.* Walney, Sheppey, Ramsey, Bardsey, Guernsey, Jersey, &c.

at them, strikes them in the crop with his hooked beak, and makes the frightened birds disgorge their spoils; then the Skua swerves in the air and catches the half-digested fish before they drop into the sea. They are fierce fellows, and the Drángey fowler has to be cautious how he approaches their haunts, as they will dash down upon him and strike him on the head, or tear his face with their powerful beaks, in a paroxysm of rage, till the blinded or bewildered man lets go the rope by which he is suspended, and is dashed to pieces among the rocks.

Drángey was said, in old times, to have been haunted by Trolls, who stretched out their arms from the hollows of the clefts as the fowler swung down, and clutching him, flung him to the bottom; but it is now believed that these Trolls were nothing more nor less than Skuas.

The bird will descend with such velocity, that he has been known to impale himself on a spiked staff which the fowler holds in his hand. He is a sociable bird among his own kith and kin, but never associates with the birds of another family, except for the sake of robbing them. He will follow a boat throughout a day, to catch the fish which are rejected by the fishermen, and he will pursue a flight of gulls with untiring patience, waiting for them to catch the fish, which he intends to secure for his own supper. The common and pomerine Skuas are brown, but the Arctic Skua has a white chin, neck, and breast; the upper part of the head is black, the primaries and tail feathers are also black, the latter very long, about ten inches from the roots; the back, wing, and tail coverts are of a soft brownish grey. The Icelanders call the Skua after his cry, *Kjói*.

The cinereous Eagle (*Falco albicilla*) fishes in the lakes and fjords of Iceland. He not unfrequently drowns himself, for he will descend with such velocity on a large fish, that he buries his beak and claws in it, and is unable to extricate them. The fish dives, and carries the royal bird along with it to Davy Jones' locker, and his body is sometimes brought up by the fishermen in their nets.

CHAPTER XV.

The Outlaw's Isle.*

(A.D. 1029—1033.)

After life's fitful fever he sleeps well.
Macbeth.

SKAGAFJORD.

POOR Grettir! hustled from pillar to post, hunted from one retreat to another, he had spent fifteen years of hardship, such as few men have undergone, yet the hatred of his deadly foe, Thorir, had not expended itself.

At length, finding that no corner of Iceland was safe, he asked Gúthmundr the Wealthy to advise him whither he should flee, to be safe from his pursuers.

---

* *Gretla*, chaps. 67, 69—87. I have curtailed the Saga very considerably.

" There is only one spot that I know of where you can be in perfect security."

Grettir replied that he had hitherto found no such spot.

Gúthmundr continued, " There is an islet in the Skaga-fjord, hight Drángey, abounding in fish and fowl, and no one can ascend it except by a rope ladder which hangs down on one of the sides. If you can reach that spot, then you may be assured that it is in no man's power to touch you, so long as you are safe and sound and able to guard the ladder."

" I will venture out there," said Grettir; " yet I am so timorous in the dark, that to save my life, I cannot abide alone."

Gúthmundr answered, " May be, but I advise you to trust no one but your own self."

Grettir thanked him for his advice, and then hastened to his mother, at Bjarg, in the Middle-frith. The fear of the dark to which he alluded had come on him ever since his wrestle with Glámr, but had increased considerably of late. No sooner did darkness set in, than the terrible eyes of the vampire seemed to stare at him from the gloom. He slept lightly, starting in his dreams, and waking repeatedly during the night. This was undoubtedly brought on by the un-ceasing strain on his mind and the excitability of nerves, caused by the hourly peril in which he had been living for so many years.

On his arrival at Bjarg his mother greeted him affec-tionately, and told him that she would indeed be glad if he could remain with her, though she feared it would be too venturesome to do so, as Thorir would certainly discover his retreat before many days had elapsed.

The outlaw replied that he would give her no inconve-nience : " For," said he, " I care to take no more trouble about preserving my life. I can bear my solitude no longer." He then told his mother of Gúthmundr's advice, adding, that he would try his best to reach Drángey, but that he must endeavour to secure some trustworthy companion to be with him.

Illugi, his brother, now fifteen years old, a fine, noble boy, was present during the conversation, and at these words of Grettir he started up, caught his hand, and said—

"Brother! I will go with you if I may, though I fear you will look on me as but a feeble helpmate; yet I will be faithful to you, and stand by you to the last."

Grettir answered, "Of all men, my brother! I would rather have you with me, and willingly will I consent to your joining your lot with mine, if our mother has no objection."

"Sorrows never come singly," replied the aged woman; "I can hardly bear to part with Illugi, yet I know how dire is your necessity of a comrade, son Grettir! therefore, I will not be selfish and keep him. It costs me a bitter pang to part with both my sons in one day."

Illugi was delighted at having thus easily obtained that on which he had set his heart, and he thanked his mother cordially.

The mother provided her sons with money and such chattels as they would require on the island, and then she accompanied them outside the farmyard, and, before parting with them, said, "Farewell, my two brave boys! I know that I shall never see you again, but what will befall you in Drángey I know not. Only of this I am certain, that there you will die, for many will resent your occupation of that island; my dreams have long forewarned me, that you will not be divided in your deaths. Beware of treachery, shun any dealings with sorcery, for nothing is more powerful than witchcraft. My blessing be upon you both!" She could speak no more, for her voice was choked with sobs; so, sitting down on a stone, she covered her eyes with her hands and the tears trickled between her fingers, falling in bright drops on her lap.

"Do not weep, mother!" said Grettir; "what though we both die! It shall ever be said of you, that you bore sons and not daughters. Long life and health attend you."

Then they parted, and the brothers went north and visited their kinsmen. So passed autumn, and with the approach of cold they went towards Skagafjord, crossed the Vatns-skarth and Reykjaskarth to Langholl, and reached Glanmbœr at the close of day. Grettir had flung his hood over his shoulders, though the wind was piercingly cold, for it was not his wont, fair or foul, warm or cold, to wear anything on his head.

Near the little farm just mentioned, the brothers stumbled upon a tall, thin man, dressed in rags, and with a very big head. They asked each other's names, and the fellow called himself Glaum; he was a bachelor out of work, and with all, a gad-about, fond of strolling through the country picking up and retailing news. He was a terrible boaster, but most people thought him both a coward and a fool. He amused the brothers by his continual chatter, and by the fund of gossip which he possessed. Grettir was especially pleased with him, and when Glaum offered to be his servant, Grettir accepted him gladly, and the man became thenceforth his constant attendant.

Says Glaum, "It is a wonder to all the people hereabouts, that you wear nothing on your head in such weather as this, and, i' faith ! it is no marvel that you are the man they take you for, if you do not mind the cold. Why, there were two of the bonder's sons down yonder going after the sheep, and they could not get clothes enough to put on them, so benumbed were they ; and yet they are plucky fellows too ! "

After this they went to Reynines ; thence they proceeded to the strand, where there is a little byre, Reykir, with a hot spring in the tún, belonging to a man named Thorwaldr. Grettir offered him a bag of silver if he would flit him across to Drángey by moonlight, and to this the man agreed.

On arriving at his destination, Grettir was well pleased with the spot, for it was covered with a profusion of grass, and was so precipitous, that it seemed impossible for any one to ascend it without the aid of the rope-ladder, which hung from strong staples at the summit. In summer the place would swarm with sea-birds, and at that time there were eighty sheep left on the island for fattening.

One of the principal chiefs in the Skagafjord was Thorbjorn, nicknamed " The Hook," a hard-hearted, ill-disposed fellow. His father had married a second time, and there was no love lost between the step-mother and Thorbjorn. It is said that one day as The Hook was sitting at draughts, she passed, and looking over his shoulder, noticed that he had made a foolish move, so she laughed; whereupon Thorbjorn retorted angrily. She instantly snatched up a draught-man, and, laying it against his cheek-bone, pressed it into his eye, so that the ball started out of its socket. He sprang up, with a curse, and dealt her such a blow, that she took to her bed and died of the injury. Thorbjorn went from bad to worse, and leaving home, he settled at Vithvik, the little farm which appears in the right of my sketch of Skagafjord.

As many as twenty farmers had rights of pasturage on Drángey, but The Hook and his brother had the greatest share.

About the time of the winter Solstice, the bonders busked them to visit the island and bring home their sheep. They rowed out in a large boat, and on nearing the island, were surprised to see figures moving on the top of the cliffs. How any one had reached the islet without their knowledge was a puzzle to them, and they had not the slightest suspicion who these occupants could be. They pulled hard for the landing-place where hung the ladder, but Grettir drew it up before the boat stranded.

The bonders shouted to know who those were on the crags, and Grettir, looking over, told his name and those of his companions.

The bonders asked who had flitted him across to the island; Grettir answered, "If you wish particularly to know, I will tell you, it was a man with a good boat and strong arms, and one who was rather my friend than yours."

"Let us get our sheep," cried the bonders, "and you come to land with us, we will charge you nothing for those of our sheep which you have eaten, and we will let you go from us in peace."

"Well offered," answered Grettir; "but he who takes keeps hold, and a bird in the hand is worth two in the bush.

Believe me, I never leave the island till I am carried from it dead."

The bonders were silenced; it seemed to them that they had got an ugly customer on Drángey, to get rid of whom would be no easy matter; so they rowed home, very ill-pleased at the result of their expedition.

The news spread like wildfire, and was talked about all through the neighbourhood, but no one could devise a plan for getting rid of the outlaw.

Winter passed, and at the beginning of spring the whole district met at the "Thing," or Council of Hegraness, an extensive island at the mouth of the Heradsvatn river, just showing on the left of my sketch. The gathering was thronged, and the litigations and merry-making made the Thing last over many days. Grettir guessed what was going on by seeing a number of boats pass the head of the fjord. He became very restless, and at last announced to his brother that he intended being present at the Council. Illugi thought this sheer madness, but Grettir was resolute; he begged Illugi and Glaum to watch the ladder and await his return.

Then he crossed to the mainland and hastened in disguise to the Council, where he found that sports of all kinds were going on among the able-bodied young men. Grettir was dressed in an old-fashioned suit, very dirty, and falling to tatters. He had on a fur cap, which was drawn closely over his eyes and concealed his face, so that no one recognized him. He sauntered among the booths till he reached the spot where the games were taking place.

Among the wrestlers no man surpassed Thorbjorn Hook in skill and prowess. He threw all the strongest men of the neighbourhood, and when he had cleared the ground of antagonists, and found that there was no one to oppose him, he stood still and cast his eyes round him. Suddenly, they rested on a tall fellow in the shabbiest and quaintest of suits, but who looked so strongly built that Thorbjorn walked up to him and caught him by the shoulders. But the man sat still, and he could not move him from his seat. "Well!" exclaimed

The Hook; "you are the first fellow I have seen for many a
day whom I couldn't pull off his stool. Come now and wrestle
with me—yet, tell me first, what is your name."

"Guest!" answered the stranger.

"A welcome guest too!" quoth the bully; "if you will
wrestle with me."

The man replied that they would not be fairly matched, as
he was little skilled in athletic sports.

Several men now chimed in, begging the stranger to try
what he could do with Thorbjorn, or, at all events, with one
of the others.

"Long, long ago," quoth he, "I was able to throw my
man, as well as the best of you, but those days are gone by,
and now I am out of practice."

As he only half refused, the bystanders urged him all the
more.

"Now mark you!" said he; "I yield on one condition,
and that is, that you take your oath to let me go free to my
home, without one of you lifting a hand against me."

There was a general shout of acquiescence, and Hafr, one
of the number, recited the peace-oath in the following legal
form :—

"Here set I peace among all men towards the man Guest,
who sits before us, and in this peace I include all the priest-
hood-holders, and well-to-do bonders, and all the young
weapon-bearing men, and all the men of the Hegraness district,
whether present or absent, named or unnamed. These are to
leave in peace, and give passage without let or hindrance to
the afore-named stranger, that he may sport, wrestle, make
merry, abide with us and depart from us, without stay,
whether he may need to go by land or flood. He shall have
peace in all places, named or unnamed, as long as is necessary for him to reach home with ease; so long only shall
peace last.

"I set this reconciliation between him and us, our relations,
our friends, and kinsmen, male or female, free or thrall, child
or full-grown. May the breaker of this peace, and breaker of

this oath, be cast out of the presence of God and good men, from the heavenly kingdom, from the company of the saints and just men. Let him be an outcast from land to its farthest limits, far as men chase wolves at furthest, as Christians seek churches, as heathens sacrifice in shrines, as flame burns, earth produces, as baby calls its mother, and mother bears baby, as fire is kindled, ships glide, lightnings flicker, sun shines, snow lies, Finns slide on snow shoon, fir grows, falcon flies in the spring day with a fair breeze under its wings, far as heaven bends, earth is peopled, winds sweep waters to the sea, churls grow corn; he shall be banished from churches and the company of Christian men, from heathen folk, from house and den, from every home—save Hell! Now let us be at-oned and agreed, each with each, in good will, whether we meet on mountain or shore, on ship or on skate, on ground or glacier, at sea or in saddle; as friend meets friend on the deep, as brother meets brother abroad, let us be at-oned one with another, as father with son, as son with father, in all our dealing. Lay we now hand to hand, and hold we now true peace, and keep we every word spoken in this our peace-telling, before God and good men, and all those who hear my words and stand around."

After a little hesitation the oath was taken by all.

Then said Guest, "Now you have done well, only beware of breaking your oaths. I am ready on my part, without delay, to fulfil your wishes." Then he flung aside his hood and almost all his tatters.

The assembled chiefs looked at each other, and were rather disconcerted, for they saw that there stood before them the redoubted Grettir Asmund's son. They were silent, and Hafr thought that he had acted somewhat rashly. The throng broke up into knots, and began to discuss whether the oath should be kept or not.

"Come now," shouted Grettir; "let me know your purpose, for I shall not long sit naked. There is more danger to you than to me, in the breach of your oaths."

He got no answer, but the chiefs moved away to discuss

the question. Some wanted to break the truce, others wanted
to keep it. Then Grettir sang :—

> " Many trees of wealth,* this morning,
> Failed the well-known, well to know,
> Two ways turn the sea-flame branches,†
> When a trick on them is tried.
> Falter folk their oath fulfilling,
> Hafr's talking lips are dumb."

Said a man hight Tongue-stone : " You think so, do you,
Grettir? Well, you are a man of dauntless courage, I will
say that for you. Look now ! the chiefs are in deep consulta-
tion about what is to be done with you."

Then Grettir sang :—

> " Lifters of shields‡ rub their noses,
> Shield-tempest gods‡ shake their beards,
> Fierce-hearted serpents' lair scatterers §
> Go on their way, much regretting,
> Peace they have made,—now they know me !"

Then out spake Hjalti of Hóf, brother of Thorbjorn Hook :
" Never let it be said of us, that we break an oath, even
though it were inconsiderately taken. Grettir shall be at
full liberty to go to his home in peace, and woe betide him
who lays hand on him to do him an injury. But should he
venture again ashore, we are free from our oath."

All, except Thorbjorn Hook, agreed to this, and were glad
that Hjalti had spoken out as became a chieftain.

The wrestling began by Grettir being matched with Thorbjorn, and after a short struggle, Grettir freed himself from his
antagonist, leaped over his back, caught him by the belt of
his trousers, lifted him off his legs, and flung him over his
back.

It was next proposed that Grettir should be matched
against the two brothers together, and he readily agreed to

---

* Periphrasis for men.
† Sea-flame = gold ; and sea-flame branches = warriors.
‡ Periphrasis for warriors.
§ Serpents' lair = gold ; serpents' lair scatterers = men.

this. The wrestling continued with unabated vigour, and it was impossible to tell which side had the mastery, for, though Grettir repeatedly threw one brother after the other, yet he was unable to hold them both down at the same time. After that all three were covered with blood and bruises, the match was closed, by the judges deciding that the two brothers conjointly were not stronger than Grettir alone, though they were each of them as powerful as two ordinary able-bodied men.

Grettir at once left the Thing, rejecting all the entreaties of the farmers, that he should leave Drángey; and, on his return to the little island, he was received by his brother Illugi with open arms.

The smaller bonders began to feel seriously their want of the island for autumn pasture, and, as there seemed no prospect of their getting rid of Grettir, they sold their rights to Thorbjorn Hook, who set himself in earnest to devise a plan by which he could possess himself of the island.

When Grettir had been two winters on the island, he had eaten all the sheep, except one ram, a piebald fellow, with magnificent horns, which became so tame, that every evening he came to the hovel which Grettir had erected, and butted at the door till he was admitted. The brothers liked their place of exile, as there was no dearth of eggs and birds, besides which, a considerable amount of drift timber was thrown upon the strand, and served as fuel.

Grettir and Illugi spent their days in clambering among the rocks and rifling the nests, and the occupation of the thrall was to collect drift-wood, and keep up the fire in the cottage. The churl lost his spirits, and became idle, morose, and reserved. One night, notwithstanding Grettir's warnings to him to be careful, as they had no boat, he let the fire go out. Grettir was very angry, and told Glaum that he deserved a sound thrashing for his neglect. The thrall replied that he was heartily tired of the life he had been leading on the island, being scolded or beaten whenever anything went amiss.

Grettir asked Illugi what had better be done, and his

brother replied that the only thing for them to do was to
await the arrival of a boat from the friendly farmer at Reykir.

"We shall have to wait long enough for that," said
Grettir; "our only chance is, for me to swim ashore and
procure a light."

"For heaven's sake!" exclaimed Illugi, "do not attempt
anything of the kind, for we are undone if anything happens
to you."

"Never fear for me," said Grettir; "I was not born to
be drowned!"

From Drángey to Reykir is about four miles. Grettir
prepared for swimming, by dressing in loose thin drawers and
a sealskin hood ; he also tied his fingers together, that they
might offer more resistance to the water when he struck out.

The day was warm and fine. Grettir started in the
evening, when the tide was in his favour, whilst his brother
anxiously watched him from the rocks. At sunset the outlaw
reached Reykjanes, after having floated or swum the whole
distance. Immediately on coming to land, he went to the
warm spring, and bathed in it before entering the house.
The door of the hall was open, and Grettir stepped in. A
large fire had been burning on the hearth, so that the room
was very warm ; Grettir was so thoroughly exhausted with his
swim, that he lay down beside the hot embers, and was soon
fast asleep. In the morning, he was found by the farmer's
daughter, who gave him a bowl of milk, and brought her
father to him. Thorwald furnished him with fire, and rowed
him back to the island, astonished beyond measure at his
achievement, in having swum such a distance.

The inhabitants of Skagafjord were angry with Thorbjorn
Hook for not having rid the island of its tenants, notwithstanding all his fine promises, but Thorbjorn was sorely
puzzled to know what measures to take.

During the summer, a ship arrived in the frith, commanded by a young active fellow, Hœring by name, who was
famous for his skill in climbing. He lodged with Thorbjorn
during the autumn, and was continually urging his host to

row him out to Drángey, that he might escalade the precipitous sides of the islet. Thorbjorn required very little pressing; and, one fine afternoon, he flitted his guest out to the island, and put him stealthily ashore, without attracting the notice of those on the height.

On reaching the usual landing-place, which was on the opposite side of the island, Thorbjorn shouted, and brought Grettir and his brother to the verge of the cliff. The old arguments were repeated to persuade Grettir to come to the mainland, and with the usual success. The Hook, however, succeeded completely in his attempt to withdraw the outlaw's attention from the farther side of the islet, up which Hœring was clambering.

The young merchant reached the top by a way never attempted before nor since; then, pausing only to take breath, he advanced towards the brothers, who were leaning over the verge of the cliff, little dreaming of danger in their rear.

Grettir was engaged in angry altercation with The Hook, but the young brother took no part in the conversation, and, beginning to feel weary of his position, he turned on one side to relieve his elbows, which had rested on the rock. In so doing, he caught sight of Hœring.

" Brother, brother ! " exclaimed he, " here comes a man towards us, brandishing an axe, and bent on mischief."

" Go after him yourself, lad ! " said Grettir; " I will guard the ladder."

Illugi sprang up and rushed towards the young merchant, who at once took to flight, ran to the edge of the crag, leaped over, and was dashed to pieces among the rocks. That spot is called Hœring's leap to this day.

" Now, Thorbjorn ! " shouted Grettir, when Illugi returned and told him what had taken place; " you had better row round to the other side of the isle, and gather up the remains of your friend ! "

The Hook pushed off from the strand, and returned home, ill enough pleased with what had taken place,

17

and Grettir remained at Drángey, unmolested through the
winter.

At this time died Skapti, the lawgiver; and in the following
spring, Grettir's relations and friends moved for a repeal of
his sentence of outlawry, but his enemies opposed this vehe-
mently, declaring that he had committed many crimes since
he had been pronounced an outlaw.

A new lawgiver, named Steinn, was elected, and when
the case came before him, he gave his opinion that after
twenty years the sentence became null and void; so that
Grettir's kinsmen had every reason for hoping that in two
years, when he would have completed the twenty, he would
be restored to their society.

Thorbjorn Hook was exasperated beyond measure at the
prospect of Grettir slipping through his fingers after all, and
he returned from the Thing, brooding over fresh schemes
against the outlaw.

It happened that he had an old feeble foster-mother, a
woman of malicious disposition; and when Thorbjorn could
get help nowhere else, he came to her, as in her youth she
had dabbled in sorcery, but had long ceased to practise it,
when after the introduction of Christianity it became illegal,
and was punishable with banishment. However, as the old
saw has it, "What is learned in youth is remembered in
age;" and though the old woman was believed to have forgotten her witchcraft, yet it remained stored up in the
chambers of her mind.

"Ah!" said she, when Thorbjorn came to her, "I see
that as a last resource you come to me, a bed-ridden old
woman, and ask my help! Well, I will assist you to the
best of my power, on one condition, and that is, that you
yield me implicit obedience."

The Hook answered her that he was quite willing to consent, as he had long since learned to rely on his foster-
mother's advice, as being most salutary.

When the month of August came round, the hag said to
her foster-son one beautiful day, "The sea is calm and the

sky bright, what say you to our rowing over to Drángey, and stirring up the old quarrel with Grettir. I will accompany you, and listen to what he says; I shall then be able to judge what lot awaits him; besides, I can death-doom him as I please."

The Hook answered, "I am tired of going to Drángey, for I never return from it a whit the better off than when I started."

"Remember your promise," said the old woman; "I shall have nothing to do with you, unless you follow my advice."

"Well then, foster-mother," quoth Thorbjorn; "let us go, though I vowed that my third visit should be the death of Grettir."

"Have patience," said the hag; "time and trouble are needed before that man is laid low; and what the result will be I know not; it may be your gain, and it may be your ruin."

Thorbjorn ran out a long boat, and entered it with twelve men; the hag sat in the bows, coiled up amongst wraps and rugs.

When they reached the island the brothers ran to the ladder, and Thorbjorn asked whether Grettir was yet tired of his island.

Grettir replied as he had replied before: "Do what you will! in this spot I await my destiny."

Thorbjorn saw now that his journey was likely to be without avail. "I see," said he, "that I have to do with the worst of men; one thing is clear enough, it will be a long time before I pay you another visit."

"So much the better," answered Grettir; "I shall not count it as a misfortune if I never see you again."

At this moment the hag began to stir in the bows of the boat. Grettir had not previously observed her presence. Now with a shrill voice, she cried: "These men are sturdy, but luck has deserted them; see what a difference there is between folk! You, Thorbjorn, make them good offers which

17—2

they foolishly reject; those who refuse good when it is offered
them, always come to a bad end. Grettir! I wish you to
be lost to health, wisdom, luck, and prudence! May these
blessings be constantly on the wane the longer you live, and
may your days henceforth be fewer and sadder than those
preceding them."

As she spoke, a cold shudder ran over Grettir's limbs, and
he asked what fiend that was in the ship. Illugi replied that
she must be the foster-mother of Thorbjorn.

"Since an evil fiend is with our foes, we can expect
nothing but the worst," said Grettir. "Never before have I
been so agitated at words spoken as whilst the hag was pouring forth her curses on us. I know now that evil must befall
me from her witchcraft; but she shall have a reminder of her
visit to me." Then he snatched up a large stone and flung it
into the boat, so that it fell upon the bundle of rugs, among
which lay the aged woman. As it struck there rose a wild
shriek from the witch, for the stone had fallen on her leg and
snapped it asunder.

"Brother, you should not have done this!" said
Illugi.

"Blame me not!" answered Grettir; "I only wish that
the stone had fallen on her skull, and that her life had been
sacrificed instead of ours."

On the return of Thorbjorn to the mainland, the hag was
put to bed, and The Hook was less pleased than ever with his
trip to the island.

"Be not down-cast," said his foster-mother; "this is the
turning point of Grettir's fortunes, and his luck will leave
him more and more. I have no fear of not having my revenge,
should my life be spared."

"You are a resolute woman, foster-mother!" said Thorbjorn Hook.

After a month the old woman was able to leave her bed,
and limp across the room. She one day demanded to be led
down to the shore. Her wishes were complied with, and on
reaching the strand she hobbled up and down till she found a

large piece of drift-timber, just large enough for a man to carry upon his shoulder.

Then she ordered it to be rolled towards her and turned over. She examined it attentively. The log seemed to have been charred on one side, and this burned portion she ordered to be planed away; then, taking a knife, she cut Runes on it, and smeared them with her blood, chanting over them, as she limped round the beam, a wild spell, that it might be borne to Drángey and there work Grettir's ill. The piece of timber was then pushed into the waves and thrust off from shore. A fresh northerly wind was blowing, but the beam swam against wind and tide, and held on its course direct for the outlaw's isle.

The old witch returned to Vithvik. Thorbjorn did not think that anything would come of what she had done; but she bade him be of good cheer, and wait till she gave him fresh orders.

In the meantime Grettir, his brother, and the churl were on Drángey catching fish and fowl for their winter supplies.

The day after that on which the hag had charmed the piece of timber, the two brothers were walking on the strand to the west of the island looking for drift-wood.

" Here is a fine log ! " exclaimed Illugi ; " help me to lift it on my shoulder, and I will carry it home."

Grettir spurned the beam with his foot, saying, " I do not like the looks of it, little brother ! Runes are cut on it, and they may betide us ill ; who knows but this log may have been sent hither for our destruction ! "

Then they sent it adrift, and Grettir warned his brother not to bring it to their fire.

They returned in the evening to their hovel, and did not mention the matter before the thrall.

The next day they found the same beam washed up, not far from the foot of the ladder; Grettir thrust it out to sea again, saying that he hoped he had seen the last of it.

The weather began to break up, and several days of storm and rain succeeded each other, so that the three men remained indoors till their stock of firewood was nearly expended.

Then they ordered Glaum to search the shore for fuel. The fellow started up with an angry murmur, and left the room muttering that the weather was too bad for a dog to be sent out in it. Then he went to the rope-ladder, descended it, and found the same beam cast up at its very foot.

Rejoiced at having so soon obtained what he wanted, he threw it over his shoulder, strode with it to the hut, and flung it down by the door.

Grettir heard the sound, and springing up, he exclaimed, "Glaum has got something at last! Let us see what he has found."

Then taking his axe he went outside.

"Now," says Glaum, "you chop it up, as I have had all the trouble of bringing it."

Grettir was angry with the fellow, and without paying much attention to the log itself, he brought his axe down on it with a sweep. The blade struck, glided off and cut into Grettir's right leg below the knee with such force that it stuck in the bone.

Grettir looked at the beam, and recognizing it at once, said, "The worst is at hand! Misfortunes never come singly! This is the very log which I have rejected twice. Glaum! you have done us two ill turns; first, in letting out the fire; secondly, in bringing home this accursed beam; and if you commit a third, it will be the death of you."

Illugi bound up his brother's wound with rag; there was but little flow of blood, but it was an ugly gash.

Grettir slept well that night. For three days and nights he was without pain, and the wound seemed to be healing nicely, and skin to be forming healthily over it.

"Well, brother!" said Illugi, "I think this cut will not trouble you long."

"I hope not," answered Grettir; "yet I have my fears!"

On the fourth evening they laid them down to sleep as

usual. Towards midnight the lad, Illugi, awoke, hearing Grettir tossing about in his bed, as though in pain.

"Why are you so restless?" he asked.

Grettir replied that he felt great anguish in his leg, and that he thought some slight change must have taken place in the wound.

The boy blew the embers on the hearth into a flame, and by its light examined his brother's leg. He found that the foot was swollen and purple, and that the wound had re-opened, and looked far more angry than when first made.

Intense pain followed, so that the poor outlaw could not remain quiet for one moment, and sleep no more visited his eyes.

Illugi remained by him, continually holding his brother's hand, or bringing him water to slake his unquenchable thirst.

"We must prepare for the worst," said Grettir; "this sickness is the result of sorcery. The hag is revenging on me that stone which I cast at her."

Illugi replied, "I ever thought evil would come of it!"

"What is done cannot be undone," said Grettir; and then, sitting up, supporting himself against his brother's breast, he sang —*

> "I fought with sword in bright old days,
> In days when I was young,
> When gladsome song and roundelay
> From happy heart I flung.

---

* The five verses given in the Gretla are, indubitably, only a poor remnant of a great poem. Two of these five refer to events in Grettir's life, of which the Saga writer knows nothing. The allusions in them are, con-sequently, difficult to understand. The subject of the extant verses is a review of the hero's past life. On account of their fragmentary nature, I have not translated them, but have taken the liberty of a story-teller, and have put a song into Grettir's mouth which may possibly embody the meaning of the lost poem, though it is my own wording. I know that I am taking a great liberty by so doing; but it seems to me that the story would be injured by omitting the incident of Grettir singing his dying lay, or at all events by leaving his song—a blank. I love Grettir too dearly for that! Besides, in telling these Sagas, one should throw oneself heart and soul into their *spirit*, and not allow oneself to be bound down and chained too closely to the *letter*.

I fought with sword in bright old days,
  When earth to me was fair,
And, fresh as heart, the lightsome breeze
  Did toss my yellow hair.

I fought with sword in bright old days,
  I loved the merry clang!
When brand met brand, and shield met shield,
  And axe on helmet rang.

As now I chant of youthful days,
  In fitful, broken, rhyme,
I seem to hear from my blue blade,
  A wild war-music chime!

I lowly laid the robber band,
  I rescued wife and maid!
My haft and hilt were purple dipped,
  And purple was my blade!

And when my friends for fire did pray,
  I sought it past the wave;
Though 'neath me gaped the fjord dark—
  Dark as an open grave!

When I returned to seek old home,
  I found my kinsmen dead;
I was a banished, outlawed man,
  A price was on my head!

A hunted man, by night and day,
  On mountain, moor, and fen;
For eighteen years to shun and flee,
  The face of fellow men!

For eighteen bitter years, to bear
  Fasting, and cold, and pain,
And never know, when I lay down,
  If I should wake again!

And now coiled up, with fevered blood,
  A grim old wolf I die;
Whilst dripping skies above me spread,
  And winds sob sadly by.

O'er tired heart and drowsy head,
  Does welcome slumber creep;
As little babe on mother's knee
  Will softly drop asleep.

With folded feet, and closèd palms,
  I will not stir nor wake :
But, hushed in happy dreaming, lie
  Till the last morning break.

And if men ask who lieth thus ;
  Say, 'tis a tired breast,
Now finding peace, finding calm,
  Finding rest !"

"Let us be cautious now," said Grettir; "for Thorbjorn
will make another venture. Glaum, do you watch the steps
by day, and draw them up at dusk. Be a faithful servant to
us, for much depends on your fulfilling your duty ; and I forewarn you, that if you betray your trust, it will cost you your
life."

Glaum promised well.

The weather daily became worse, and a fierce north-east
wind blustered over the country, bearing with it cold and
sleet, and powdering the highlands with snow. Grettir asked
nightly whether the ladder had been drawn up. Glaum
answered churlishly, "How can you expect people to come
out in such a storm as this? Do you think that folk are
so anxious to kill you, that they will be crazy enough to
jeopardize their own lives in the attempt? No, no! You
have lost all your pluck and manliness since you have been a
little unwell! You are now scared and frightened at the
merest trifles."

Grettir answered, "You have none of our pluck and
manliness yourself! Go now, and guard the ladder as you
have been bidden, instead of standing here reproaching us
with cowardice !"

So Illugi and his brother drove the churl from the house
every morning, notwithstanding all his angry remonstrances.

The pain became more acute, and the whole leg became
inflamed and swollen ; signs of mortification appeared, and
wounds opened in different parts of the limb, so that Grettir
felt that the shadow of death was upon him. Illugi sat
night and day with his brother's head on his shoulder,

bathing his forehead, and doing his utmost to console the fleeting spirit. A week had elapsed since the wound had been made.

Thorbjorn Hook was at home, ill pleased at the failure of all his schemes for dispossessing Grettir of the island. One day his foster-mother came to him, and asked whether he were ready now to pay the outlaw his final visit. Thorbjorn replied that he had no wish to do so, as it would come to nothing; and asked his foster-mother whether she had any desire to seek out Grettir again, or whether she had been satisfied with the success of her former visit.

"I may not seek him myself," answered the hag; "but I have sent him my greeting, and by this time it has reached him. Speed now to Drángey as swiftly as you can row; for, if you delay, he will be beyond your reach."

The Hook had come off so ignominiously on every former occasion when he had visited the island, that he did not much relish the notion of making another attempt, especially on a day when it would be dangerous to venture on the water in a boat.

"You're a helpless fellow!" exclaimed his foster-mother, when Thorbjorn told her his objections to her scheme. "Do you think that I, who have called up this storm, cannot refrain it from doing you injury?"

Well! in the end, the man allowed himself to be persuaded; so he beat up the neighbouring farmers, asking them to assist him in manning a large boat. None of them would come with him, but The Hook brought twelve of his own men; his brother Hjalti lent him three more; Eirik of the Good-dale sent him one man; Tongue-stone furnished him with two; Halldorr let him have six of his house churls: and these were all he could get. Of these, the only two whose names I need mention were Karr and Vikarr. Thorbjorn went with his party to Haganess, where he borrowed a large sailing-boat. None of the men were in good spirits, as the weather was so bad, and they had no confidence in their leader. By dusk, they got the vessel afloat, spread sail, and, with a lurch, she ran out to sea.

As the wind was from the north-east, they were under the lee of the high cliffs, and were not exposed to the violence of the gale.

Heavy scuds of rain and sleet swept the fjord; the sky was overcast with dense whirling masses of vapour, and, beneath their shadow, the waters of the frith were black as ink. For one moment the clouds were parted by the storm, and the rowers looked up, to see the heavens barred with the crimson rays of the Northern Light; then the vapours, dense as volcanic smoke, swept across the gap. A flame ran along the cordage, and finally settled on the mast-head of the boat, swaying and rocking with the motion of the vessel. It was that electric spark which Mediterranean sailors call S. Elmo's light, and Icelanders, Hrœvarelldr.

A line of white foam marked the base of Drángey; and, now and then, a great wave from the mouth of the fjord thundered against the crags, and shot in spouts of foam high into the air.

Along the western shore of the frith, which was exposed to the full brunt of the gale, the mighty billows were beaten into white yeasty heaps of water, rolling onwards and recoiling, shivering against the rocks, and falling back in lashing spray, booming down long caverns till they choked them, and then bursting out with a roar, in steam-like jets. Upon the top of Drángey, one ruddy spark shone from the window of the hovel in which lay the dying outlaw, and it was reflected as a streak of fire on the tossing deep.

Gulls cried and wheeled around the smack as it ploughed its way through the water, and the stormy Petrel fleeted past in the trough of the waves, the Kittiwakes and Tern wavered and dipped before the boat, uttering their melancholy scream, kreeah! kreeah! The Diver passed, dancing like a cork, or rushing through a wave to appear on the farther side, with a fish in its beak. Seals rose out of the water and watched the boat, floating with only their round black heads above the surface—heads, in the gloom, appearing so fearfully like those of human beings, that it seemed to the shuddering rowers that the drowned men of the fjord had risen to greet them on their desperate errand.

Now let us return to Grettir!

He had been in less pain that day: Illugi had not left him, but remained faithful to his post.

The churl had been sent out as usual to watch the ladder and draw it up at nightfall. But instead of doing as he was bid, the fellow laid himself down at the head of the steps, in a sheltered nook, and went to sleep. As dusk set in, the thrall partially awoke, and looked drowsily at the ladder. "Humph!" said he; "I see no use in taking the trouble of pulling this up to-night, when there is such a sea running that no boat could venture out on it. I'll just take another snooze, then saunter home, and say that all is safe." So he turned on his side, and was soon snoring.

When Thorbjorn and his party reached the shore, they found that the ladder still hung down.

"We are in luck's way!" exclaimed The Hook; "now, my men! perhaps you will think that our journey will not prove as bootless as you expected. Up the ladder with you! and let us all be of good courage!"

Then they ascended, one after another, Thorbjorn taking the lead. On reaching the top, they noticed Glaum, asleep under a rock, snoring loudly. Thorbjorn recognized the man at once, and struck him over the shoulders with his sword-hilt, bidding him wake up, fool that he was! and tell them truly all that he knew about those whom they sought.

Glaum turned over on his side, rubbed his eyes, and growled forth—"Cannot you leave a poor wretch alone! assuredly never was man so ill-treated before; you won't even let me sleep out here in the cold!"

"Idiot!" exclaimed The Hook; "look up and see who are come! We are your foes, who purpose slaying every man of you."

Glaum started up, and screamed with terror when he saw the black figures around him.

"Silence!" cried Thorbjorn; "I give you your choice of

two things,—answer the questions I put to you, or die on the spot."

The churl was silenced, and stood trembling before The Hook, with great drops of perspiration rolling off his face.

" Are the brothers in the house ?" asked Thorbjorn ; " or shall we find them out of doors ? "

" Oh ! " cried Glaum, " they are both within, Grettir sick to death, and Illugi watching and never leaving him ! "

The Hook asked for particulars, and Glaum told him all the circumstances of Grettir's being wounded. Then the Hook burst out laughing, and said, " The proverbs come true, ' Old friendships are the last to be broken,' and, ' Woe to him who has a thrall for his only friend ! ' especially if he be such a fellow as you, Glaum ! for shamefully have you betrayed your master, bad though he be ! "

Some of the men caught Glaum by the throat, and beat him till he was nearly senseless ; then they flung him down, and pushed on towards the hovel.

In the meantime, Illugi had been sitting near the fire with his brother's head on his lap, whilst Grettir lay on some sheepskins beside the hearth. All that evening, the sick man's eyes had been wandering among the rafters, watching the light play among them, as the firewood blazed up or smouldered away. Presently, he turned his head towards his brother, saying that he thought he could sleep : and in a few moments he closed his eyes.

Illugi watched his face kindled by the scarlet glow from the embers. It was more tranquil than he had seen it for many days ; the muscles were relaxed, and the wrinkles furrowed on the brow by the intense pain which the poor outlaw had suffered, were now smoothed quite away. Grettir's face was not handsome, but it was grave and earnest, tanned dark by continual exposure to the weather. His breath came evenly in sleep ; one hand lay open, palm uppermost, on the floor, the other played with the tassel of his spear, which stood ever by his side. Suddenly, there was a crash at the door, and the sleeper opened his eyes dreamily.

"It is only the old ram, brother! he wants to come in," said Illugi; "and is butting at the door."

"He butts hard, he butts hard!" muttered Grettir, and at that moment, the door burst open. They saw faces looking in.

Illugi sprang to his feet, grasped a sword, flew to the doorway, and defended it valiantly, so that none could come within a spear's length of it, for the lad brought down his weapon on their lances and smote off the heads.

Then some of the men clambered up on the roof, and began to rip off its covering of turf. Grettir tried to rise to his feet, but could only stagger to his knees. He seized his spear and drave it through the roof, among those who were tearing it down. It struck Karr in the breast and pierced him.

"Be careful!" cried The Hook; "be careful, and no harm can happen to you!"

Then the men pulled at the gable ends, heaved the ridge-piece aside and broke it asunder, so that a shower of rafters and turves fell into the chamber.

Grettir drew his sword and smote at the men, as they leaped upon him from the wall. With one blow he struck Vikarr, the servant of Hjalti, over the left shoulder, as he was upon the point of springing down. The sword sliced through him and came out below his right arm, and the corpse dropped upon Grettir. The blow was so violent that Grettir fell forward, and, before he could raise himself, Thorbjorn Hook struck him between the shoulders, and made a fearful wound.

Then cried Grettir, "Bare is man's back without brother behind it!" and instantly Illugi threw his shield over him, planted a foot on either side of him, as he lay on the floor, and defended him gallantly, so that all were amazed at his courage.

"Who showed you the way to the island?" asked Grettir, of The Hook.

"Christ showed us the way!" answered Thorbjorn.

"Nay, nay!" muttered Grettir; "it was that hag, your foster-mother, who directed you hither!"

The mist of death was in his eyes, he attempted to raise himself, but sank again on the sheepskins, which were now drenched in blood. No one could touch him, for the brave lad warded off every blow that was aimed at his brother. Then The Hook ordered his men to form a ring around them, and to close in on them with shields and beams. They did so, and Illugi was taken and bound, but not till he had wounded the majority of his opponents, and killed three of Thorbjorn's churls.

"You are a brave fellow!" said The Hook; "and never have I seen one of your age who fought so well!"

Then they went up to Grettir, who lay in a state of unconsciousness, without being able to make any resistance.

They dealt him many a blow, but little blood flowed from the wounds. When all thought that he was dead, Thorbjorn tried to disengage the sword from his cold, damp fingers, saying that Grettir had wielded it long enough.

But the strong man's hand was clenched around the handle so firmly that his enemy could not free the sword from his grasp.

Several of the men came up, and endeavoured to unweave the fingers, but they were unable to do so. Then The Hook exclaimed, "Why should we spare this vile outlaw? off with his hand!" and they held it down, whilst he hewed it from the arm, at the wrist. Then the muscles of the fingers relaxed, and The Hook was able to loosen them, and possess himself of the sword. Standing beside the body, and grasping the hilt with both hands, he smote at Grettir's head; the edge of the blade was notched with the blow. "See!" laughed Thorbjorn; "this mark will be famous in the history of my sword. I shall show the notch and say—'This was done by Grettir's skull!'" He smote twice and thrice at the outlaw's neck, till the head came off in his hands.

"Here have I slain a famous warrior!" exclaimed Thorbjorn; "this head shall come with me to land, that I may

claim the price that has been set upon it, and that none may
be able to deny that I slew the redoubted Grettir."

The rest of the party told him to do as he chose, but they
did not think much of his act, for they believed Grettir to
have been dead before Thorbjorn smote at his head, and
they suspected that he had wrought his foe's sickness and
death, by unhallowed means.

Then The Hook turned to Illugi, saying—"It would be a
pity that a brave lad like you should die, because you have
associated yourself with outlaws and evil-doers."

Illugi answered, "At Al-thing, you shall be summoned
to give an account of this cursed deed, and answer to the charge
of witchcraft, which I shall bring against you, if I live."

"Listen to me, boy," said The Hook, "Lay your hand
to my hand, and take a vow never to revenge that which has
taken place to-night, and I will give you life and liberty."

"And listen you to me, Thorbjorn," replied Illugi. "If
I survive, but one thought shall occupy my heart, night and
day, and that will be, how I can best avenge my brother.
Now that you know what to expect from me, choose whether
I shall live or die."

Thorbjorn took his companions aside to ask their advice,
but they shrugged their shoulders, and replied that, as he had
planned the expedition, he must carry it through as he
thought best.

"Well," exclaimed The Hook, "I have no fancy for having
the young viper ready to sting me wherever I tread. So he
shall die."

Now when Illugi knew that they had determined on slaying
him, he smiled, and said—

"You have chosen that course which is most to my mind."

As the day began to dawn, they led him to the east side
of the island, and slew him there. It is said that they neither
bound his hands nor eyes, and that he looked fearlessly at them
as they smote him, and neither winked nor changed colour.
Then they buried the brothers beneath a cairn, but they took
the head of Grettir and bore it with them to land.

As they rowed home, the thrall, Glaum, made such out-cries that they were tired of his noise, and on reaching the mainland they slew him.

One morning, Thorbjorn Hook rode with twenty men to Bjarg, in the Middle-frith, with Grettir's head hanging at his saddle-bow. On reaching the house he dismounted, and stalked into the hall, where Grettir's mother was seated with her servant. Thorbjorn flung her son's head at her feet, and sang :—

> " Flitted I from the island,
> With me the head of Grettir ;
> That yellow head, which women
> Weep ; with it I am standing.
> Look you ! the peace-destroyer's
> Head lyeth on the pavement ;
> Look you ! it cannot moulder
> Now that it well is salted."

The lady sat proudly in her seat, and did not shed a tear ; but lifting her voice in reply, she sang :—

> " Milksop ! no less than sheep
> Flee before the fox,
> Would you have fled before
> Grettir strong and hale ! "

After this the Hook returned home, and folk wondered at Asdisa, saying that none but she could have borne such sons as those twain who slept in Drángey.

I must detain the reader for awhile with a few remarks I wish to make on this story—a very touching one to me, who have followed the brave man from cradle to grave, and have watched his character unfolding, in all its stern beauty. The death of Grettir is mentioned in the Icelandic annals, under the date 1038, as also in the Landnama, or Icelandic domesday book.

Some incidents in Grettir's eventful career are related in other Sagas, and the brave outlaw is mentioned in several genealogies. The persons spoken of in the Gretla are heard

18

of, again and again, in numerous other Sagas, and in no case
is there an anachronism.

Grettir was once captured by several boors, and would
have been hung, had not a rich lady interposed to save his
life. This incident is also related in the Fóstbrœthra Saga;
and there is a curious incidental expression in a fragment of
the Thorthar Saga, printed in Nordiske Oldskrifter XXVII.,
which seems to bear on this point. Thorth is related to have
blessed the Middle-frith in these words: "This I say, that
the man who grows up in this vale, shall never be hung!"

The Gretla mentions a certain Gisli, a boaster, who vowed
that he would slay Grettir. The outlaw stripped the man,
whipped him, and sent him home with a flea in his ear. The
Viga-styr Saga, the earliest of all Icelandic Sagas, casually
speaks of this very Gisli, and, without relating this incident,
gives him precisely the character which is attributed to him
in the Gretla. The murder of Atli, brother of our hero, and
the consequent revenge, are spoken of in the Bárthar Saga.
The circumstance of Grettir having lived in a cave near the
farm of Bjorn Hitdœlakappi, is alluded to in the Bjarnar
Saga. In the Gretla, mention is made of the squabbles which
took place between Bjorn and a certain Thorth. The Bjarnar
Saga gives an account of this feud. In our Saga, Grettir
is spoken of as meeting Viga-Barth wounded and exhausted,
after a hard fight in which he had revenged the death of his
brother: the Heitharvigum Saga gives the details of the
murder and the expedition to revenge it. Thus one Saga
explains and supports another.

I have mentioned only a few instances out of many, to
show the reader the authority we have for regarding these
stories as historical. But these are not the only testimonies
to the truth of the narrative; there are geographical details
in the Saga which will bear the closest scrutiny. In the
account of Grettir's life among the Eagle tarns, his hut
is said to have been situated near a spit of land which projects
into the Great Arnarvatn, and he is spoken of as swimming
along the side of this spit without being observed by a hired

assassin, who was on the look-out for him, standing on the end of the promontory.

Now the appearance of the lake itself bears out the statement of the Saga in the most emphatic manner. The only grassy spot along the shore of the lake faces the south, and is close to a tongue of land covered with herbage, rising about nine feet abruptly from the water's edge with an almost overhanging scarp; so that any one swimming close along the side would necessarily be hidden from the person on the top. Near the same lake is a cleft in the rock, in which Grettir defended himself against a band of men, led against him by Thorir of Garth. A friend of the outlaw, named Hallmund, stationed himself behind him, and guarded his back, without being himself visible. The description in the Saga of the way in which this support was given is puzzling; but on seeing the rift where the conflict took place, all the difficulty vanishes. There is a nook like a sentry-box in the side of the cleft, and it was in this that Hallmund ensconced himself, so that he could hew down any one who attempted to pass through the chasm, whilst he remained completely screened from observation.

There remain, it must be allowed, some incidents in the narrative which are undoubtedly fabulous, and which have been inserted to fill up and adorn the story; but these are easily distinguishable from the facts. For instance, it is told of Grettir that he broke open an old vikings-cairn, and, after a hard struggle with the tomb-dweller, despoiled him of his sword and treasures. This same adventure is found in the Flóamanna, Holmverja, Hrómundar, Bárthar Sagas; also in those of Olaf Geirstafa-alfs and the elder and younger histories of Olaf the saint. It is unquestionably a myth which has suffered anthropomorphosis, and represents the descent of the sun into the grave of winter, to return after having despoiled it of its prey—the fruits of the earth.

The belief in vampires is so widely spread, that I hardly like to dismiss the story of Glámr as a fable; it is, moreover, susceptible of an explanation. The Vale of Shadows was always famous for being a stronghold of robbers and men-

18—2

slayers, so that Glámr may have been simply a freebooter, who was invested with a supernatural character by the fears of the inhabitants. The witchcraft in the account of Grettir's death is easily explained away. It was the interest of Grettir's friends, and of Thorir his enemy, to circulate the story that Grettir came by his death through unhallowed means; thus the outlaw's kindred were able to bring an action against his slayer, and Thorir was freed from the necessity of paying the price he had offered for his head.

The Saga writer is careful to tell us that the thrall informed Thorbjorn's companions of the cause of Grettir's sickness, before he was killed. That the outlaw wounded himself whilst chopping up wood is by no means improbable, and that his condition of body was so unhealthy that the wound mortified, is likely enough, as his food had been confined to sea-birds and fish.

This sickness was connected with Thorbjorn Hook by the fact of Thorbjorn having a foster-mother who had always been accounted a witch; and thus the supernatural portion of the story was engrafted on the main body of fact.

The Saga has undergone embellishments, but the main facts are indisputably true. Perhaps no better idea of its relation to pure unadulterated history can be obtained than through an incident related to me by the priest of Hvammr in Northrárdalr. He said that there had lived an old farmer at Lángadalr within the memory of man, the same who had transcribed the volume of Sagas which I saw at Underfell in the Vatnsdalr. This farmer had been enthusiastic in his admiration of the Gretla. One night he saw, or dreamed that he saw, a tall figure approach his bedside, and at once he recognized Grettir the Strong.

The old man's face brightened up, and rising in his bed he held with the hero a long conversation, of which only this is reported.

Said the farmer: "Tell me, who was the strongest man in Iceland: was he Orm Storolfson?"

"Orm Storolfson!" exclaimed the apparition, with contempt; "I tell you, I have seen stones in Norway which Orm had lifted, and I could toss every one of them."

"Then, who was Iceland's greatest hero?"

"Egill Skallagrimson! He was unsurpassed."

"Tell me next, Grettir! whether the story of your wrestle with Glámr is true?"

"That was a bad business," answered Grettir, shaking his head. "Ask me no more on that subject."

"And about your descent of Gothafoss, after a tussle with a Troll woman?"

"It is true that I strove with *something* there, but what that thing was I cannot tell, as the struggle took place at night. I did swim beneath the falls."

"Is the Saga about you quite true?" asked the bonder.

"It is *touched up*," answered the strong man; "but it is for the most part quite trustworthy."

To his dying day the farmer protested that this interview really took place.

CHAPTER XVI.

FROM SKAGAFJORD TO MITHFJORD.

Svatha-stathr—Heathen Charity—Church of Vithimyri—An odd Cow—Borg,
an old Castle—Queen Victoria's Cousin—Icelandic Duck—The long-
tailed Duck—Red-necked Phalaropes—Deep Bogs—Melr—A happy
Meeting—A Wedding—Female Dress—Music—National Anthem—
Michaelmas Hymn—Mr. Briggs in Love—Church Service—Sacred Music
Ecclesiastical Position of the Icelandic Establishment.

AFTER having turned our backs on the Skagafjord, our course
lay up a broad valley, through which flows the Herathsvatn.
This river breaks into a multiplicity of streams, divided by flat
low holms and insulated ranges of low hill.

This district of islands is called Hegranes, and is the place
where the provincial Thing, or Council, was held in olden
times.

The view over it from Svatha-stathr was very beautiful,
the whole extent of islet and stream being bathed in sunlight.
Countless flocks of wildfowl sailed, sun-kindled, through the
clear air, filling it with their piercing cries.

Svatha-stathr receives its name from a certain Svathi, who
lived in the tenth century.

It happened during that period that Iceland was suffering
severely from a bad year, so that there was a large amount of
destitution over the country; and unless something were done
by the wealthy bonders to relieve it, there was a certainty of
many poor householders perishing during the approaching
winter.

Svathi, heathen though he was, stepped forward and libe-
rally undertook to provide for a considerable number of suf-

Pl. 12.

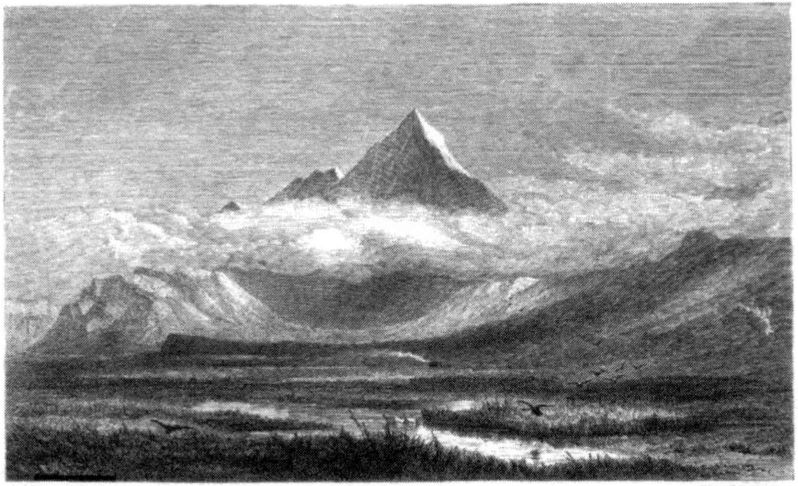

S. Baring Gould, delt. Edmund Evans, sc.

MÆLAFELLSHNUKR.

ferers. Accordingly, the poor starving wretches assembled at his door, and were ordered by him to dig a large pit in his tún. They complied with alacrity, and in the evening they were gathered into a barn, the door was locked upon them, and it was explained to them that on the following morning they were to be buried in the pit of their own digging.

"You will see at once," represented Svathi, "that if twenty or thirty of you be put out of your misery, the number of mouths wanting food will be reduced, and there will be more victuals for those who remain."

There certainly was something in what Svathi said; but, unfortunately, the poor wretches did not see it in the same light as he, nor appreciate the force of his argument; and they spent the night howling with despair.

Thorwaldr of Asi, a Christian, who happened to be riding by towards dawn, heard the outcries, and went to the barn to inquire what they signified. When he learned the cause of the distress, he liberated the captives, and bade them follow him to Asi. Before long, Svathi became aware that his birds had flown, and set off in pursuit. However, he was unable to recover them, as Thorwaldr's men were armed. Thus the golden opportunity was lost, and he was obliged to return home, bewailing the failure of his benevolent scheme. As he dashed up to his house, blinded with rage, and regardless of what was before him, the horse fell with him into the pit which his protégés had dug, and he was killed by the fall. He was buried in it next day, along with his horse and hound.[*]

From Svatha-stathr we obtained a guide to show us across the river; and here I parted with the faithful Jón, paying him twelve dollars for the time he had been with me.

We were a long while traversing the streams of the Herathsvatn, which seemed innumerable. The view from the grassy flats was particularly striking. To the south rose the magnificent Mœlafellshnukr, or the Measuring mountain, with a belt of cloud along its base. To the right lay a moun-

_____

* _Younger Olaf S. Tryg._, chap. 225.

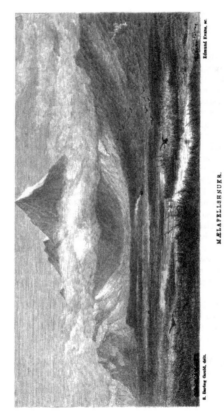

S. Baring Gould, delt.

MÆLAFELLSHNUKR.

Edmund Evans, sc.

London: Published by Smith, Elder, & Co., 65, Cornhill. 1863.

After a rest of an hour, at a little byre, with the euphonious title of Gauks-myri, situated amongst the worst bogs I ever saw, we started again, guided by the farmer through morasses which were perfectly impassable to all but those who knew them thoroughly. We were a long while traversing them, as we had to follow a serpentine course, and as the horses floundered in repeatedly, and could only be extricated with difficulty. One of them strained himself in his efforts to reach firm soil, and limped for the rest of the way. We descended into Mithfjord, by the small farm of Reykir, and, in the evening, reached the river which winds through the vale.

After wading through it, we rode to the little church and parsonage of Melr,* which lies on the western slope of the hill, in a large, grassy tún.

Hurrah for the wedding! We were in time for the festivities, that was clear, for our arrival was hailed by the house disgorging its contents. A motly crowd of men and women came out to stare at us; the former in white shirts, and turned

---

* *Melr* means a sandy hill, *mól* a sand flat. I take the word to be derived from a root signifying to grind. This appears in several Icelandic words, and is applied to sand as crushed or pounded stone, also to meal as comminuted corn. The words are—*mala*, to grind; *melja*, to crush. Hence, Thor's hammer is called *mjölnir*, or the shatterer. *Mylna* is a mill, *malana* a grinding, *moli* a particle, *mjöl* meal, *mjöll* snow, which, to a certain extent, may be said to resemble a sprinkling of meal: *melur* is the wild corn which produces meal, and which grows on sandy flats.

The same root is found in many of the Aryan languages.

In English we have the verb to *mill*, substantives *mill* and *meal*.

In Danish there is the verb *at mala*, substantives *mölle* and *meel*, also *smule*, a particle.

In Gothic, the verb *malan*, and the substantives *melo* meal, and *melm* dust.

In Lithuanian, the verb *málti*, and the substantive *miltai* meal.

In Russian, the verb *molótj* to grind, *mélivo* a grinding, *molj* meal, *melj* a sand bank.

In Latin, *molere*, *mola*, *malleus*.

In Greek, μύλλω, μύλη.

In Sanskrit, *malana*, a grinding; and in Hindustani, *malnd* to grind, crush.

In old Persian, *marad* to grind.

Holmboe, however, connects *melr* with *mold*, loose earth, and the verb *myldja* to dig, which I think is a mistake.

down collars, showy ties, and blue jackets; the latter in black
gowns and silver-worked bodices. Hold! surely I see a round,
merry, well-known face, peering at me from the door. I
sprang to my feet, and was locked in the embrace of Mr. Briggs.
"Oh, my friend!" I exclaimed; "how came you to be here?
Where are Martin and the Yankee?"

"They have gone to fish in Skoradals-vatn, and are tired
of this valley; but I—I could spend my life here!"

"Is it so very attractive?" I looked round; the vale
was broad and well-watered. It lay between hills, nowhere
rising above a thousand feet, and was by no means as picturesque as the Vatnsdalr.

"Attractive!" exclaimed Mr. Briggs; "it is Paradise,
Elysium, Valhalla!"

"I see," said I, patting him with an air of commiseration,
"the corn-brandy has been too strong for you."

"No, indeed!" answered my fat friend; "but, Padre,
you shall hear to-morrow all about it. She has such eyes,
such a mouth, such a nose! such a heavenly nose!"

"She!" I almost screamed; "my dear Mr. Briggs, I am
all in a quiver; this marriage, is it yours?"

"Alas, no!" he moaned; "would that it were!"

Any further communications were broken off by the Archdeacon of Melr coming up to me, shaking hands, and inviting
me indoors. I thanked him, and apologized for having arrived
at such a time; but he replied that he was delighted that I
should have the opportunity of being present during the
festivities.

I made myself as tidy as circumstances would permit, and
then entered the guest-room, which was thronged with people.
The middle of the chamber was occupied by a long table
covered with eatables and drinkables; these consisted of ship's-
biscuits, skonrogs, or hard biscuits, which can only be broken
with a hammer or a stone, cakes of flour and sugar, wild
corn pancakes, cold boiled ducks'-eggs, and bottles of corn-
brandy. A bowl of curd was brought in for Grímr and me,
and we made a hearty supper off it, the biscuits and the eggs.

I had now an opportunity of seeing the festive Icelandic female costume. This consists of a black cloth skirt, white sleeves, a green or black velvet bodice, worked over with silver flowers in the most beautiful and tasteful manner, and fastened in front by a silver bodkin which laces the sides together by passing through silver rings with tinkling balls and flowers of the same metal attached to them.

On the head is worn a tall white cap, fastened to the hair by pins with beautifully wrought silver-gilt heads. This cap is, however, set aside in the evening for the more convenient, coquettish black skull-cap with long silk tassel, which is in common use.

The hair of Icelandic ladies hangs to the shoulder, and is then cut off. Round the neck is worn a small black or coloured handkerchief, tied in a bow. I had brought some brooches with me to Iceland as presents, but these I found were quite useless, as the ladies never wear shawls or kerchiefs folded over their bosoms, which could be fastened by a brooch.

I learned, to my regret, that I was late for the religious portion of the wedding; however, I was in time for all the convivialities.

These lasted till two or three in the morning, and consisted in eating and singing. There was no dancing, to Mr. Briggs' great mortification, neither was there the hard drinking which is popularly believed to characterize all Icelandic merry-makings. Everything was in moderation, except the singing, and of that I thought the good people would never tire. Every one of the guests joined in, Mr. Briggs included, who thundered forth a reckless bass which suited one tune as well as another. All the rest sang in unison, except one little girl who in the most ear-piercing notes shrieked out a part, just a fifth above the melody.

All the songs were in Icelandic, but the melodies were mostly Danish. However, I recognized our "God save the Queen," and "Beautiful Star," both set to Icelandic words, descriptive of the glories of the island. One melody of a character quite distinct from the rest, riveted my attention;

it was grand, stately, and composed in one of the old church modes.

I requested to have it repeated, and the singers willingly complied, telling me that both words and tune were old as the hills, literally, fire-old (elld-gömll váru lagit ok hljóth). I at once wrote down the melody, and here it is :—

ICELANDIC NATIONAL ANTHEM.

Harmonised by W. A.

NATIONAL MELODY.

Harmonised by W. A.

V. S.

The tune will only be admired by those whose ears have been accustomed to ancient ecclesiastical music. The second version of the melody was given me by a lady at Reykjavik; it is evidently a modern corruption of the old tune. As a little variety to these pieces, I transcribe one pretty little air which in style is quite different from those melodies which are unquestionably of Icelandic origin.

W. B.

Alleluiah! sons of morning,
Walking in triumphal might,
Gleams of the last Easter dawning,
Kindling this our wistful night!
Alleluiah! Alleluiah!
Sing ye Alleluiah!

Alleluiah! trumpet pealing,
  Hosts of light to battle fly!
Fell Abaddon, downward reeling,
  Drops like lightning from the sky:
    Alleluiah! Alleluiah!
  Shout ye Alleluiah!

Alleluiah! at the clashing
  Of your arms he quakes with fear;
Michael beats him, binds him, gnashing,
  Chains him to his dungeon drear.
    Alleluiah! Alleluiah!
  Sing ye Alleluiah!

Alleluiah! victory gained!
  Shout, the battle's fought and won!
Satan and his host restrained,
  Upward flash like rays of sun!
    Alleluiah! Alleluiah!
  Sing ye Alleluiah!

Alleluiah! upward streaming,
  White as sunlit flakes of snow,
All in Heaven's glory beaming,
  Through the golden portals flow!
    Alleluiah! Alleluiah!
  Thund'ring Alleluiah!

Alleluiah! throng in singing
  To the Royal city blessed!
All Jerusalem is ringing,
  All with flowers of Eden dressed!
    Alleluiah! Alleluiah!
  Ceaseless Alleluiah! *

It was morning before the party broke up, and the guest-
room was converted into a bed-chamber for Mr. Briggs, Grímr
and me.

The following day was Sunday. After breakfast I walked
outside the farm to look out upon the sea. The great Húnaflói, a mighty bay, out of which branch numerous fjords, lay
to the north, calm and blue, with a line of snowy mountains

---

* The Icelandic words are of a sacred nature, so that I have not chosen to
adapt secular words to this graceful melody. The verses which I have given
are original, and are added to facilitate the performance of the little piece.

19

rising out of it, far off on the horizon. These I take to have
been Sandfell and Burfell, distant between forty and fifty
miles. At first, I mistook them for soft white clouds, lit by
the morning sun, but, on examining them through my tele-
scope, I was able to distinguish the pearly range—

> Faintly, flushing, phantom fair,
> A thousand shadowy pencilled valleys,
> And snowy dells in golden air.

Two little fishing boats skimmed over the sea, like gulls,
and were reflected in the azure surface. The scene might
have been taken for a Mediterranean peep—had it not been
for the icy wind which puffed up from off the Arctic Ocean.

"Padre!" said a sepulchral voice near me, which made
me start in alarm; "Padre! come and pour balm into a
lacerated bosom."

I looked round, and saw that the portly Mr. Briggs was
reposing in the lee of a cowshed, wrapped in a warm rug.

"Well!" said I; "let me hear now that story which
you would not confide to me last night."

"That story is comprehended in four words," answered
my fat friend; "I am in love!"

"I judged so by the hints you gave me, and also by the
languishing glances which I saw you casting at one of the
bridesmaids, over your stock-fish and butter."

"Is she not an angel, a houri?" exclaimed the enraptured
Mr. Briggs.

"Your Ariadne is certainly very beautiful," I replied.
"How does the love-making progress?"

"Very badly," sighed the fat man; "I only know one
Icelandic expression; that is, 'ver-thu sœl!' (may you be
blessed!) I have said it to her at least fifty times, till I think
she is tired of that remark, and quite thirsts for a new one.
Now, Padre! you have arrived in the nick of time. Tell me
the Icelandic for—Adored angel, I fling myself at your feet,
a grovelling victim of your charms!"

I stopped my friend and advised him strongly to join
company with me, and try to forget the eyes.

"Not the eyes so much!" interrupted the enamoured
swain; "but the nose, the adorable nose!"

If the flame could not be smothered, I recommended him
to return to Iceland next summer and propose.

Mr. Briggs agreed to my suggestion, and so the matter
was settled.

In the meantime, horses had been brought round to the
farm door; and the guests, including the lady whom Mr.
Briggs adored, departed. Scarcely were they gone than the
church-bell rang for service.

It would not have been etiquette to have chimed for
worship till the guests had departed, lest they should think
that the parson wished to take an unfair advantage of their
presence in his house, and force them to listen to his Sunday
discourse.

Mr. Briggs, Grímr, and I went to church, and were
accommodated in the chancel near the altar. At Melr the
holy table stood free from the east wall, and was not enclosed
within rails. Against the east wall was a poorly painted
triptych, and on the altar two glass candlesticks, not matching
each other, one being intended for three lights, the other for
a single candle. The holy table was vested in a blue cloth,
with a triangle surrounded by flames, in the centre.

The men sat in the choir, and the women in the nave.
The clerk occupied a seat, like a returned stall, against the
screen. This man, an odd old fellow with large horn spectacles,
lighted the altar candles, consisting of two tallow dips, and
snuffed them at intervals during the service with his previously
wetted fingers.

Enter the Archdeacon in cassock without girdle; the clerk
proceeds to vest him, throwing an alb over his head, then a
chasuble of crimson velvet, with a gold cross down the back.
In doing this he so ruffles the parson's hair, that he licks his
hand, and plasters the hair down with it.

The priest advanced to the altar and took up his position between it and the east end, facing the congregation,
in the manner prevalent in the Basilica churches at Rome.
This is unusual. Grimr told me that he had never seen

19—2

the priest in this position elsewhere, and that it was occupied
in this case only because there were no communion rails.
The appearance of the minister looking towards his congregation over a gaudily-draped table, between glass candlesticks,
reminded one unpleasantly of a conjuror.

Raising his powerful but not musical voice, the priest sang,
"Latum us bithja !" (Let us pray); then he chanted the
collect for the day. This was followed by the Epistle and
Gospel, not said from different sides of the altar, as in English
and Roman churches, but from the middle. Before the Epistle,
was said "Pistilinn skrifar postulinn Pál" (the Apostle Paul
wrote the Epistle), and before the Gospel, "Guthspjallith
skrifar guthspjallamathurinn Marcus" (the Evangelist Mark
wrote the Gospel). This was followed by "Guthi se lof ok
dyrth fyrir sinn glethilegan bothskap !" (Praise and glory be
to God for His joyful message !) sung by the clerk. Both
Epistle and Gospel were chanted to a plain tune. The Lord's
Prayer, which came next, was similarly chanted.

The people remained seated during the prayers till the
Paternoster was said, when they rose, but sat down again, to
sing the following hymn : —

W. B.

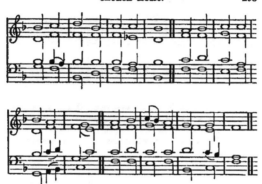

After another prayer, the priest gave out a second hymn, which I copied also, as a specimen of a genuine Icelandic melody :—

W. A.

Then his chasuble and alb were removed, and he ascended the pulpit, where he said a bidding prayer, and preached a sermon on a text from the Gospel. The discourse ended, another hymn was sung, whilst the minister returned to the altar.

W. A.

The pastor then said a prayer, answering to our "Church militant," after which he dismissed the people with a blessing, making the sign of the cross over them in the air. This service resembles the Roman "Missa sicca," and that form of the Communion office which is so prevalent in our own church; the faithful coming for the Blessed Sacrament, and being dismissed with the sermon, like the children in the Gospel, who ask for bread and are given a stone. The service being ended, the priest went round and kissed all the congregation, beginning with the visitors. There are more women than men at an Icelandic service.

In the Icelandic liturgy there is no creed. I asked Grímr how that was. He told me that a creed was to be found in some of the old prayer-books, and that it was used only in out-of-the-way churches—never at the Cathedral of Reykjavík. I found, on reference, that the creed to which he alluded was a metrical version from which the ninth article has been expunged; and one looks in vain in the Icelandic office books for the three Catholic symbols. This is natural enough, as a profession of faith in the article on the Holy Catholic Church would be unmeaning in a Lutheran community. Communion takes place twice or thrice in the year, and before presenting themselves at the altar, the communicants have to be shriven. If at any other time a parishioner signifies to the priest that he wishes to communicate, the pastor is bound to celebrate.

Much of the dogma of pre-Reformation times lingers among the Icelanders; they believe implicitly in baptismal regeneration, the real presence, the power of the keys as vested in the priesthood; but all notion of sacrifice, as connected with the "mass," has been practically lost, though they retain the name of priest for their ministers, and though the Augustan Confession admits the sacrificial views of the whole Primitive and even later Greek Church ; so that Molanus, Abbot of Lokkum, together with the Hanoverian theologians, were able to satisfy on this head the eminent Gallican, Bossuet.   Danish and Icelandic orders are derived in direct and unbroken succession from a certain Buggen-hagen, and not, as in our Church, from the apostles of Christ.

It is much to be desired that the Icelandic and Danish establishments should be restored to the unity of the Church, by receiving the succession through our own bishops.

The importance, I should rather say the absolute necessity, of there being an unbroken chain from the Apostles in order that the Sacraments may be valid, is now beginning to be felt in Denmark, and a move on the part of the English Church might prove of incalculable advantage to the Lutheran societies in Denmark and its dependencies.

It is surely better that by securing the succession from our bishops, their orders and sacraments should be rendered valid, rather than that individuals from those societies should be restored to participation in the merits of Christ through the mutilated communion of the Roman Church, and be brought again into worse than Egyptian bondage to the Papal throne.

S. Baring Gould, del.

Edmund Evans, sc.

BJARG.

CHAPTER XVII.

THE MIDDLE FRITH.

Glacial Action—Bjarg: the Home of Grettir—Cairn—Spear-head—Carved Stone—Magical Characters—The Story of the Banded-Men: a Saga.

On Sunday evening I galloped with Grímr and the son of the Archdeacon to Bjarg, the farm where Grettir was born and spent his childhood. To reach it, we rode up the vale, passing the church of Statharbakki, and crossing the river where the banks contract and the hills become steeper. Our path lay over the shoulder of a hill, the rocks completely smoothed and grooved by glacial action. The striæ ran from N.N.E. to S.S.W. There are, however, no mountains near from which glaciers could have descended. Above Bjarg is a long hill with the rock exposed in several places. I found it everywhere polished. It is strewn with large blocks of stone, one of which, a Grettis-tak measuring 46 feet in circumference and 15 feet 4 inches in height, rests on a pivot, like a logan stone. This huge mass, in being moved along the smooth rock, has scored a furrow about twelve feet long. This is only one among numerous blocks which lie thus resting at the end of ruts in the polished surface. The larger of these stones rest on the top of the hill, and they diminish in size towards the bottom of the valley.

Here is Bjarg! This little farm with its red gables and grass-grown roof, and its green tún in a dell full of buttercups! I stood on the rocky platform in front of the house to survey the landscape over which Grettir's eyes must have roamed so often. Below us was the river plaintively murmuring over a floor of pebbles. Beyond it swelled up the heithi, over which lies the road traversed by Grettir, when he rode to avenge his brother's murder. Beyond it rises a snowy mountain head. Turning south, the eye ranges over the wild deserts of Tvidœgra and Arnarvatns-heithies, to the faint white cupola of Eireks Jökull, thirty-seven miles distant, beneath which the poor outlaw spent many years of his exile.

I inquired for the spot where rested Grettir's head, and was shown a green mound in the tún. The saga states that the head was buried in the church at Bjarg, but there is no church there now, nor any trace of one.

I obtained permission to examine the mound at the cost of a dollar, and dug into it till we came to a large stone which we were unable to move without levers. As no iron bars which would serve the purpose were to be discovered in the house, I was obliged to leave the stone where I found it. It may possibly cover the spot where lies the skull, but it is more probably an erratic block which the possessors of Bjarg have been unable to remove, and have accordingly covered with earth.

No remains of antiquity have been found at the farm, nor are they likely to be found, as Icelanders never dig. The only vestige of olden times, which I could procure in the island, was a spear-head of the Sturlúnga period, which was discovered in the sand near Myvatn, the day before my arrival there. It is shaped like a knife tapering to a point, and is of highly wrought steel. The blade is twelve inches long, and an inch and a third broad at the broadest part. It was originally fastened into its pole by five nails, one of which has a head turned in a scroll to allow of a bunch of feathers being attached to it.

On our way back to Melr, we visited the priest of Statharbakki, who showed us a handsome brass woman's saddle, covered with embossed griffins, cherubs, and flowers. It was of seventeenth-century workmanship, and had the date, and the name of the maker, in Icelandic characters on the side.

The priest informed me of the existence of a large stone lying on its side, and curiously carved, in the tún of Thorfastathr on the opposite side of the river. I at once crossed the river with Grímr, and visited the farm. We were shown the stone in question, which is about twelve feet long. The only marks on it are these :—

The larger of the two is certainly intended for Thor's hammer, a magical character. Whether this stone were used in heathen times, for sacrificial purposes, or at a later period, for the incantations of witchcraft, I cannot say.

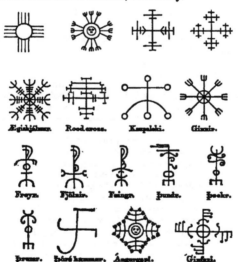

This table of Icelandic magical signs, I give for the benefit of those of my readers who dabble in the black art.

Below Melr is the cairn of Kormak; it has been opened, the Archdeacon told me, but was found to contain nothing.

The Middle-frith is full of historic interest, for not only does the Gretla relate to it, but also the Sagas of Kormak, Thorth hretha, and the Banded-men. The latter contains an account of a law-suit, so spirited, so full of curious interest, and giving such a lively picture of the manner in which litigation was carried on in the eleventh century, that I shall epitomize it now for the benefit of the reader, as I did at Melr on that Sunday evening for the amusement of my friend, Mr. Briggs.

The Story of the Banded-Men. *
(Eleventh Century.)

At Melr, in Mithfjord, lived a worthy, straightforward sort of a man, Oddr by name; he was rich, for he had been much in commerce, and every cruise had been attended with success.

Oddr was, unfortunately, on no very good terms with his father, who lived at Reykir, beyond the river; and it is difficult to say which was in the wrong. Certainly Oddr had left home without saying good-by to his father; but then, the old gentleman had treated him shabbily, and had invariably kept him at a distance.

Now that Oddr was rich, pride kept the old man from claiming relationship, and Oddr's resentment had not yet cooled, so they never met, though they lived within sight of each other. It happened that Oddr was involved in strife with a vile fellow named Uspak, who had murdered Vali, their mutual friend, in mistake for Oddr. Uspak had influential kinsmen, yet there was little chance of their taking up his cause,—at least, so thought Oddr, when he commenced legal proceedings against the culprit, with a view to getting him outlawed. Having summoned witnesses to the murder, Oddr went to Althing, the general council, as summer drew nigh,

---

* Banda-manna Saga. I use the word banded-men instead of confederates, as it is a more exact rendering of the Icelandic name.

hoping to get his action through court without difficulty. All went on smoothly for a time, and he came at last to summoning the defendant to take legal exception to the suit.

Now it happened that two relations of the accused, Styrmir and Thorarinn, were sitting near the Doomring, chatting. Quoth the former, "Do you hear, friend? There goes the summons for exception to be taken! what do you intend doing?"

"Letting the action proceed, of course," answered Thorarinn; "it is perfectly just. Oddr has excellent reason for desiring a villain like Uspak to be punished."

"That is all very well! but remember the delinquent is a relation."

"I care naught about that!" said Thorarinn.

"Perhaps not," spoke Styrmir; "but you must consider that all sorts of ill-natured things will be said of you, if you let a cousin be outlawed, without lifting your little finger to help him. There is a flaw in the action, and that you must see as clearly as myself."

"I have noticed it all along," broke in Thorarinn.

"Then why on earth don't you answer the summons! This Oddr tosses his head, as though no one dare oppose him. In a little while you will find that he has become so powerful and arrogant, that he will tread us all underfoot."

"As you like then!" answered Thorarinn; "but I have a presentiment that our meddling with a bad cause like this will only bring us into trouble."

"Not it!" exclaimed Styrmir, starting from his seat, and walking to the doomring. "Oddr!" called he, "I take exception! The action is wrongly set on foot. You have summoned your ten witnesses at home, that is against the law; they should have been summoned at the Thing, and not in the district where you live. Thus your action breaks down."

Oddr was silent, and thought the matter over, then, finding that the exception was legitimate, he withdrew, and returned to his booth.

As he was passing through an alley between two of the booths, he nearly ran against an old man, dressed in a black-sleeved cloak, which was torn; and he had one of the sleeves dangling behind his back. The aged man had a spiked staff in his hand, and a broad-brimmed hat on his head, from under which his eyes wandered restlessly to and fro. He was bent nearly double, and hobbled onwards, groping his way with his staff.

This old man was Ofeigr, the father of Oddr.

"You are quick in coming from court!" quoth he, in a harsh, grating voice; "things have gone on swimmingly of course, and now, that knave Uspak is proclaimed guilty, ay?"

"No, indeed," answered Oddr; "he has got off."

"Ah, ha!" giggled the old fellow; "you fine gentlemen are fond of poking fun at old people. Pray how is it possible that the man should be allowed to escape? is he not really criminal?"

"There can be no doubt on that score," said Oddr.

"What!" exclaimed Ofeigr; "and he not declared guilty?"

"A flaw was found in my summons, and the case broke down," answered Oddr, with some irritation.

"It is quite possible," said the old man, "that you are more in your element when bartering or steering a ship than when engaged in an action at law. Yet I still think that you are trying to hoax me!"

"Well!" answered Oddr, with considerable asperity; "I don't care a straw whether you think so or not."

"Perhaps not," retorted Ofeigr; "but this I can tell you, that if you had only asked me, I could have told you at the outset that your case was illegally set on foot. You were too high and mighty to ask advice, or you need not have come on this wild-goose chase."

"I don't see that your meddling with the matter would mend it!"

"The case is not hopeless yet, and I do not mind assisting you; but you must let me know whether you are ready to pay well."

" I shall not spare my money," answered Oddr.

" Then hand over to me a tolerably heavy bag of silver. I tell you what," quoth the old man with a leer; " the eyes of many a man hanker after hard cash! We shall see what effect this money will have when it is brought into court!"

Oddr gave him a large purse, and then walked home to his booth.

Ofeigr, however, hobbled along the plain, till he came to the Northern quarter court, and there he asked, how the actions were going on.

He was told that some had been settled, others had broken down, and others, again, were ready for the final summons.

" Pray how ended my son's case? Is it legally concluded?" asked Ofeigr.

" It is concluded after a fashion," answered the judges, who were winding up matters for the day, as it was becoming late in the evening.

" And, of course, Uspak is doomed guilty!"

" Nay, nay!" was the answer; " he is not indeed."

" How comes that?" inquired the old man.

" There was found a flaw. Your son had set the action afoot illegally."

" Ha!" exclaimed Ofeigr; " have you any objection to a poor old body, such as me, coming within the doom-ring?"

" None in the world!" So he tottered into the circle and sat down.

" Well!" quoth the old gentleman; " I can't quite understand what you have told me. Actually, Uspak is acquitted—so I understand you—though his guilt was as clear as daylight."

" There was legal exception taken!"

With an expression of amazement, he exclaimed—

" How could you allow a trifle like that to interfere with the course of justice! so that you have acquitted a depraved

scoundrel, who is both a thief and a murderer. This is what
I call giving verdict against all right and equity."

They replied that there was a difference between equity
and common law, and that though morally Uspak might be
criminal, yet legally he was guiltless.

"May be!" said Ofeigr, with an incredulous shake of his
head; "but your oaths, dear life! To think of the oaths you
have taken!"

"What of them?"

"Why, did you not swear, on your admission to the seat
of justice, that your finding would be according to truth,
equity and law; was it not so?"

The judges answered in the affirmative.

Then Ofeigr said, "And what could be more according to
equity, than that he who slew an innocent man, who is moreover a deliberate thief, should be sentenced to outlawry.
Think! Is it not better to follow the dictates of conscience,
to judge truly and righteously, than according to the letter of
the law. A very serious responsibility is incurred by those
who undertake to judge in a court of law, for they are not
only bound by the laws under which they act, but they have
to answer to a much higher tribunal—that of their own consciences. Conscience is a tender plant, which should not be
tampered with. I am an old man, my hair is grey, and I feel
daily more keenly the necessity of obeying conscience above
all things. I am pained for you, dear friends! as I know
what a heavy load your consciences will have to bear, not
only for having given wrong judgment this day, but for having
thereby violated your solemn oath to judge according unto
right."

The old man had allowed the bag of silver to slide from
beneath his cloak during this address, and had then hitched it
up again. He noticed that the eyes of the judges had followed
the purse; so he continued—

"It would be perhaps as well to reconsider your verdict,
and to judge righteously and with justice, and thus secure the
friendship of all good and upright men."

As he spoke, he poured the silver out before him, and began counting it into little heaps.

"It is quite pleasant to find myself here among old friends," said Ofeigr; "and really, I cannot resist the temptation of making you all a little present, just by way of keeping up old acquaintance. So please, my judges, accept an ounce of silver apiece, and you who break up the action, half a mark each. Dear friends and relatives, oaths are too sacred and solemn to be tampered with. Excuse my anxiety for your welfare, in thus pointing out to you the path of duty."

The judges looked very solemn, and conscience began to gnaw like a worm. The sin of violating an oath struck them in a way it had never struck them before. They came round to Ofeigr's way of thinking, pocketed the silver, reviewed the suit, sent for Oddr, hurried through the legal forms, declared Uspak guilty, and then the court broke up.

Nothing transpired during the night; but, next morning, Oddr stood up and called aloud: "Hear all! a man, hight Uspak, was outlawed last night in the northern quarter court, for the slaying of Vali. As a mark by which he may be recognized, know all that he is a large-built man, has brown hair, high cheek-bones, large hands, and a sturdy pair of legs."

People stared, for few had heard of the proceedings during the night, and it was thought that Oddr had acted with wonderful skill in having got his case through. Styrmir and Thorarinn were much disconcerted at the action having terminated thus, and at their becoming the butt of every one's jokes. Matters could not stop thus, that was certain; so they assembled their kinsmen, Hermundr, Gellir, Egill, Jarnskeggi, Skeggbroddi, and Thorgeir, and these eight took an oath that they would hold together, till they had driven Oddr to self-doom or outlawry. It was called self-doom when a defendant threw himself on the mercy of the plaintiff, and paid any fine which his antagonist chose to demand.

Next spring they rode to Melr, and summoned Oddr for having used bribery in the law court. Oddr took things

20

easily, and did not trouble himself much about the impending
suit, till his father warned him of its serious nature, and of
the impossibility of his escaping the sentence of banishment,
unless he followed his advice.

"And what is that?" asked Oddr.

"My advice is," said the old man, "that you get your
chattels on board ship, as soon as all men go to the annual
Thing, and that you cruise about, waiting to hear the end
of the action. I will go to the council, but you must provide
me with a good bag of silver, in case of emergencies."

Oddr obeyed; and no one was aware of the arrangement
except himself and his father. The Thing met, and the
eight confederates were in high glee, for they were satisfied
that no one would venture to oppose this action, so that they
could easily carry it through court.

One day the old man, Ofeigr, in a brown study, limped
out of his booth, shaking his head sadly, for no man of consequence could he find, who would lend his support to Oddr
against the banded chieftains, for love or money. The aged
man's knees shook under him, as he lurched from side to
side, moving among the booths. At last he came to that
of Egill, one of those who were banded together, and hung
about the door, weighed down by his infirmities, whilst some
men were talking within to Egill. When they walked out,
Ofeigr hobbled in and greeted Egill, who looked at him, and
asked his name. He told it.

"Then you have come to speak about your son's case:
that won't do. I have no leisure for it, and it is useless your
thinking of talking me over; I am only one among many,
who are involved in this suit."

"No, no," croaked Ofeigr; "I am but a stupid old man,
who dearly loves a bit of gossip with clever and courtly gentlemen like you, so you will oblige me, I know, by letting me
listen to you, whilst you talk."

"Well, I have not the least objection! Come and sit by
me, and we can chat on indifferent matters; only indifferent
ones, mind!"

"Quite so," answered the old rogue. "So you are a landowner, Egill, I hear, and have got a venerable mansion, out at Borg."

Egill nodded.

"I hear all sorts of things about you, Egill. You are just a man to my fancy. You don't stint any man of his food, but keep open house. I so like old-fashioned hospitality. And you are fond of company, and love to have a house full. That is so very nice, so prince-like, I must say!"

"I am glad you think so," quoth Egill, smiling; "for I have always heard that you are a man whose opinion is worth something, and you are a man of good family, too!"

"Ah! it is all very civil of you to say that!" answered Ofeigr; "but I'm a man of little consequence, and make no pretention to a place among the landed gentry of Iceland. Whilst you—*you* are quite one of our leading aristocrats. More's the pity that you should be short of cash, and unable to keep up your former hospitality."

"Ay! but the tide will turn! I reckon on getting some pretty good pickings soon," quoth Egill, slily.

"You do!" exclaimed Ofeigr; "I am positively delighted to hear you say so. But where will they come from, tell me?"

"Well," answered Egill; "since you ask me, I answer that we confederates reckon on getting Oddr's money; and he is so wealthy that the eighth part, which is to be my share, will quite set me up."

"I can give you every information on that head; perhaps, you may wish to know, with precision, what you may expect to obtain."

"You are an excellent, shrewd old man: I long to hear!" answered Egill.

"Then you may calculate on receiving precisely one-sixteenth portion of the land of Melr, a few acres of irreclaimable bog: that will be your share, nothing more! The law officers have a right to half of all confiscated property, the other half will be divided among you who are banded together."

20—2

"Nonsense, man!" exclaimed Egill, with vehemence; "how can you make that out, when your son is one of the wealthiest bonders in the island? There are his goods and chattels——!"

"Do you really think that Oddr would be such a fool as to let his moveable goods fall into your hands? Between ourselves—but don't breathe a word of this to any one else—my boy has shipped everything which could be got away, and is now cruising *rather* near to your home at Borg. I suppose that outlawry will be a matter of indifference to him at sea, and with all the world before him. My son knows, of course, that you are opposing him. He is not far from the Borgarfjord, rather near your house, unpleasantly near perhaps, for I presume that the venerable mansion is left undefended."

Egill became very red and uncomfortable; he took a turn up and down the booth, and then, reseating himself, said,—

"The case is altered certainly. I wish to goodness that I had never become involved in it! We confederates shall become the joke of the whole island; confound it! I have been reckoning on making a pretty penny out of this! My purse is drained, and, confound it! I shall not be able to scrape through another winter without some windfall dropping in. I might have expected that Oddr would never suffer us to pounce on his moveable goods! A nice mess I am in, to be sure!"

"Come, come!" said the old man; "things are not yet at their blackest. You are not yet the laughing-stock of all Iceland, not yet starved out of house and home, not yet constrained to sell the venerable mansion. I suppose that you must be uncommonly badly off for cash just now, well——!" and the big money-bag began to dance up and down on the old man's knee. Egill's eyes lighted on it, and Ofeigr at once drew it from under his cloak, saying,—

"I should quite like to make you a little present, Egill!" Then he untied the bag, and poured two hundred pieces of silver into Egill's lap.

"There now!" quoth Ofeigr; "accept this from me, and

if you have a chance of showing a kindness to me or my son, pray show it."

Egill answered sorrowfully: "There is my oath to the other men of the gang in the way; take back the silver, I cannot break it!"

"Who wants you to break an oath?" asked Ofeigr, with virtuous indignation; "not I certainly! an oath is a solemn and holy thing! All I ask of you is that you will observe the conditions of your oath. These conditions are that you will stand by your comrades till Oddr is brought to self-doom or outlawry; are they not?"

"Quite so."

"Well, then, supposing that we accept self-doom, and *you* are appointed one of the umpires to decide what fine Oddr is to pay, you will not be hard on him, will you?"

"You are a crafty old fox, on my word!" exclaimed Egill; "but I cannot stand alone in this. Get some one to back me up, and perhaps I may do as you propose."

"Whom will you have?" asked Ofeigr, sharply.

"Let me see;" and after running through the list, Egill pitched on his friend Gellir.

"Humph! I shall have hard work with him: but I shall try what can be done. Now, farewell! and keep your own counsel."

The old man trudged off to the shanty of the second confederate. His steps were tottering, his back was bent, his thin hands tremulously clasped the staff which supported him, and he was regarded with contemptuous pity by all who passed.

On reaching Gellir's door, he stood cringeing at the entrance, and begging to be admitted.

"Not if you want to talk about your son's business," answered Gellir.

"Dear life! such a thought never entered my head. I was but passing, and being tired, I wished for rest: besides, I am old, and dearly love hearing men of genius discourse. You, according to all accounts, are remarkably talented. I don't mean to flatter, but people *will* talk, and one can't help hearing what they say."

"Well, what shall we talk about?" asked Gellir.

"I care not; any subject which will suit you—the pretty girls of Iceland, ay?"

"As you like, old gentleman!" laughed Gellir.

"You have got some charming creatures in your neighbourhood, haven't you? Who are the marriageable girls with money, down in your parts?"

Gellir mentioned the daughter of Snorri the priest, and that of Steinthor of Eyri.

"I hear that you have the most beautiful daughter; people *will* talk, you know, and I hear that, in personal charms, she excels every maiden in Iceland."

Gellir replied that the damsel was handsome, certainly.

"How comes it that she was not included in your list? Is she married yet, the pretty pearl?"

"No!" answered the bonder.

"Dear life! why, how comes that?"

Gellir answered that hitherto no rich and noble suitors had offered. "You see," he added, "I am not a man of great means myself, though I am pretty well connected, and I am on the look-out for a husband for the girl, who may have a fortune of his own, and not expect much of a dower from me. Come! you have asked me plenty of questions, now I shall take my turn as questioner. Who are the rising young men in your northern parts?"

"Oh! there are several," replied the aged man; "there are—let me see!" and he began to count on his trembling fingers. "There is, first, Einar Járnskeggi's son, and there is Hall Styrmir's son, and there is—well, you know people *will* talk, and I believe they say that my boy, Oddr, is a promising young chief. Do you know, I was actually going to ask, this summer, for your daughter Ragnhildr as his wife!"

"Ha!" exclaimed Gellir; "if you had only asked a little earlier you might have had a ready answer: now it is out of the question."

"I do not quite see wherefore ? " said Ofeigr.

"Why, because Oddr is under a cloud at present, and that a pretty threatening one, too ! "

"Nevertheless, the best thing you can do is to accept him as your daughter's betrothed. He is prodigiously wealthy, is of good birth, and is a kind-hearted fellow to his friends. You said just now that your means were small, ay ? "

"This action is an insuperable obstacle," said Gellir.

"Not insuperable," contradicted the old man, decidedly.

"I fear that it is," quoth Gellir ; "if I could only see my way out of it, I should be glad enough to come to terms with you, about the match which you propose."

"Probably, you have been calculating on obtaining a good share of my son's property," said Ofeigr ; "I may as well let you know exactly what you may calculate on receiving." He then explained to Gellir how paltry would be his gain, if he remained in the confederacy. "Not only will you gain little by proceeding with this suit, but you will stand a chance of losing a great deal ; for my son is at present cruising near Eyjafjord, to look up Járnskeggi's farm ; then, if he gets a favourable wind, he purposes calling at Skeggbroddi's house. Your farm is on the coast, also, I believe ; unfortunate, certainly, for you confederates, that you should live so near the sea, and be away from home when the man you are hunting down by law is making little visits with fire and sword at your homes. I tell you all this because you are a favourite of mine, and I am sorry to see a chief like you come to such an utter and irrevocable break-down."

Gellir was much agitated. "I am vexed for the sake of my friends, rather than for my own," said he.

"Look here ! " spoke the old man, and his voice was now wonderfully firm. "You have just two courses open to you—either step into the utter hobble which is before you, or else marry your daughter to Oddr, and accept this bag, containing two hundred pieces of silver, as a token that my son will deal handsomely by you."

"I can't do it ; I can't, indeed ! " burst forth Gellir, in

intense agitation: "I never have deceived those who have trusted me, and I never will!"

"You are an honourable man!" exclaimed Ofeigr; "I myself am the soul of honour; but who the deuce asked you to deceive those who rely in you? Not I, for one! Only should it happen that you are appointed to award the fine which my son is to pay, you will, I am sure, be lenient towards him, considering that he is to be your son-in-law."

"You are desperately keen, on my word!" said Gellir; "but I am not strong enough to bear the brunt of this alone."

"Whom will you have to back you up?"

"Egill is my nearest kinsman."

"Then have a chat with him to-night on your way to evensong. See what he says!"

As Ofeigr left the booth, he flung aside his staff, and walked with a firm step, no longer bent double, but upright, and with a triumphant light in his wild gray eyes.

The day arrived when judgment had to be given, and the Hill of Laws was thronged.

Egill and Gellir had gathered their clients together. Ofeigr stood forward on his son's behalf, and said, "I have not been mixed up with the law proceedings hitherto; however, I must now step in at the last moment, and try to effect a compromise. You men, who are banded together, have instituted proceedings in a most unusual manner, and I cannot deal with eight plaintiffs at once. I shall therefore select Hermund to be your mouth-piece, and demand of him what compensation he will take for wrong done."

"Compensation, forsooth!" exclaimed Hermund; "we will receive nothing less than self-doom."

Ofeigr answered, "Who ever heard of a defendant handing himself over to the tender mercies of eight men? That is monstrous, and against all legal precedent. Such a thing cannot be claimed! However, I do not object to handing the matter over to one or two of you, and letting them be awarders of the doom."

"As you like! Pick any two of us; no matter whom."

" I claim the right of selection," said Ofeigr.

" You are welcome to it," answered Hermund.

But Thorarinn exclaimed, " Stop, Hermund! do not say yes to-day, to what you may rue on the morrow."

" I have said it, and it can't be helped; not that it matters much," said Hermund.

Then Ofeigr took securities, and they were easily procured, for the defendant was known to be in a position to pay. Then it was formally agreed that Ofeigr should pay whatever the two awarders decided, and that the confederates on their side should abide by the decision, and let the action terminate.

The banded chieftains now gathered their men round the doomring. Those of Egill and Gellir sat together. Ofeigr stepped into the circle, turned up the flap of his hat, stroked down his arms, straightened himself, and stood before all, perfectly upright, rolling his eyes from side to side. " Now I shall choose those whom I please," said he, " to be awarders of my son's fine. Whom shall I select first? There you sit, Styrmir! and it would be strange if I were to pick you, for I am one of your clients. More's the token that you have received presents of considerable value from me, and have never given me anything. I think you a very shabby fellow, to be always receiving and never giving! So I reject you. There you sit, Hermund! a mighty man you, who are fond of dabbling in dirt and getting hold of the wrong end of the stick. You have poked your nose into this suit, simply from a love of mischief; for that alone I reject you. Ah! Járnskeggi, there you sit, puffed up with arrogance; you had a banner carried before you at the Vathla-thing, as though you were a king, so please your Royal Highness we'll have no kings to judge in this case—I reject you." Then Ofeigr turned round and said, " There you sit, Skeggbroddi! Is it a fact that King Harald Sigurth's-son said you were the man best suited to be King of Iceland? Ay! "

" The king often said things which he never really meant," answered Skeggbroddi, becoming scarlet in the face.

" Pray be king in your kitchen, but not in this judgment

ring," said Ofeigr; "I reject you. There you sit, Gellir, and
nothing has drawn you into this action but dire poverty, which
is to a certain extent excusable. You eight men are a bad
lot, all of you, and I hardly know whom to select. However,
you, Gellir, are not the worst of the batch, and so I accept you.
Come forward into the ring. There you sit, Thorgeir! with
just wits enough to talk about bullocks, but such a blunder-
head as you would never make an umpire, so I reject you.
As for Thorarinn, he was one of the first to set upon Oddr, so
I will have nothing to say to him!"

Then Ofeigr stared round him, "Well!" he exclaimed;
"I am like the wolf who eats and eats at the lamb, till nothing
is left but the tail, and that is hardly worth the munching. I
have had to select among you chieftains, and I have rejected
one after another, and so I am obliged to content myself with
this piece of fag-end, Egill."

Egill burst out laughing, and said, "Honour is forced on
one sometimes, so come along, Gellir! Let us step aside and
discuss what the fine is to be." So they went into a place
quite apart from every one, and Egill said, "Half measures
will never do; better let us offend one party than leave both
half satisfied, and disposed to grumble."

The two now agreed on the fine, and walked back to
the doomring.

There was a murmur of expectation through the Thing.
The confederates looked triumphantly at Ofeigr, who, with
arms akimbo, and upturned hat-rim, glanced defiance in return.
A hush fell on the assembled crowd, as Egill and Gellir
stood up.

"We have carefully weighed the matter," said Gellir;
"and we have come to the decision that the price imposed
should be thirteen ounces of silver."

"What, what!" snapped out Hermund. "Thirteen
hundred of silver, of course!"

"I saw you put your hand to your ear, Hermund; so you
must have heard perfectly well," answered Egill. "We said
only thirteen ounces of silver, and that such silver as no

gentleman will accept: old shield scrapings, and broken bits of rings, any rubbishy silver will do, to make up the weight."

"Egill," hissed Hermund, through his teeth; "you have deceived us."

"One may well deceive a man who trusts no one," retorted Egill.

"You are a liar!" shouted Hermund, furiously. "And that every one shall know. Here's my proof! Last winter I invited you to my home, as your own is a crazy tumble-down sort of place, letting in wind and weather. You jumped at my invitation. Shortly after Yule you began to grow terribly down-in-the-mouth at the prospect of returning to starvation in your old shanty; so I begged you to remain through the spring with me;—you were only too glad, of course. Well; after Easter, you returned home to Borg, and told every one that I had fed my guests on thirty old baggage horses."

Egill answered: "There might have been a little exaggeration, perhaps, for we couldn't eat the horses, they were too tough for that! *My* family and visitors never lack food, I beg to tell you! though I am a poor man, compared with you. At all events, we have never been driven to picking the bones of old broken-down mares!"

"On my word, I hope that we shall not meet here next summer or I——"

"I hope not, with all my heart!" interrupted Egill, with a laugh; "for it has been foretold of me that I should die of old age. May the devil take you long before then!"

Styrmir rose and said: "Egill, the best that can be said of you is, that you are an intriguer."

"You had better not call me names," answered Egill; "for in your drinking bouts, when you have been pretty well boozy, you have often said that we two are a pair. By the way, I understand that you have got a blot on your fair fame, which none are supposed to know of but yourself. Some one is said to have taken to his heels when an axe was lifted, some one to have starved his servant; but I won't mention names."

Styrmir was silent, then up stood Thorarinn.

"You had better hold your tongue, Thorarinn," said Egill, "or you will get into such disgrace that you will be thankful to escape. It must be an odd sight to behold you squatting over the fire, with your legs curled up, and all the maid-servants giggling at you behind your back."

Thorarinn said: "One must take advice, from whatever quarter it may come," and sat down.

"Thirteen ounces of silver!" growled Thorgeir.

"Ay," cast in Egill; "thirteen is a capital number; it will remind you of the thirteen bruises which you had, when a country clod banged you about the head at Rangárhlith."

Thorgeir was quiet, and neither Skeggbroddi nor Járnskeggi would exchange a word with Egill.

"Well," quoth Egill, turning to Ofeigr; "there's not another man but you in Iceland who could have weathered a gale like this, and sailed in the teeth of such a storm!"

After this the Thing broke up, and all returned to their homes. Oddr married the daughter of Gellir and gave presents to Egill.

CHAPTER XVIII.

THE HRUTAFJORD.

Leave Melr—Icelandic Etiquette—A beautiful Frith—Mr. Briggs' Story—
Thorodds-stathr —Row across the Fjord—Merchant Vessels—Hospitality
—Mr. Briggs recovers his Heart—Trouble with the Pack-saddle—
Holtavörthu Heithi—Sclavonian Grebe—Arctic Foxes—Icelandic Tra-
ditions concerning the Fox—Icelandic Mice—Travellers' Tales—Swans—
Terns—Difficult Pass—Baula—MSS.—Church of Hvammr.

ON Monday, July 21, we started tolerably early, after having
drunk the bride's health, in what the archdeacon called port,
but which seemed to me to be a composition of black currant
jam, treacle, and water. We paid the pastor in English gold,
with the request that he would hammer it into a keeper for
his daughter's bridal ring.

The son of the archdeacon accompanied us over the heithi,
past Burfell, to the scene of the fight which took place between
Grettir and Kormak. The hill-top is strewn with stones,
deposited during the glacial period, but nowhere did I see any
traces of morraines.

One large block, a Grettis-tak, mentioned in the Saga,
marks the scene of conflict. Below it is a little pool, on
which floated a diver. Here we parted with our young guide;
and Mr. Briggs, our guides, and I struck over the hill, S.
by S.W.

In a little dell, filled with bog, stood half a dozen shaggy
ponies, leisurely cropping the rank grass, whilst the men to
whom they belonged lay on their backs fast asleep in the

sun, with their caps drawn over their eyes. Grímr stopped
his horse, jumped off, stepped up to the men, lifted their caps,
and awoke each with a kiss.

"May you be blessed!" said the men, starting up and
scratching.

"And may you be blessed!" replied Grímr. Then he
remounted his horse, the men lay down for another nap, and
we rode on.

"Could you not leave those poor wretches to sleep in
peace?" asked I.

"Certainly not," replied Grímr. "It would have been
against all laws of etiquette, to pass men on a journey without saluting them."

We descended suddenly upon the wildly beautiful Hruta-
fjord, a narrow strip of water extending twenty-three miles
inland, and only a mile and a quarter broad at the point where
we descended upon it. The frith is hemmed in between stony
wastes, and the only grass visible is at its head, and in the
swamps which fill indentations of the hills on either side.

Anchored near the farther shore, in front of the wooden
store of Bortheyri, were two merchant vessels. On the map
of Iceland, Bortheyri is marked in large type, as though it
were a capital town, and I had expected to find that it consisted of at least half a dozen cottages, and not of a wooden
shed only, which is locked up all the year round, except during
the fortnight in the summer when the merchant ships lie
off it.

"Thank goodness!" exclaimed Mr. Briggs, when he
heard that these vessels were floating shops. "Now I may
chance to get what I have been wanting for many a day,—a
bottle of strong essential oil, by means of which I hope to
keep my tormentors at bay."

"My dear fellow!" said I; "I provided myself with
camphor and oil of lavender, before leaving England. But
Icelandic vermin have no noses, and set at defiance all precautions taken against them."

"These creatures always out-manœuvre me," said Mr.

Briggs. "Once, when I was in Wales, I was just as un-successful in keeping them off. Shall I tell you what I did ? I was stopping at a little inn, and had not been in bed five minutes, before I became sensible that it was alive. I lit a candle, but the light only served to show me the whole room was swarming. I rang the bell. Up came a servant. ' Mary ! a pot of treacle ! ' The treacle was produced. I made a ring of it round each leg of a chair, then folded myself up in my rug, and sat complacently on the seat, thinking that I had defeated my inveterate foes at last. But no ! they were too clever for me ! What do you think they did ? They crept up the walls, and dropped on me from the ceiling."

We reached the base of the hill at Thorodds-stathr, the farm which had belonged once to Thorbjorn Strong-as-a-bull, the murderer of Grettir's elder brother. It is a neat farm, with a large tún, enclosed within high turf walls, with a gate, a rare sight in Iceland !

Between this and Reykir is a swamp, in which, according to the Saga, Grettir lost his spear-head. This was found about two hundred years afterwards, and the marsh is called " Spear-swamp " to this day.

The farmer of Thorodds-stathr was absent at the ships, so that we had to ride to the next farm before we could obtain a boat in which to row across to the vessels. The object I had in visiting the ships was to procure small change, as I was unprovided with any smaller coin than dollars, and I could get none exchanged at the farms, as the people live by barter, and use money only in occasional transactions.

The larger of the vessels belonged to M. Sandhop, a mer-chant, who, if I remember right, was a Dane by extraction, though he had been born in Iceland. He received us with every civility and insisted on our dining with him.

He had come from the I'safjord, and was going to visit two or three other stations, and then sail for England.

The hold of his vessel was fitted up like a shop, with counter and desk. Round the sides were ranged sacks of rye

meal and coffee, canisters of sugar and snuff, kegs of
brandy and rum, suits of secondhand clothes, whips, bridles,
saddles, ranges of pottery and hardware.   Above the entrance
to the lower hold were heaped up fox and swan skins, and
bales of wool and eider-down, which had been received from
the natives.

Unfortunately for Mr. Briggs there was neither camphor
nor oil of lavender among the stores. M. Sandhop let us over-
haul the ship's medicine-chest, but we could find nothing
which would avail us as a specific against "jumpers,"
as the Danes designate a disagreeable form of insect life.
We then visited the second ship, which belonged to a kind
old Dane with white hair, who was bent on showing us hospitality, and was distressed beyond measure at being unable
to provide us with what we wanted.

Surely Iceland is a glorious field for the operations of
Mr. Harper Twelvetrees!

On this vessel we found the whole of the upper deck converted into a shop under canvas. A steady traffic was going
on, bags of wool were being hoisted up the ship's side from
Icelandic boats, and meal, coffee, and brandy barrels were
being swung down in exchange. The merchant laughed when
we told him of our discomforts, and assured us that he was
compelled to swab down his deck, and wash the boards which
constituted his counter, every night, so as to purify them
from the loathsome creatures, which had been left on board
by his customers.

The old gentleman brought us into his cabin and insisted
on our toasting Denmark with him, in bumpers of raw brandy.

On our return to the larger vessel, we found that dinner
was ready in the cabin.

The merchant and his captain, Mr. Briggs, Grímr, and I
sat down to a capital repast of hot roast mutton and black bread.

After having lived for so long on curd, stockfish, and
occasional junks of cold, semi-putrid mutton, the fresh roast
meat was most delicious, and I never enjoyed a dinner so
thoroughly as that on board the merchant vessel.

We drank bottle after bottle of foaming Bavarian beer, and glasses of good claret; after which an "Alexandra pudding" was brought in. This consisted of a very light raisin pudding, floating in egg, brandy, and flour sauce. With this we drank port and champagne, and the meal concluded with a steaming bowl of punch, very hot and strong.

"Well! has either of you lost his heart in Iceland?" suddenly asked the merchant. Mr. Briggs dropped his head, and became red as a peony.

"Why, what is the matter?" asked M. Sandhop; "there is something in the wind, I can see!"

"What glorious Jökulls there are in this island," said my fat friend, making a clumsy attempt to turn the conversation.

"Yes! but that is neither here nor there. What do you think of the fair maids of this icy clime?" Grímr burst out laughing, and looked at my friend.

Poor Mr. Briggs! his confusion became terrible.

Neither Mr. Briggs nor I answered, but Grímr maliciously told the story of my friend's *affaire de cœur*, to the great amusement of the merchant.

"Well!" said M. Sandhop, "you have certainly chosen the prettiest of all Icelandic belles. The lovely Thorney has not only got the most beautiful eyes——"

"And nose," interpolated Mr. Briggs.

"But she is also as good as she is beautiful; and she is well connected too! for she is the grand-daughter of an Arch-deacon, daughter of a Sysselman, and niece of a Thing-man. I do not know that a more eligible match is to be found in the whole island, but there is a drawback."

"A drawback!" echoed Mr. Briggs, with a groan.

"Yes," answered the merchant; "though no daughter could behave better to her widowed mother—still there is a drawback!"

"Tell me, oh! tell me, what that is!" pleaded my fat friend, with an expression of agony on his usually cheery countenance.

21

"I hardly like to mention it," answered the merchant;
"but——she has got the sheep's disease." *

Poor Mr. Briggs collapsed with a moan, and since that
moment, he never alluded in my presence to the angelic
Thorney.

It was late in the evening when we left the merchant-ship,
and rowed back to the farm from which we had borrowed the
boat. We paid for the hire of it the trifling sum of three
marks (1s. 1½d.) Guthmundr, the guide of Mr. Briggs, had
been sent on with my friend's baggage horses, to Melr—
another Melr, not that from which we had come; but my
ponies remained at the farm.

Our progress up the fjord was somewhat impeded by the
condition of Grímr and Mr. Briggs, who, of course by the
merest succession of accidents, repeatedly tumbled off their
horses, and nearly came to blows by charging each other
with being drunk. Our track led across numerous swampy
gills, and deeply rutted streams. Grímr led the way in a
series of zigzags, between the fjord and half-way up the hill.
This may have been all right, but I put it down at the time to
a certain confusion of mind, produced by the merchant's
champagne and punch.

In two hours we reached Stathr church and parsonage,
where every one was asleep: and then began a series of
troubles which hindered our journey, so that it was past midnight before we reached Melr in Hrutafjord, which was to be
our destination for the night.

The crupper of the principal baggage pony broke, and pack
and boxes slipped over the brute's head. It was a quarter of
an hour before these were again adjusted. A horse ran loose,
with only a saddle on his back, and I suggested that his crupper
should be brought for the sumpter-pony. Grímr thought
otherwise. "This will not happen again!" said he, and he
drove the horse on without one. In five minutes the load was
over his head again, causing another delay of a quarter of an

---

* See page 101.

hour. Mr. Briggs and I set off after the horse with the spare saddle, but the discordant yells of my friend so alarmed the pony that we could not catch him. Grímr strapped on the pack as before, and would not assist us.

At the next hill over went the boxes again; and this actually happened six times, Grímr protesting on every occasion that it would be the last, till Mr. Briggs and I completely lost our temper with him. Will it be believed that on opening one of the boxes at Melr, we found a bran-new crupper that Grímr had put at the top—one which we had brought with us in case of emergencies? Yet Grímr would rather undergo the trouble of saddling and loading the horse six times than use a crupper, simply because it was my proposal that he should.

The bed at Melr was so dirty that I preferred sleeping on the floor, with a saddle for my pillow, and relinquishing the bed to my guide.

The farmer, next morning, gave us nothing to eat, and charged us exorbitantly; so that I advise all future travellers to shun lodging at Melr.

We started early, and wound over a desolate upland tract of stone and mud, following a stream which brawled and tumbled in beautiful falls over the black rock, tearing itself a course and burrowing among crags as hard as iron. At one spot it leaped over a jet black glossy shelf into an inky pool, on the side of which was seated a Sclavonian grebe (*Podiceps cornutus*) gravely watching the bubbles. It was a handsome bird with rich auburn tufts on its head, red neck, and silvery white breast. The back and forehead, crown and ruff, were of a deep glossy brown, the beak black with a white tip. This grebe builds a large floating nest of rushes and slime, and ties it to a couple of strong reeds. The nest is so strong and so carefully plastered and plugged with clay that not a drop of water can get into it. The bird lays three or four eggs of a pale greenish white colour, one inch and three quarters long by one inch and one quarter in breadth. When the mother leaves her nest, she carefully covers the eggs with reeds and twigs, to conceal them from the skuas which are on the

look-out for them. After that the young are hatched, she
takes them under her wings and dives with them to give them
their first lessons in swimming and dipping. The sportsman
will find the little ones still under the wings of the bird he has
shot, as it rises from the water into which it has plunged in
alarm on his approach. The little things support themselves
under the maternal wing by their bills, which rest on the back
of the mother towards her tail. M. Preyer, in his tour in
Iceland, speaks of the eared grebe as most common, whilst
the Sclavonian grebe is rare except on Myvatn. Neither Mr.
Fowler nor I saw the former, whereas the latter is met with in
most of the Icelandic lakes.

The heithi we were now crossing is called Holtavörthu
heithi; it is more tolerable than most of those I had passed,
as the views from it are peculiarly fine. We jogged over it at
a good pace, seldom falling into a walk.

We disturbed a blue Arctic fox, and sent it darting behind
some rocks. The Icelandic fox (*Canis lagopus* L.) is very shy
in summer, and seems to know that his fur is much in request
then; whereas, in winter, he is full of audacity, and will venture
close to the farms in search of food. The foxes then go about
in troops and attack the sheep with the ferocity of wolves,
flying at their throats and hanging on till the frightened
animals drop from exhaustion. Then a number fall upon them
and the sheep are soon devoured. Every farmer loses several
of his flock, as many even as twenty or thirty sometimes
during the winter by this means, and he consequently does his
utmost to kill down these mischievous creatures. When the
snow is on the ground, the fox loses his blue grey hue, and
becomes perfectly white. His fur is then worth from two to
three dollars, whilst it would fetch from six to nine in its
summer colour. When the fox happens to be caught by the
foot in a gin, he gnaws off his leg without scruple, and then
limps away to his den. As soon as one has been captured and
killed, he is at once skinned, and his skin is bartered for rye or
brandy at the merchant stores. I purchased a great many of
these skins, but could not obtain a single specimen with the

bones of the legs and head entire, fit for stuffing. The Icelanders tell many stories of the cunning of this proverbially
crafty animal; of course, among the number is that world-wide
story of his ridding himself of fleas by retreating slowly into
water, letting the vermin crawl first from his tail as that is
immersed, then from his body as the water closes over it, till
they are all congregated on his nose, when he dips that, and
sends the whole company adrift.

Another story told of him is that he walks with erect and
nodding brush towards a flock of sea-fowl, which mistake the
dancing tail for one of themselves, and only find out their
mistake when Reynard turns sharply on one of them and
makes a meal of it. Once upon a time, however, the sea-bird
got away even after it was in the mouth of the fox; and by
this means: "For shame!" exclaimed the bird, "you godless
creature, you are going to eat me without even saying grace!"
Reynard abashed, folded his paws, turned up his eyes and
opened his mouth;—out flew the bird. "Bother!" said the
fox; "henceforth I shall only say grace after meals!"

Henderson relates that in the vicinity of the North Cape,
where the precipices are almost entirely covered with various
species of sea-fowl, the foxes proceed on their predatory expeditions in company; and previous to the commencement of
their operations, they hold a kind of mock-fight upon the
rocks, in order to determine their relative strength. When
this has been fairly ascertained, they advance to the brink
of the precipice, and taking each other by the tail, the
weakest descends first, while the strongest, forming the last
in the row, suspends the whole number till the foremost has
reached the nests and eggs. A signal is then given, on which
the uppermost fox pulls with all his might, and the rest assist
him as well as they can with their feet against the rocks. In
this way they proceed from rock to rock, until they have provided themselves with a sufficient supply. Credat Judæus non
ego! This looks much as though poor Henderson had reproduced from schoolboy recollections the famous story of the
men of Gotham, who descended in a similar manner to the

water's edge from Gotham bridge. The wise men, however, it will be remembereed, got a ducking, for the mayor, who was holding to the parapet, let go for a second that he might spit on his hands.

Olafsen and Povelsen give a wonderful account of the sagacity of the Icelandic mice. These little animals are often obliged to cross rivers to make their distant forages. On their return with the booty to their magazines, they are obliged to repass the stream. This they effect by selecting a flat piece of dried cowdung, on which they heap the berries they have collected; then they push their original boat to the water's edge, launch it, surround the mass of dung with their little heads resting on it and their bodies in the water, and so they paddle across, steering with their tails. Henderson brings forward two eye-witnesses to the truth of this most wonderful story. I found that our guides knew of mice thus crossing lakes and streams, but they had neither of them seen it done themselves.

" I'll tell you what, Padre!" exclaimed Mr. Briggs, when I related to him these anecdotes; "travellers *must* tell wonderful stories—it is expected of them, it is a duty! and if you are unprovided with them, you had better not write a book. Now, I will tell you a little fact in Arctic zoology, just as authentic as those stories you have just been telling me. It was given me as a fact, and you may believe it if you like. Well! you know that all the bears there are in Iceland came over from Greenland on the ice, don't you?"

I assented.

" Very well, then," he continued—" two or three years ago—I am not quite certain of the date, a small trading vessel ran on the ice north of the I'safjord, and was deserted by the whole crew. A company of Polar bears coming over from Greenland took up their quarters in the vessel, and, I have no doubt, found themselves very pleasantly accommodated. In spring the ice melted, and the vessel floated off with all the bears on board. There was at that time a succession of northerly gales, so the ship was driven south, passed Rockall,

sighted the west coast of Scotland, left the Mull of Cantire
in the wake, doubled the Calf of Man, and stood right into
Liverpool harbour. The custom-house officers thought her
a queer-looking craft, and came alongside in a boat, climbed
up the side, saw nobody, and walked to the cabin-stair, when
—imagine their dismay!—a procession of Polar bears marched
up the ladder with the utmost gravity and composure, headed
by an august maternal bear bowed down with years, and in
the rear half a dozen cubs, which had been born on the voyage.
Fact, Padre!"

We halted to change saddles at a small lake on which
were floating several wild-fowl, enjoying the brilliant sunshine
and rocking at their ease on the crisp wavelets which flowed
before the fresh northerly breeze. They seemed to be perfectly
indifferent to our presence, and made no attempt to escape,
with the exception of a common scoter (*Oidemia nigra*), which
rowed off at a great rate, and appeared only at ease when in
the shade on the farther side of the tarn.

The other birds consisted chiefly of teal and pintails
(*Anas crecca* and *Anas acuta*). Besides these, a pair of swans
floated in a dreamy majesty on the blue water. They were
Hooper swans (*Cygnus ferus*); another species, Bewick's swan
(*Cygnus Bewickii*), breeds in Iceland, but is not common; so
that Brehm was unfortunate in naming it Cygnus Islandicus.
M. Preyer never saw the bird, and Mr. Fowler doubts its
being a native of the island, as the Icelanders whom he questioned were very positive that only one variety frequents
their lakes. On the other hand, my friend J. W. R., who
has contributed some notes for sportsmen at the end of this
volume, shot a specimen, and has the head in his possession
at present. Neither M. Preyer nor Mr. Fowler found the
little grebe (*Podiceps minor*) in the island, yet one was shot
by J. W. R., who gave the skin to Mr. Briggs. I have a
suspicion—I cannot say that it is more than a suspicion—
that I saw a red-necked grebe (*Podiceps rubricollis*) on Myvatn,
but I could not get near enough to the bird to thoroughly
convince myself.

The Icelanders are tolerably unanimous in their assertion
that only one species has been seen in the island, but, as
in the case of the swans, their testimony is open to question.
The terns again cannot surely be represented by one species
alone, and that the Sterna arctica. I am convinced that a
naturalist will find other varieties if he looks for them among
the islets and along the coast. As I have already mentioned,
we shot what I believed at the time to be the common tern
(*Sterna hirundo*), on the Thingvalla heithi, above the Allmen's
rift. Mr. Martin, who brought them down with his gun, has
written to me in answer to my queries, and given his unhesi-
tating opinion in accordance with mine.

As the day began to decline, we descended into the vale of
the Northrá, and passed the little farm of Fornihvammr.
In one place the track lay over a narrow ridge of rock, not
two feet wide, descending to the river on one side, and to a
brawling torrent on the other, in abrupt precipices. The
pack-horses refused to advance over it, and we found that they
were frightened, and could not be driven by blows. Consequently, I rode past them, and taking one by the bridle,
walked my little piebald across, hoping that the rest would
follow lead. This they did, till they reached the middle, and
then they halted, and stood trembling on the ridge. If their
feet had slipped, and they had fallen over on the river side,
they would have been killed; if they had slipped over on the
torrent side, they would have certainly broken their legs; so
that it was a moment of anxiety to us all.

Grímr was behind with Gúthmundr, and neither of them
could pass to the front, so that I was left to do what I could.
I pulled at the bridle of the foremost pony, but he would not
move a step, neither could he retreat, as there was not room
on the ledge for him to turn. I drew off my comforter, and
bound it round his eyes, then caressingly urged him to advance. This he did, still trembling violently, and pawing the
ground in front, before he planted each foot. So, with much
trouble, I got him completely across, and the others followed
in his steps.

In the meantime, Mr. Briggs had made a considerable détour, having gone back some way till he found a spot where the hill admitted of being ascended, and the torrent crossed without difficulty.

Before us rose the cone of Baula, a tall grey mass of trachyte about 3,000—3,500 feet high, and so precipitous that snow can never rest on its head or flanks. Near its base is little Baula, a singular crater, containing in the centre of its bowl a sugar-loaf of red cinder, considerably higher than the walls of the crater. The old myth, that at the top of Baula there is an opening to the land of the elves, has been exploded, for it was ascended, two years ago, by some German naturalists.

The mountain is composed of a pyramid of pale grey trachytic columns, three or nine sided, and arranged with the most beautiful regularity. From the top of Baula can be counted thirty-seven lakes, and innumerable chains of snowy mountains.

We stopped the night at Hvammr, a little parsonage planted under a precipice of dark rock. The old priest was an enthusiast on the subject of Icelandic history, and was able to give me some curious information corroborating the statements in some of the Sagas.

He had seen the stones in the Hitará, which mark the spot to which Grettir and Björn had swum, and had sounded the stone in the Hvitá to which Thangbrand the Christian missionary had attached his ship. The Saga speaks of it as giving out a musical note when struck. It does so still, and is called the Glöckustein. It is egg-shaped, of a yellowish tinge, and about six feet high.

The old man showed me a parchment MS. history of his parish, written in the sixteenth and seventeenth centuries, also a MS. volume of Sagas, containing those of Asmund viking, Jasone Bjarta, Thorstein forvitna, Florus oll sonum hans, Dynuse Drambláti, Eirek Artussyni, and Halfdane Eysteinssyni. The church contains little of real interest except a font basin of brass stamped with the Annunciation, and a fine

brass chandelier. On the altar are two triple candlesticks.
In the churchyard are stone staves over the tombs, and one
forms the top of a stable door adjoining the house. This
bears an inscription which I could not decipher, as the stone
was much overgrown with the turf, of which the walls are
composed.

CHAPTER XIX.

THE VALE OF SMOKE.

Runaway Horses—Grjóthâls—Glorious View—I succeed in mastering the Horses—Cruelty to Birds—The Future of Animals—Ptarmigan—Whimbrel and Plover—A Heithi—Icelandic Way of Sleeping—Names—Tunguhver—Boiling Jets in a River—Reykholt—Snorro's Bath—The Church—Snorro Sturlason.

On the following morning, Mr. Briggs and I were lying in bed, laughing and talking, when Grimr came in, with the news that two of my horses had run away.

"Then go after them!" said I.

"Ah!" said he, "of what use is that? You would lose guide as well as horses then. When once the horses make up their minds to run, they will sometimes go for a week without stopping. There was once a man from the Skaga-fjord came to our house at Reykjavik, and left his horse standing outside our door, whilst he spoke to my father within. The pony started off, and next day was seen passing Thing-vellir; then it ran through Kaldidalr, and twelve men who were making up the way marks on the side of Ok, tried in-effectually to stop it. The horse ran on till it reached its home, and that was six days after it had left Reykjavik." Grimr had always got a dismal story on his lips, when one was at all inclined to be cheerful, so I took this anecdote at what it was worth, and waited patiently for two hours, till the horses were recovered.

The priest would take no money for our lodging; so I presented him with the *Illustrated London News Almanack*, and his wife with a pretty necklace, to which was suspended a cross of Cornish diamonds.

The almanack for 1862 contained chromo-lithographs of our English farm-stock; and as there are neither pigs nor donkies in Iceland, the representations of these animals roused the interest of the old people.

"What is that?" asked the priest's wife, pointing to a pollard willow in one of the pictures.

"That," said Grímr, "is a forest tree."

"Tree!" exclaimed the old lady; "how wonderfully trees must grow in southern climes! Why, it is twice my height, and as thick round as my waist. Trees must be fine things, indeed, if they grow as big as that!"

Our road lay over the Grjóthála, or "Stony neck," from which we obtained a magnificent view of Baula, starting up in its strange isolation, out of an elevated plateau above the Northrár-dale.

From the other side of the ridge, we beheld one of the most enchanting landscapes I have ever seen. The sun was out in all its noontide glory. Below us lay a wooded plain, spangled with pools and streams of water, and stretching to the Borgarfjord, which flashed in the sun, like a mirror. Far away to the south lay the purple chain of Skarths-heithi, crested with snow; to the left rose Ok Jökull, above a range of low brown hills.

We descended to the plain and rode for an hour through a coppice of brightly glistening birch, among which darted the redwing (*Turdus iliacus*), and the white wagtail (*Motacilla alba*). The soil was sandy, and sprinkled with the orange Alpine cinquefoil (*Potentilla aurea*). Passing North-tunga, we were involved in bogs, and had to scramble up an almost precipitous hill to escape from them. Presently Grimr exclaimed—"We have lost the track!" Ten minutes afterwards I found it, but Grímr kept resolutely from it, as I, and not he, had been its discoverer.

Up to this time the pack-horses had held me in sovereign contempt, and would not advance a step when I shouted to them to go on.

It is tedious work driving a troop of horses in Iceland, unless one has a man to every three, for, instead of continuing in the track, the wretched creatures separate, and stand quietly cropping the grass a hundred yards apart. The driver has to ride from one to the other, shouting and cracking his whip. As soon as one horse is urged on, it advances half a dozen steps and then stops till the rest of the troop have been whipped, and the driver returns to it again. It often takes an hour to get over a couple of miles, and the trouble and worry are incessant. With my little switch, I used to ride after some quietly grazing pony, uttering shouts like a fiend. The animal would look round at me with indifference, crop a mouthful, look again, take another mouthful, and, as soon as I came within reach, would saunter off, with perfect nonchalance. Now, however, I made myself redoubted, having buckled a leathern strap, a yard and a half long, to the end of my whip; and with this I dealt such frightful blows that the ponies fled like the wind before me. As we rode over a long rolling hill, the sun was at our backs, and threw our shadows before us. The appearance of my shadow, running over the pasture on which the ponies were grazing, was now sufficient to strike a panic into their hearts, and send them on at a gallop.

Two or three times we disturbed families of ptarmigan, the mothers starting up between the legs of our horses and running away with a terrified cluck, whilst their numerous progeny darted hither and thither in the wildest alarm. I dismounted once and caught a fledgling, a droll little ball of yellowish grey down. The poor chick remained perfectly tranquil in my hand, its tiny heart beating very fast, and its little head turning right and left in search of its mother. Presently I heard a harsh croak close by, which was immediately followed by a feeble "Cheep!" from my captive and a struggle to be free. On looking down, I found that the

mother bird was only a few yards off, with her family about her,
eyeing me with unmistakable anxiety. I at once released
the little fluffy ball; out started the rudimental wings—quaint
flaps they were—and the diminutive creature scampered off to
its mother in great glee. The old bird seemed now to be quite
satisfied as to my peaceable intentions, and let me remount
my horse and ride off without making any further attempt to
escape.

What a lovely sight is a nest with the parent bird seated
patiently on her eggs, or a mother gathering all her little ones
around her! I daresay that Clement Brentans is right when
he says:—

> " Engell, die Gott zugesehen,
> Sonne, Mond und Sterne bauen,
> Sprechen: ' Herr, es ist auch schön,
> Mit dem Kind ins Nest zu schauen ! "

" Angels who see God's face, who sustain sun and moon
and stars, say: ' Lord ! how goodly it is also looking into a
nest with a child ! ' "

It always goes to my heart to kill a bird. The feathered
creation are so wonderful in their perfection and beauty, that
it gives me real pain to rob them of the precious gift of life
which God has bestowed upon them. It may be necessary
sometimes to despoil them of it in the cause of science, but
it inspires me with a sickening disgust to see the wanton
manner in which some take pleasure in destroying these
precious pieces of mechanism; these glorious little bodies, so
matchless in their beauty, so lovely in their motions, so
buoyant in their joy of life. It is small satisfaction, too, to
stuff a poor little skin and set it up under a glass shade.

The bird is the same; true, but never can the most skilled
hand restore the gloss to the draggled feathers, which the
living bird's bill could polish in dainty pride.

The bird-stuffer does his best: he inserts glass eyes,
gives the head a correct twist; places the body in a suitable
attitude by the means of wires; supplies camphor to keep
away insects; and tries to think that the stuffed creature

before him is as good as a living bird. No, no! Life and
beauty are inseparable; the beauty of the bird has returned
with its feeble spirit to the Father who gave them both, and
who is the author of life and perfection, and the enemy of
death and decay.

Look at the stuffed specimen; at its dull eyes which see
not, sparkle not; its feet, which run not now among sweet
heather; its wings which quiver not now in the blue expanse
of heaven: and can you do otherwise than regret that the
happiness of that little being is over, and that what its Maker
has pronounced very good, should be marred and destroyed.
How wanton and irreparable an injury it is to rob the simple
creature of the life which is its all! In one moment to
change the active, restless, sturdy little frame into a senseless,
soulless heap; and then, having spoiled, to remould and
vainly attempt to restore to it a semblance of that life which
it once possessed in full perfection.

Of all orders of living beings, that of the birds is the
most beautiful; perhaps it may have been less affected by the
Fall, in that the curse fell on the earth and its inhabitants,
rather than on the air and its denizens.

Among quadrupeds there is often something to be found
which is offensive, and the imagination can picture them in
a higher state, when that which is uncomely shall be done
away; but among birds there is little that displeases—their
form, their colour, their order, their intelligence, are all
exquisite. Birds are never alluded to reproachfully in
Scripture; beasts get an occasional word of rebuke, but birds
never. Birds, beasts, and fishes, were faithful to their Master
when man deserted Him; turtle doves redeemed Him, a bird
woke Him on Easter morning;* fish came into the net
obedient to His call; a fish brought Him tribute by the lake;
ox and ass knew Him in the stable of Bethlehem,† wild
beasts were His companions in the desert; the ass and colt

---

* Eccl. xii. 4.   † Isa. i. 3.   I take these texts to refer to Christ on
Patristic authority.

bore Him to His Passion ; and, in return for their obedience,
when Christ refers to these simple creatures, it is with
tenderness and love.

The animals have been hardly dealt with by man, and I
cannot but believe that justice will be done. them hereafter,
and that as they suffered by man's fall, so his restoration will
be their restoration also : when also I hear of a covenant
being made, after the resurrection, with the beasts of the field
and with the fowls of heaven, and with the creeping things of
the ground ; when I hear, too, that in the renovated earth the
beasts will lie down in peace together, and that an angel will
summon to high feasting all the fowls of the air, and that
man will be but the first fruit of *all* creatures, I cannot but
consider that these words are to be taken in their plain and
literal signification. But to return to the ptarmigan.

There can be no doubt that the Icelandic ptarmigan is a
distinct species from that found in Scotland and Norway, and
Faber has accordingly named it Tetrao Islandorum. It is
an interesting bird of a hardy nature, frequenting the highest
and bleakest heithies, and only descending to lower ground
to lay its eggs and educate its young. The probable reason
is that there is more shelter from the rapacity of the hawk and
falcon among the willow and blaeberry twigs than on the exposed moorland ; besides, the tender buds and flowers are the
food with which the little ones are nourished. The ptarmigan
sits motionless till one almost treads on it, and, from its
umbreous brown plumage, it is scarcely distinguishable from
the soil. When alarmed it runs, but does not fly, and that
with great rapidity ; a horse cannot keep up with it over
the broken hassocky ground. In autumn the bird becomes
grey, and in winter perfectly white. This is an example of
the care of Providence for the brave bird, which is exposed
more than any other to the cold ; for, as the plumage becomes
a bad conductor of heat, there is less evaporation and waste
from the body, and the bird fattens upon a far smaller quantity of food than it requires in summer when in its rich brown
costume. In all probability, moreover, the ptarmigan feels

as warm whilst burrowing in the snow as it does when basking in the sun.*

I have passed through a whole flock of these birds, crouching down and hardly distinguishable from the ground, without their attempting to escape, for they are not shy birds, and seem to have little dread of man. The ptarmigan cannot bear to be long silent, and it discovers itself by its peculiar cry "Rjö-rjö-rjö-rjö!" whence it has obtained its Icelandic name of *rjupa*.

The ptarmigan either builds no nest at all or constructs one in the rudest manner of a few littered twigs, on which she lays sixteen or seventeen eggs; Icelanders assert positively that they meet with as many as twenty sometimes: these are of a reddish yellow colour spotted with rich brown, but less marked than those of the red grouse.

The male bird remains constantly on the watch beside the hen during incubation, only leaving her to bring her blaeberry flowers.

The Icelanders tell that when a falcon has slain a ptarmigan and torn her open, he utters bitter piercing cries, as he finds then that he has slain his sister, whom he did not recognize till he reached her heart.

Another bird which is found in great plenty on the heithies is the golden plover (*Charadrius pluvialis*), always in company with the whimbrel (*Numenius phæopus*). Almost any number of these birds may be shot by a sportsman on the moors, so that he is always certain there of finding his dinner. We lived for many days on nothing but these three kinds of birds: the ptarmigan far surpasses the other two in flavour, and there is very little eating on the plover for hungry men; the whimbrel is a tasteless bird as we cooked it, and there was always a scramble amongst us to bring up a bit of ptarmigan out of the pot in preference to the stringy whimbrel.

I can scarcely imagine a more drearily wild situation than an Icelandic heithi towards evening, the ground broken into huge humps of ash-grey moss; a little dwarf willow here

---

* Mudie's *British Birds.*

and there cowering on the ground, apparently clinging to it
as a stronghold against the wind, its white roots straggling
out of the soil, or completely laid bare by the fury of the
gales, looking much like blanched bones; the grey sky is
mirrored in little patches of stagnant water; a wan glare is in
the north; contorted blocks of lava, or splintered fragments
of trap cut sharply the cloud-covered horizon; and through
the solitude ring from all sides the weird mournful call of the
whimbrel and sad pipe of the plover—high pitched, clear, in-
expressibly melancholy, whilst, from a distance, out of the
sky, comes the dreary, ghost-like laugh of a wandering gull.

We reached the little farm of Sithumuli, where the people
were making hay—the women without their bodices, the men
in their drawers. These poor bodies considered the weather
pleasantly hot; I wore a fur coat and a pea-jacket over that,
as some protection against the piercing wind. In summer
Icelandic men present the most ludicrous appearance, as they
step out of their outer clothing and appear in very tight-fitting
jersey and drawers, both white once upon a time; as the men
are all thin, they look like skinned rabbits skipping about in
the hay. When I arrived late at a farm, and had to rouse the
people from sleep to take me in, a man invariably stalked out
to the door in this costume, so that I fancied at first that it
was the usual night gear. I was soon undeceived, however,
for on passing through the long sleeping apartment, the bathstófa, where the whole household roost, great and small, male
and female, old and young, I found that the locker beds along
the walls were filled with people in no night gear at all, tightly
packed together, lying two, three, and even four in a bed, the
head of one at the feet of the other. The ordinary custom
is for an individual to get into bed with his clothes on, and
undress beneath the feather bed. Towards midnight the
chamber becomes intolerably close and stifling, so that the
eider-down coverlets and feather beds are kicked aside; and
the appearance of the room with twenty or thirty sleeping
bodies in ranges along the walls is sufficiently overwhelming
to a bashful intruder.

As we rode into the tún of the farm, a man stepped forward and asked my guide's name. He told it. The man nodded, and then turned to me.

"Hvat heitith thèr?" (What is your name?)

"Sabine Jatvartharson" (Sabine Edward's son), I replied, for Icelanders have no surnames, and are called So-and-so's son, after their father. My guide was named Grimr, his father Arni; consequently he went by the name of Grimr Arnason, but his son would be Grímsson.

"Ok hvat heitir móthir thín?" (What is your mother's name?)

"Sophia," I replied. This was a name new to him, so he exclaimed "Svóinna! Ok fathir hennar?" (Indeed! and what was her father's name?) This he asked, because a wife in Iceland does not take her husband's name, but is called So-and-so's daughter. Thus if Grimr had a daughter, married or single, to the end of her days she would be Grímsdottir.

We had no time to waste, so Grímr called the farmer, and asked him whether it were possible to cross the Hvítá, or White river, a rapid and deep stream a quarter of a mile below the farm.

"No!" said the farmer. "You will want the ferry-boat. I will row you across for a dollar."

"That is not true," said one of the farmer's maids, who was making hay. "There is a ford, only it is not very safe."

The man now acknowledged that there was a ford, but said that it was dangerous, and he undertook to show it us. We crossed the river, and then scampered over a hill till we came out above the Valley of Smoke. Immediately below us there rose a dense white cloud, and we trotted towards it. This we found to arise from Tunguhver, the spring which Sir G. Mackenzie calls the alternating Geysir. From the hill-side starts up a mound about fourteen feet high, composed of brick-red clay. From the side of this, jet more than sixteen springs of water, all boiling furiously, fizzing, and puffing off steam, pouring down the side of the hillock in scalding rills, forming

22—2

pools of hot clear water, and then gurgling between banks
of green moss to meet the river.

I ascended the mound, and looked at the jetting spluttering fountains hard at work below me; the sun flung my
shadow on the ascending masses of steam with perfect dis-
tinctness, surrounding my head with a bright rainbow.

TUNGUHVER.

The spring which Sir G. Mackenzie describes as squirting
up about fourteen feet is now choked with stones, which
thoughtless travellers have rolled into its bore. It can now
only throw up water to the height of three feet, and all signs
of alternation have disappeared. Close to it I found a specimen of crimson Alga* growing in the almost scalding spray

* The Rev. M. J. Berkley, to whom I sent the specimen I collected at
Tunguhver, writes to me : " Had I received your Icelandic cryptogam without
any note as to its habitat, I should have said at once it is a barren Fusis-
porium. As, however, it is a production of hot-springs, it must be an Alga,
and is nearer to Kützing's genus Hypheothrix than anything else that I know.
I have not been able, however, to refer it to any species described in his
species Algarum. The most like it is H. Teukeri. It has some resemblance
to Leptothrix Kermesina, but if I mistake not, the threads are vaginate, and,
if so, it is not a Leptothrix."

from the jet, and overflowed by a boiling ripple at every explosion.

Having satisfied our curiosity, we remounted our horses, and ascended the valley, crossing the river repeatedly. Ten or twelve steam clouds rose from different sides of the dale. One we noticed in the middle of the river, where the boiling water had heaped up a mound of red, black, and purple deposit about ten feet high. On the top of this are three boiling jets, the largest of which plays to the height of three feet; the others boil briskly, but do not erupt. I rode my horse into the river, and tried to reach the mound, but he reared and snorted in such manifest alarm that I was obliged to conduct him to the bank, and wade through the river myself without shoes and stockings. I found now that the main cause of the horse's alarm had been a line of little hot springs, rising in the bed of the stream; these had undoubtedly scalded his feet as they sank into the mud and gravel through which the hot water rises.

On reaching the heap I was obliged to put on my shoes, as the stone was too hot and too covered with rills of boiling water for the naked foot to rest upon it. The volumes of steam which rose from the three orifices were blinding; however, I was able to mix a glass of hot brandy and water at the main jet, and then I jumped down into the river with my coat, cap, and hair, drenched in the condensed steam.

There are other interesting springs in the valley, which need not be described, as careful accounts of their phenomena have been given by Sir George Mackenzie, Henderson, and Captain Forbes.

At nine o'clock we reached the parsonage of Reykholt, near which are situated Snorro's bath and castle. The former consists of a circular pool enclosed within walls of hewn stone, about fifteen feet in diameter, and cemented with clay; the floor is paved with slabs, and a stone bench runs round the inside of the bath; water is conveyed to it through a stone-coated drain from Skrifla, a furiously boiling spring, about 150 yards distant, which is surrounded by a mud and stone wall.

The bath is now used only for washing clothes, the idea of corporeal ablutions being quite foreign to the Icelandic mind. The water can be raised to the height of four feet, above which there is an escape pipe, the whole depth of the bath being ten feet. Snorro, who constructed the bath, supplied it as well with a stream of cold water, so that the temperature could be regulated at pleasure. This conduit has fallen into decay, and in the 500 years which have elapsed since his death no Icelander has arisen of sufficient enterprise to clear out or re-dig the old channel, a labour which might occupy an English workman two or three days.

Skrifla is about fourteen feet in diameter, and discharges a considerable amount of water. A little north of this fountain is another boiling spring, into which I amused myself by plunging my shirts. The farm-servants undertook to wash my stockings and the rest of my clothes, and I was obliged to ride next day without a shirt, as it was not dry, but in an unwonted condition of jubilation at being free at last from my tormentors.

The castle of Snorro is simply a large tumulus covered with grass. It is situated in the tún, so that there is no prospect of its being dug into for many a day. The church is interesting, as it contains several relics of antiquity. Over the altar is a triptych, containing a crucifix with SS. Mary and John beside it. There are other saints on the doors. These figures are probably old, and are not wanting in spirit, but they have been reset and freshly coloured.

There is a magnificent brass basin serving as font, and a chandelier of the same metal, of excellent design, but late. A modern chasuble of violet cloth, embroidered with yellow and crimson flowers, is the work of the farmers' daughters in the neighbourhood. Outside the church door is a stone slab, on which are engraved Runes.

This the old man who unlocked the church door for me assured me was the tomb of Snorro Sturlason, the author of the *Heimskringla, or World's Circle,* a history of Norway, and the compiler of the Younger Edda, a composition giving an account of ancient Icelandic mythology. I believe, however, that the statement of the old gentleman is not to be relied upon.

Snorro was born in the year 1178 at Hvammr, where we slept last night. He was a fierce, turbulent chieftain, and is accused of having betrayed the independence of his country, and contributed to reduce Iceland to the state of a province of Norway. In 1241 he was murdered by his sons-in-law, at Reykholt, in the sixty-third year of his age. Both the *Edda* and *Heimskringla* will make the name of Snorro famous as long as the world lasts. His style is pure and nervous; he introduces episodes with singular discretion, and as a graphic historian will never be surpassed.

CHAPTER XX.

OK.

Ascend the Side of Ok—Flowers—Strange Sight—Skogkottr—Meet old Friends—The Skrimsl—Mermaids—Francesco de la Vega—Smoking in the Tent.

ON the following morning we started early for Thingvöllum, over the rough mountain pass of Ok. The horse which had been lamed in the bogs of the Middle frith was now unable to proceed, and I was obliged to leave him behind, much to my regret, as he had cost me 8l. 10s. However, one must make up one's mind to such losses in Iceland, and it was a matter of daily astonishment to Grimr that we had met with so few accidents.

We toiled for some hours up the steep flank of the Jökull, over soil spangled with golden Marsh Saxifrage (S. hirculus), the flowers one inch and an eighth in diameter, and with the beautiful white Tufted Saxifrage (S. cæspitosa), the large flowers of which, an inch and five-sixteenths in diameter, blossomed on stems four inches high.

On the farther side of Ok we rested at a little patch of turf and low willow, where grows in abundance the sickly flesh-coloured Water Avens (Geum rivale).

We were now approaching the road along which we had come when entering Kaldidalr. In front of us, behind the snowy head of Hlöthufell and Bláfell rose columns of red sand, high above the mountain tops, and forming dense lurid clouds, which sweeping over Skjaldbreith, tarnished its silver bosom.

R. Baring Gould, delt.

NEAR BRUNNEN

London: Published by Smith, Elder, & Co., 65, Cornhill   1865

Edmund Evans, sc.

We had heard at Melr and Hvammr confirmation of the report that Skapta was in eruption, and the appearance of these coppery pillars in the direction of that mountain led us to believe that the volcano was in full activity. We watched the red bursts of what seemed to be smoke rise up in curling eddies, and then dissolve into a cloud which completely obscured the mountains over which it was carried. We were mistaken, however, in supposing that these red pillars arose from the Skapta. They were in fact caused by the wind sweeping down the gullies of the Jökulls, and carrying up the sand which lay in the deserts between them.

Of course one can only form a rough approximation of the height of these pillars from where I stood; but estimating Hlöthufell at 2,500 feet above the sand district, and the Biskupstungna desert whence the columns rose at thirty-seven miles distant from me, I being twenty-six miles from Hlöthufell, I suppose the altitude of the sand column to have been 2,808 yards, as it appeared to rise at the least as high again as the Jökull.

We baited at Brunnum, where we had slept on our way north, and I had a delightful bathe in the ice-cold lake.

Towards evening we reached Skogkottr, branching off from the Thingvellir track below Armannsfell. A gleam from the setting sun kindled up the long cinder range of Tindarskaggi, the tooth-like peaks of Klukku-tindar, and the precipitous Hrafna bjarg, or Raven's Castle.

The bed of the valley, which is two or three miles wide, was bathed in the deepest purple gloom, out of which rose the tilted slabs of curdled lava.

To our great delight, we met at Skogkottr with Martin and the Yankee, who had come from Skoradalsvatn, where they had been spending a week in fishing. They were full of the appearance of a skrimsl, a half fabulous monster, which is said to inhabit some of the Icelandic lakes, but which has generally been considered the offspring of the imagination.

It so happened that my two friends had arrived at the lake only the day after the monster had been seen disporting itself

on the surface, and they had been able to obtain some curious
information with regard to it.   One morning, the farmer and
his household had observed something unusual in the lake,
and presently they were able to descry a large head like that
of a seal rising above the water, behind this appeared a
back or hump, and after an interval of water, a second hump.
The creature moved slowly, and seemed to be enjoying itself
in the sun.

The following account of the skrimsl has been sent me
by one of the party who visited Skoradalsvatn.   "The
skrimsl measures forty-six feet long; the head and neck are
six feet, the body twenty-two feet, and the tail eighteen feet
long, according to the estimate of the farmers on the shores
of the lake.   The monster was seen the day before we arrived
at Grund, by the farmer of the place.   His story and description of the fish were so very remarkable, that we instituted
inquiries, which resulted in our hunting out several individuals
who had seen the monster.   On one occasion it was observed
by three farmers who reside on the banks of the lake, two of
whom I met and questioned on the subject.   One of these
men produced a sketch of the creature, which he had made
whilst it was floating and playing on the surface of the water
for half an hour.

"I should have been inclined to set the whole story down
as a myth, were it not that the accounts of all witnesses
tallied with remarkable minuteness, and that the monster is
said to have been seen not in one portion of the lake only but
at different points.   The annexed sketch is taken from the
drawing alluded to."

THE SKRIMSL.

I beg to call the attention of the reader to the remarkable
coincidence between the description of the creature seen in

the Skoradalsvatn with that of one or more observed in the
Lagarfljót, which is on the farther side of Iceland, as given in
the following accounts.

In the *Icelandic Annals* there is mention of the appearance
of a similar monster during the summer of 1345, in these
words,—" There appeared a wonderful thing in the Lagarfljót,
which is believed to have been a living animal. At some times
it seemed like a great island, and at others, there appeared
humps several hundred fathoms apart, with water between
them. No one knows the dimensions of the creature, for none
saw its head or tail, consequently there is no certainty as to
what it was." There is a curious description of the animal
given in Jón Arnason's *Thjóth-Sögur*, Vol. I., taken from the
accounts of eye-witnesses. The skrimsl is there said to have
appeared in the same lake during the years 1749–1750. It
was seen by Peter the lawyer and two other men; they
described it as having been the size of a large vessel, and to
have been moving rapidly. These men, after having watched
it for some time, came at dusk to Arneithar-stede, where they
mentioned what they had seen. Whilst they were speaking the
monster rose to the surface, in front of the farm, and seemed
to be thirty or forty fathoms long with a large hump on its
back. One of the men present believed that he could distinguish a line or filament from behind the animal, as though
there were a tail submerged. All the farm people without
exception, at Arneithar-stede, saw the creature.

After this, the inhabitants of Hrafngerthi saw three humps
rise out of the water, and remain above it all day, with a
hundred fathoms of water between each. One morning Hans
Wíum was at Arneithar-stede with the priests Hjörleif, Magnús
and Grímr, when they observed rising out of the water jets
like those thrown up by a whale, when blowing off. After
that the monster seemed to have advanced to the lake head,
and was there observed by several persons. It again appeared
in 1819. Dr. Hjaltalin, at Reykjavik, told me that for many
years he had believed the story of the skrimsl to be a fable,
till he had been shown a mass of skeleton and monstrous

bones which had been washed up on the beach of the
Lagarfljót; these bones, he said, were very different from
those of the whale, and he was unable to identify them with
the bones of any marine animal frequenting the northern
seas.

The Grímsey fishermen are very positive that the skrimsl
frequents the shores of their lone isle, and that it comes
ashore and leaves traces on the turf where it has reposed.

It is said that a skrimsl haunts the Thorska-fjord, and is
often dangerous to vessels, as it swims at them, staves them
in and sinks them. It is frequently seen in that frith, rocking
on the surface of the water, like a large boat floating keel
uppermost.

Now that I am on the subject of monsters, let me give the
reader an account of a merman, which was found on the
north-west coast in the belly of a shark. Wernhard
Guthmund's son, priest of Ottrár dale, gives the following
description of it :—

" The lower part of the animal was entirely eaten away,
whilst the upper part, from the epigastric and hypogastric
region, was in some places partially eaten, in others completely
devoured. The sternum, or breast bone, was perfect. This
animal appeared to be about the size of a boy of eight or
nine years old, and its head was formed like that of a man.
The anterior surface of the occiput was very protuberant, and
the nape of the neck had a considerable indentation or sinking.
The alæ of the ears were very large, and extended a good
way back. It had front teeth, which were long and pointed, as
were also the larger teeth. The eyes were lustreless, and
resembled those of a cod fish. It had on its head long,
black, coarse hair, very similar to the Fucus filiformis;
this hair hung over the shoulders. Its forehead was large
and round. The skin above the eyelids was much
wrinkled, scanty, and of a bright olive colour, which was
indeed the tint of the whole body. The chin was cloven,
the shoulders were high, and the neck uncommonly short.
The arms were of their natural size, and each hand had a

thumb and fore fingers covered with flesh. Its breast was formed exactly like that of a man, and there were to be seen on it something like nipples. The back was also like that of a man; the ribs were very cartilaginous. In places where the skin had been rubbed off, a black coarse flesh was seen, very similar to that of the seal. The animal, after having been exposed about a week on the shore, was again thrown into the sea."

In the Korungs-skuggsjá, or King's mirror, an Icelandic or Norse work of the twelfth century, is the following description of a mermaid :—

"A monster is seen also near Greenland, which people call the Margygr. This creature appears like a woman as far down as the waist, with breast and bosom like a woman, long hands, and soft hair; the neck and head in all respects like those of a human being. The hands seem to people to be long, and the fingers not to be parted, but united by a web like that on the feet of water birds. From the waist downwards, this monster resembles a fish, with scales, tail and fins. This prodigy, along with that already mentioned (the Haf-stramba), is believed to show itself especially before heavy storms. The habit of this creature is to dive frequently and rise again to the surface with fishes in its hands. When sailors see it playing with the fish, or throwing them towards the ship, they fear that they are doomed to lose several of the crew; but when it eats the fish, or turning from the vessel flings them away from her, then the sailors take it as a good omen that they will not suffer loss in the impending storm. This monster has a very horrible face, with broad brow and piercing eyes, a wide mouth and double chin."

The Landnama speaks of a *marmennill*, or merman, having been caught off the island of Grimsey. The Icelandic annals relate that a *margygr* was seen off the East Friths in 1305, and 1329.

With regard to the appearance of the merfolk in other countries, I may state that one was fished up on the coast of Suffolk in 1187, and was kept by the governor for six months.

It closely resembled a man, but was not gifted with speech. One day when it had an opportunity for escape, it fled to the sea, plunged in, and was never seen again. In 1430, after a violent tempest which broke down the dikes in Holland and flooded the low land, some girls of the town of Edam in West Fruzeland, going in a boat to milk their cows, observed a mermaid in shallow water and embarrassed in the mud. They took it into their boat, and brought it into Edam, dressed it in female attire, and taught it to spin. It fed with them, but never could be taught to speak. It was afterwards brought to Haerlem, where it lived for several years, though still showing a strong inclination for water. Parival relates that it was instructed in its duty to God, and that it made reverences before a crucifix (*Délices de Hollande*). In 1560, near the island Manar on the west of Ceylon, some fishermen entrapped in their net seven mermen and mermaids, of which several Jesuits and Father Henriques, and Bosquez, physician to the Viceroy of Goa, were witnesses. The physician examined them with a great deal of care, and dissected them. He asserts that the internal and external structure resembled that of human beings. We have another account of a merman seen near the great rock called Diamon, on the coast of Martinique. The persons who saw it gave a precise description of it before a notary; they affirmed that they saw it wipe its hands over its face, and even heard it blow its nose. Another creature of the same species was captured in the Baltic in 1531, and sent as a present to Sigismond, King of Poland, with whom it lived three days, and was seen by all the Court. Another was taken near Rocca de Sintra, as related by Damian Goes. The King of Portugal and the Grandmaster of the order of St. James, are said to have had a suit at law to determine which party the creature belonged to.

At Lierjanes, near Santander, was born in 1657 Francesco de la Vega, of poor parents, This lad had always a strong predilection for water, and this so greatly irritated his widowed mother that she bade him one day, in wrath, go to sea entirely. Francesco remained quietly at home till he was fifteen, and was then apprenticed to a carpenter at Bilbao. Here he idled his time away for two years, till one day in 1674 he went to bathe with some comrades, but disappeared. Five years after, in 1679, some fishermen at Cadiz were surprised by noticing a figure on the surface of the water, now rocking on the waves, then plunging. Thinking it to be a merman, the fishermen rowed towards it, but he dived and disappeared. Next day several other boats accompanied them to the spot, and the men had the good fortune to see it again. It again eluded their attempt to take it. On the third day the fishermen attracted it by casting pieces of bread into the water; it swam up to these and eat them, gradually nearing the boats, till it was caught in a net. The boatmen conveyed their prize to the Franciscan monastery in Cadiz. The creature was found to be a man, with a few scales on his spine; his nails were gone, his flesh was colourless and flabby, his hair was short and reddish. For two days he remained in the monastery without uttering a word, but one day he distinctly pronounced the name Lierjanes. As this was the only sound he enunciated, it was presumed that it was the name of his native village; and a friar undertook to conduct him home.

Inquiries were set on foot, and information of the disappearance of Francesco, five years previously, was obtained at the monastery, and the merman was believed to be Francesco de la Vega.

The friar conducted him to the top of the hill above Lierjanes, and then bade his strange comrade lead him the rest of the way. Francesco, without hesitation, walked down to his mother's cottage; his brothers and mother recognized him at once, but he showed no signs of affection or recollection, staring at them with chill, fishy eyes, and receiving their embraces with cold indifference.

Henceforth, he resided at home. His habits were still peculiar; he never spoke, or seemed to have any intellect, though a dull instinct remained. He disliked clothes, and would cast them off if put upon him. He would never tolerate shoes. He eat anything that was put before him without showing any preference for one food over another; and, if his dinner were forgotten, he never asked for any. He seemed to understand simple sentences, but never replied. He was employed in conveying letters as a post-boy, and was most punctual in the discharge of his duties. On one occasion, when the ferry boat was absent, he swam across a large sheet of water between Lierjanes and Santander. Thus he spent nine years at home, and then disappeared.

Some fishermen declared afterwards that they had seen a figure like him playing in the bay of Asturias; but he was never again captured.

The facts of this singular narrative have been collected and critically examined by the great German writer, Ludwig Tieck, in Der Wassermensch, 1835, and it is almost impossible to escape from the conclusion that they are authentic. I have given them here to show the reader that it is quite possible that there may be a foundation of truth to the world-wide fable of the existence of merfolk.

I cannot conclude this digression better than by translating the verses sung by a marmennill, when he was carried back to his favourite element after a brief sojourn on dry land. They are given in the Saga of Half and his knights:

> " Cold water to the eyes !
> Flesh raw to the teeth !
> A shroud to the dead !
> Flit me back to the sea !
> Henceforward never,
> Men in ships sailing !
> Draw me to dry land
> From the deep of the sea ! "

From Skogkottr we could not see the mountains on the farther side of Thingvalla lake, so dense was the cloud of dust and sand which filled the air. The head of Armannsfell was visible only through a film, and Hengill was blotted completely out of the landscape.

We slept the night in our tent, and intended riding to the

Geysir on the following day. Mr. Briggs, the Yankee, and Martin, were accustomed to doze off with their pipes in their mouths; but, to prevent accidents, these were attached to the main pole of the tent by pack thread, so that as the smokers dropped off, the pipe slid from between their lips and hung in the middle of the tent. I heard each pipe click against the pole before I fell asleep. Next morning we were awakened by Gúthmundr stepping over us with a tray of hot coffee and sugar-candy.

CHAPTER XXI.

GEYSIR.

Flowers—A Natural Chimney—Extensive Plain—Laugarvatn—An eccentric Bridge—Uthlith—Sleeping in a Church—Position of the Geysir District —Description of the Springs—The Little Geysir—Jack in the Box— Boiling Wells — Strokr—Blue Ponds—Experiments—Great Geysir— Keeping Watch—Magnificent Explosion—Mr. Briggs misses seeing it— Theory of the Geysir.

It was not till afternoon that we left Skogkottr for the Geysir, as some of the horses had escaped, and as the luggage had to be sorted, so that what was superfluous might be sent to Reykjavik under the charge of a guide.

The road lies through very picturesque country, better wooded and more grass-grown than any I had seen in the island hitherto.

The brightness of the day, the warmth of the sun, the fragrance of the birch, and the brilliancy of the flowers, made our ride most enjoyable. A change had come over the scene since our last visit to Thingvöllum; the mountains had lost much of their snow, the soil was drier, and the coppice was now radiant with the purple Wood Geranium, whose flowers are of every possible hue, from carmine to plum. My little friend the moss Campion was out of flower, and the Dryas had shed her eight cream white petals, and was busy maturing seed in the warm days of July. The Dandelion thrust its showy head out of grass nooks, the Saxifraga hirculus, like a golden star, sprinkled the shaley slopes; Thrift shook its pink tufts among the lava crevices; the sward was dappled with the rosy tassels of the Alpine Catchfly, and the golden cups of the

KÁLFS-TINDAR.

Ranunculus glacialis. Under covert of the bushes, crept a little Briar, (*Rubus saxatilis*), with bunches of white blossoms peeping saucily from under the glossy birch leaves.

We scrambled across the Hrafnagjá, a rift much like the Allmannagjá, only on a somewhat smaller scale, and passed some caves resembling Surts-hellir in conformation; but probably not so extensive. We had neither time nor inclination to explore their recesses.

A low rise of sand and cinder on the right, crowned by a slag chimney, is worth a visit, as this chimney appears to be the vent from which has been ejected the ash and dust which cover the mound. A trembling Alpine Rock-cress (*Arabis petræa*) was nestled within the lip, vibrating with every blast which rolled over the black gulf. The funnel-shaped throat is about five feet across, and stands up about fifteen feet above the sand; it is vitrified, and streaked red, yellow, and black. If a stone be dropped into the abyss, it is heard to strike before very long, and, judging from the time it takes in falling, I suppose the hole to be about seventy feet deep.

After leaving this ash vent, we rode over cinder till a rapid descent brought us to a sandy plain, beneath the gloomy but picturesque range represented in the opposite plate.

The most distant cone is reddened by volcanic fires, though not to the extent to which the charring of Hlitharfjall and other mountains around Myvatn has been carried.

On surmounting the next rise, we came in sight of a great marshy plain, extending for fifty-five miles to the sea. Two lakes lay before us, the Apavatn, and the Laugarvatn famous for being full of hot-water jets; beyond these, stretched vast morasses, out of which rose puffs of steam, which rolled away before the breeze. Heckla stood up majestically beyond, covered with snow, and flushed with the evening sun; and far away in the horizon, soft as summer clouds, and tinted like the tenderest blush-rose petals, rose the peaks of Tindfjalla and the back of Eyjafjalla, their bases lost in the bloom of evening.

The boggy tract in front of us was blue-veined with some

of Iceland's goodliest rivers, the White river, the Salmon
river, the Tongue-flood, the Bridge-stream, and the Bull
river.*

We changed horses at Laugarvatn. The lake is certainly
curious. In the green morass east of the lake rose a dense
column of dazzling white steam, and we could see by the
leaping and tossing of the vaporous whorls that the water
below was boiling savagely. Beyond the lake close to the
margin, is a jet which intermits, sending up puff, puff, puff,
like a steam engine. Vapour rises from the sheet of the lake
itself, showing that hot springs are bubbling up in its bed,
and near the farm are fountains, some throwing up water
about two feet, whilst others smoke quietly, or gurgle through
holes in the beach. The soil around these springs consists
of grey, blue, and red clay containing sulphur and gypsum.
These jets empty themselves by a little brook into the lake.
The smell of rotten eggs which issues from the water shows
that it contains sulphuretted hydrogen in solution. On
attempting to bathe in Laugarvatn, one soon finds out that
there is a stratum of hot water spread over the surface of the
lake, whilst the body of water below is intensely cold.

We skirted the lake, and our course then lay over smooth
ground, passing near farms which studded the slopes of this
smiling country. Certainly a tourist who runs to the Geysirs
and back to Reykjavik gets no true idea of Icelandic scenery.
I saw nothing so bright, fertile, and grass-grown in any other
portion of Iceland. Yet poor is the best, and inferior to an
Irish bog.

Presently we came upon the Brúará, which is crossed in
a somewhat eccentric fashion. In the midst of the river is a
fissure, into which a considerable portion of the water roars;
the rest flows down a series of shelves, till it tumbles over
a ledge into still water, and unites with the foaming stream
bursting from the rift. Over this chasm a slender wooden
bridge has been thrown, and the horses wade to it,

---

* Hvítá, Laxá, Tungufljót, Brúará and Thjórsá.

scramble across, and wade again to the opposite bank. The width of the river is about eighty yards, that of the chasm seven or eight at the point where spanned by the bridge. This bridge consists of planks laid from one wall of the crack to the other, with a hand-rail on either side, to prevent one from slipping off—an accident which might easily happen, as the floor is wet with spray.

It is a curious sight, looking up the chasm from the middle of the bridge. Where the rent terminates, the water rolls in with considerable violence, whilst minor cascades gush down on either side of the fissure, at right angles to the stream; and the torrent tumbles foaming angrily and hissing through the narrow gap below one's feet.

We spent the night at Uthlith, in the church, and supped off coffee and Icelandic moss stewed in milk, which the farmer's wife laid for us on the communion-table. We ate it, lying on our beds placed within the rails, by the light of the altar candles, which the good woman had kindled to do us honour.

The farmer and his wife were the cleanest people we had met for some time; their pleasant, cheerful faces and neat dresses were quite refreshing. The lady wore handsome gold earrings in the shape of crescents, and I have no doubt that the farmer was a man of means, for the land around was productive, his tún extensive, his flocks of sheep and droves of cattle numerous.

The church of Uthlith is quite new; over the altar is a gilt frame, containing a black board like that usually seen in schools. On this are chalked the hymns which are to be sung during divine service; but this is intended to give place to a painting when the parish is rich enough to buy one. The candlesticks are poor, and the vestments modern; but the silver chalice is peculiarly graceful and of considerable antiquity.

On the following morning we took leave of our kind host and hostess, promising to look in upon them on our return from the Geysir, and reached our destination about noon.

A thin vapour, visible from a mile off, marked the site of the springs so celebrated as one of the world's wonders.

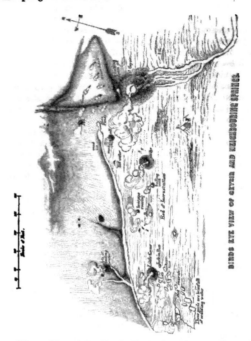

The position of the Geysir district is rather peculiar. A range of hills on the north rise out of the great plain; south of them, a node of trachyte has been forced like an island out of the morasses to the height of 600 feet; and it is on the slope of this that the Geysir jets lie.

The springs form a rude ellipse, the axis major of which is 440 yards and the axis minor 140 yards. The direction of the axis major is N.E. by S.W. ·

The springs first reached are situated in a marsh of hot mud and moss, and consist of steaming pools, with here and there a fountain boiling and hissing. North of these lies the Little Geysir, dancing and overflowing, tossing up jets to the height of three feet, with occasional bursts of steam. The fountain is in the centre of a small heap of deposit; its mouth is about two feet in diameter, and the depth of the bore is twelve feet, below which it takes a bend towards the S.W.

Curiously enough, we saw no explosion of the Little Geysir; it boiled briskly during the three days that we spent in this district of hot springs, but did not erupt, as we had been led to expect from the account of former travellers.

Close to the Little Geysir is a puddle of black mud, presenting the most ludicrous appearance. It remained tranquil for about half a minute, and then a bell rose like a thumb, to the height of four inches and sank again, without bursting, scattering, throwing out steam, or making the slightest sound. I named it "Jack in the box." But Jack was not always so demure: on my choking the throat of the Little Geysir with turf, I found that the slime puddle was converted into a jet of steam and inky water, which played with vivacity, till the Little Geysir had relieved itself of its dose.

South of these are some limpid pools, so hot that the hand cannot be borne in them for an instant, and sufficiently shallow at the edges to make them admirable places for washing clothes.

To the north, on the farther side of a scalding brook, is a noisy fountain, which may have played at one time to a considerable height when it was not choked with stones. Now the water can only escape in hot squirts, which fizz and growl among the encumbering fragments, without the power of dislodging them.

I must take this opportunity of protesting most earnestly

against the mischief which travellers do in choking the throats of boiling fountains with stones. As I have already said, the alternating Geysir at Tunguhver is spoiled by the wanton folly of visitors ; and I fear that damage will be done to Strokr by tricks of this kind being played on it. Turf can do no harm, but stones are very likely to block the throat or injure the mechanism. The following quotation is from a newly published voyage to Iceland by Carl Vogt. It arouses my indignation most thoroughly.

" In the hopes of being able to excite a second eruption, moreover with the view of forcing the Geysir (*i. e.* that at Reykir) to exert its utmost violence, we dragged great stones up and *filled the bore of the fountain* completely to the brim. However, our expectations of thus arousing the spring were disappointed, for no second outburst took place before we rode away on the following morning."

Proceeding north-east from Little Geysir, we came upon several splits and holes in the ground or floor of incrustation which extends to the Great Geysir.

Down them the water can be seen and heard, lashing and sobbing, whilst the steam blows off from the orifices. The largest of these is a tunnel with an oblique pipe, so that the boiling water cannot be seen, though it can be heard roaring angrily in its den. Though we baited it with a great amount of turf, we could not induce it to break loose. In the crevices of the pipe overhanging the water was a bunch of dog violet in blossom, seeming to enjoy its warm berth.

To the south-west again, close to Strokr, is a well of boiling water, which can be seen, chafing the sides, six feet below the surface of the floor on which we stood.

We came next to Strokr or the Churn, whose crater is like a saucer turned the right way up, whilst the Great Geysir mound is like a saucer inverted. In the middle of this bowl is a well, down which, at the depth of sixteen feet, is the water in violent ebullition. Strokr erupts of itself, but it can also be made to jet at the visitor's pleasure, by choking the gullet with sods of turf.

More of Strokr presently!

The next spring in a direct line north-east is the Great Geysir, but to the north-west are several others: first, a small steam escape; then a mud spring; and hard by this a grimly abyss, with the water out of sight, down 20 feet at least, muttering, growling, and sending up gusts of steam.

Above this are two lovely ponds of still blue water, connected by a stream which flows over a partition of silicious deposit, leaving a space for the visitor's kettle. The beautiful forms of petrifaction can be seen far down in the water, tinted green and blue, while the edge on which one stands, shelves over the water and appears but a finger-breadth in thickness. The depth of these pools varies from 17 to 24 feet, and the length of the two conjointly is 90 feet. They seem to be supplied with water from the south-west, and are always brimful.

They are at a considerable elevation above the Great Geysir and Strokr. Indeed, one of the most curious phenomena connected with the Geysir district, is the difference of level in the various pools and springs. The range is as great as 50 feet.

We raised the level of the well just south of the blue ponds 20 feet, by turning the stream from these ponds into it, and completely altered its character, converting it from a well of furiously boiling water to a pool steaming tranquilly.

Not satisfied with this experiment, we tried another, and dug into a small puddle of hot mud; it was at once converted into a bubbling pool of five jets. After this, a mud pond which had been in active ebullition, 180 feet north of it, on the farther side of the blue ponds, gave up working, and left a dry black bed, riddled with holes out of which the water had previously risen.

I mention this to show how variable the springs are, and to urge the importance of every traveller taking precise notes of their position and character.

North of the blue ponds is a well similar to that already described, the water about 16 feet down the bore briskly boiling.

S Rapi(ground, de l   Edmund Evans, sc

THE GREAT GEYSIR.

London Published by Smith Elder, & Co. 64 Cornhill 1862.

The Great Geysir is farthest east of all the springs; it is indicated by a mound of sintery deposit like a heap of dry grey leaves, piled up about 30 feet above the soil, not quite regular in shape, as on the south, where the water flows away, the deposit spreads farther down the hill, and the rise is not so steep as on the north, where a miniature ravine causes the mound to be contracted and precipitous.

At the summit is the basin, generally full to overflowing; it measures 56 feet by 46 feet, and is four feet deep, shelving gently to the bore, whose position is distinctly visible, even when the bowl is full, by the slight agitation of the surface immediately over it, and the darkness of the water.   The deposit from Geysir and the other springs is formed with great rapidity.   At every overflow it is stated to leave a layer of silex of the thickness of silver paper, and this is confirmed by the fact that I found near the stream flowing from Geysir a French newspaper completely coated, so that the printing was quite legible through a film of silex.   I found also a bunch of moss Campion, every leaf and petal encrusted in a similar manner.   Under a little fall which the Geysir water makes after it has become cool, and the deposit is less rapid, was hung a plover's wing; it was covered with a silicious film. The deposition of silex must be rapid indeed, if a newspaper, a tender flower, and a bird's feather, are overlaid before the hot water has had time to destroy them.

The finest incrustations are not found in the rill from Geysir, but in the stream from the blue ponds, where leaves, flowers, and mosses are to be found in great abundance.

It is interesting to notice the difference between deposits left by the same waters.

The grey laminated sinter, of which the Geysir mound is built up, is different from the glossy white, orange, and scarlet deposit with which the stream bed from the blue ponds, the gullet of Strokr, and the fall of the Geysir rill, are lined.

In the tiny ravine north of Geysir is a whistling steam jet and a few spluttering orifices.   At right angles to this ravine is another gully leading up the hill, and serving in spring as a

S. Baring Gould, delt.   Edmund Evans, sc.

THE GREAT GEYSIR.

London: Published by Smith, Elder, & Co., 65, Cornhill. 1863.

watercourse : the bottom of this is also full of little blubbering springs; and the red mound in the angle between
these channels is pierced with openings from which steam
rises.

On ascending the hill above the springs, which is composed
of red bolus, very similar to that of Námarfjall, I found sintery heaps covered with red clay. By means of a spade the
wash of earth was removed, and one of the mounds I found
to be nearly as large as that of the Great Geysir. I was able
to make out the central bore, but found it blocked with disintegrated sinter.

We pitched our tent between Geysir and Strokr, and cooked
our supper on the bridge between the blue pools. In the
depths of one we could descry a kettle, which some former
traveller had dropped.

It was agreed that we should take turns to watch through
the night, so as not to miss an eruption of the Geysir. These
eruptions take place now about every third day, and not once
in twenty-four hours, as was the wont of Geysir a few years
back. He who kept guard was to be employed in digging a
trench between the blue ponds and Strokr, which we purposed
filling by turning the stream into it.

Strokr went off in the night without being baited, but the
Great Geysir remained tranquil.

Next day, Saturday, we spent in filling the well from the
pools, as already mentioned. We got on slowly with the
trench to Strokr, as the ground is very hard. Ultimately we
abandoned our conduit; the future traveller will probably find
some traces of it, and I hope that he will continue it, and try
the effect of a dose of lukewarm water on the irascible Strokr.

During the day the Great Geysir gave false alarms. Before
each explosion, a sound like a gong beating in the bowels of the
earth gives warning to be on the alert.

We heard this signal thrice, and each time it was followed
by a lifting of the water over the bore to heights from two to
eight feet, after which it subsided.

Martin shot several plovers in the marshes below Laugar-

fjall, and we cooked them at a fire which we made between
Strokr and the well.   In the evening we drew lots to decide
the order in which we were to keep guard.   My watch came
first, and it was to be followed by those of Mr. Briggs, Martin,
and the Yankee.

Mine was over at two o'clock in the morning, and I had
gone comfortably to sleep on my air cushion and fox-skins,
when a violent concussion of the ground brought me and my
companions to our feet.

We rushed out of the tent in every condition of déshabille,
and were in time to see Geysir put forth his full strength.

Five strokes underground were the signal, then an overflow, wetting every side of the mound.   Presently a dome of
water rose in the centre of the basin and fell again, immediately to be followed by a fresh bell, which sprang into the
air full 40 feet high, accompanied by a roaring burst of steam.
Instantly, the fountain began to play with the utmost violence,
a column rushed up to the height of 90 or 100 feet against
the grey night sky with mighty volumes of white steam-cloud
rolling about it, and swept off by the breeze to fall in torrents
of hot rain.   Jets, and lines of water tore their way through
the cloud, or leaped high above its domed mass.   The earth
trembled and throbbed during the explosion ; then the column
sank, started up again, dropped once more, and seemed to be
sucked back into the earth.

We ran to the basin, which was left dry, and looked down
the bore at the water, which was bubbling at the depth of
six feet.

The diameter we found to be 9 feet 6 inches.   A plummet
sank about 76 feet.

We observed that Strokr and the well threw up clouds of
steam during the eruption of the Geysir, and the little
blubbering jets on the sandbank were also more active.

But a very few minutes after the explosion, Martin called
out, " Where is Mr. Briggs ? "

We looked about us, but our fat friend was not to be seen.
We shouted, but there was no answer.

"Very odd!" said Martin, "this is his watch. He ought
to have roused us, and he has vanished; I hope he has not fallen
down Strokr!"

We ran in some alarm to the tent, he was not in it, but an
exclamation from Martin brought us round to the side opposite
that which faced Geysir.

Here we discovered that "Tun of a man" fast asleep, the
spade by his side, and his head reposing on a heap of turf.

We aroused him with "Wake up, old fellow! you are a
nice sentinel, you are, sleeping during your watch!"

Mr. Briggs opened his eyes, rubbed them, and looked
sleepily round.

"Do you know what you have missed?" I asked.

"Ay, what?" with a drawl.

"Why! Geysir has erupted during your comfortable nap.
So you have lost that sight."

"Ay!" Mr. Briggs started to his feet; "Geysir gone
off? Then where are all my flannels?"

"Your *what?*"

"Why, I tied a couple of shirts, a pair of trousers and
some stockings to my fishing line and let them down the
mouth of the Geysir at the beginning of my watch. Did any
of you see them erupt with the water and the steam?"

We none of us had, and our search for the missing articles
on the following morning was without result, so that these
portions of my friend's clothing must have been torn to
minute shreds and carried off by the wind, or sucked into the
bowels of the earth, where they may still be stewing for aught
I know to the contrary.

On Sunday morning we baited Strokr twice. The bore is
8 feet in diameter at the top, and 44 feet deep. Below 27 feet
it contracts to 19 inches, so that the turf thrown in completely
chokes it. Steam then generates; a foaming scum covers the
surface of the water, and in a quarter of an hour, it surges up
the pipe, allowing one ample time for escape to the edge of the
saucer. The fountain then begins playing, sending its bundles
of jets rather higher than those of the Great Geysir, flinging

up the clods of turf, which have been its obstruction, like a number of rockets. This magnificent display continues for a quarter of an hour or twenty minutes. The erupted water flows back into the pipe from the curved sides of the bowl. This occasions a succession of bursts, the last expiring effort, very generally, being the most magnificent.

Strokr gives no warning thumps like the Great Geysir, and there is not the same roaring of steam accompanying the outbreak of the water.

Bunsen, who visited this part of Iceland in 1845, ascertained by experiment that water, long boiled, becomes almost entirely free of air; by which the cohesion of the aqueous particles is so much increased that, when it is exposed to a heat sufficient to overcome the force of cohesion, the production of steam is so instantaneous and so considerable as to cause an explosion. To this cause he accordingly attributed the eruption of Geysir and Strokr, which, being in constant ebullition for many hours, are then so freed from air, that the intense heat at the bottom at last overcomes the cohesion of the particles, and an explosion ensues.

The only objection which can be raised to this theory is that the characteristics of Geysir and Strokr are very different. The latter is in constant ebullition, not so Geysir. Now, if air were liberated in great quantities, the surface would be continually boiling; as a matter of fact, the surface, except during an explosion, is perfectly at rest, and scarcely a bubble rises through it.

Now, let an iron tube be bent to the angle of 110, keeping one arm half the length of the other. Let the pipe be filled with water, and the short arm be placed in the fire. The surface of the liquid will remain still and calm for a minute; then the pipe begins to quiver, a slight overflow takes place without any signs of ebullition over the lip of the bore, and suddenly, with a throb, the whole column of water is forced high into the air.

With a tube the long arm of which is 2 ft., and the bore $\frac{3}{8}$ inch, I can send a jet to the height of 18 feet.

The rationale of this is simple enough. Steam is generated in the short arm of the pipe, and presses down the water, causing an overflow till the steam bubble turns the angle, when it forces out the column in the long arm with incredible violence.

This experiment is merely an adaptation of Sir George Mackenzie's theory, that the Great Geysir is supplied from an underground reservoir. I suggest that a bend in the pipe is sufficient to produce all the phenomena of the Great Geysir.

As we lay in our tent, towards evening, when the weather was cold, Mr. Briggs turned to me, and asked me to tell one farewell Saga before our return to Reykjavik.

" And let it be a good one," said he.

After thinking for a few moments, I began the following story.

CHAPTER XXII.

Thorgils' Nursling.*

A SAGA ABOUT GREENLAND.

(CIRC. A.D. 986—99.)

To the south of this well-watered plain, over which our eyes reach to the horizon in full view of the white peaks of Tind-fjalla, is a farm named Tratharholt, crowning a toft which rises out of green meads and almost impassable swamps.

This hillock was occupied by farm buildings in 986, when my story begins.

It then belonged to a bonder, Thorgils, who lived on it, surrounded by his family and friends.

The farmer was a man of strong convictions. Having satisfied his own mind that the religion of the Œsir was false, and that Christianity opened a more glorious future to man, besides being more accordant with his own internal convictions of morality, Thorgils received baptism, and remained ever after a staunch adherent to the true faith. He did not exactly become a model Christian, but he acted consistently with the little light given him, and that is saying a great deal for any man.

One idea he certainly had grasped, that he owed a chival-rous obedience to Christ as his Monarch; but of the laws which govern the Kingdom of Grace, and the obligations which he had incurred on his admission into it, he was profoundly ignorant.

Thorgils' own disposition prevented him from acquiring the requisite knowledge; for he was passionately fond of

---

* Floamanna Saga, chap. 20—24, 28, 29, in Fornsögur, Leipz. 1860.

change ; and the love of visiting new scenes was so strongly
implanted in his soul, that not even the fertility of the meadows around Tratharholt could attach him to a quiet life, and
restrain him from indulging in his natural roving propensities.
However, he had spent several years at his fair farm, and
kept himself from vegetating by squabbles with his neighbours,
and by litigation with his brother chiefs.

At last, on an inauspicious day, came a message from
Eirek the Red, the discoverer of Greenland, begging Thorgils
to pay him a visit, and assist him in colonizing the new
country.

This was a proposal after the bonder's own heart : he
fired up with enthusiasm at once ; and if Greenland had been
Paradise, he could not have been more impatient of the delay
which the rigour of winter imposed on him of postponing his
voyage till the summer.

"What think you of this expedition to Greenland ?"
asked Thorgils one night of his good woman ; "people *do* say
that Eirek is becoming enormously wealthy."

"You must please yourself about going," answered Thorey,
his wife ; "but the voyage is long and boisterous enough to
make me shrink at the notion of it. I have become much
attached to Tratharholt !"

"My dear," quoth the bonder, "I can think and dream of
nothing but Greenland, with its glorious meadows and rich
mountain pastures ; so that I shall have no rest till I have
seen it with my own eyes. But there is no necessity for you
to accompany me at present. If you prefer it, you can remain at home and keep house till I return and report the
state of the country."

"No ! sweet friend !" replied the housewife ; "wherever
you go I shall follow, only I have no great forebodings of
success."

The husband added, "I have been talking matters over
with your foster-father, Jósteinn, and he is already half inclined to join me in my venture, and bring along with him
that termagant of a wife of his, and his fine strapping boys,

24

Kolr and Skarkathr. Our son, Thorleif, shall also accompany us, and he is a man of experience after his Norway voyage. So, you see, we shall be quite a family party!"

When summer was well set in, all was ready for the voyage. Thorgils had purchased a vessel conjointly with Jósteinn, and had laden it with all things necessary for farming in Greenland. Both of the bonders brought a select party of thralls to manage the vessel and tend the cattle in the new settlement. Unfortunately, Thorgils' little daughter sickened at the last moment, and had to be left behind with a friendly farmer, as it was impossible to delay sailing till her recovery.

On the eve of departure, Thorgils had a dream. A mighty red-bearded figure stood before him, brandishing a heavy mallet.

"I am Thor!" spoke the apparition; "you have renounced me for the white Christ. I shall be your foe, unless you return to my worship. Storm and tempest, or soft breezes, are at your choice. I can fan you swiftly to your new home over a blue scarce-ruffled deep, or sink you like lead with one blow of my hammer, in the green, boiling ocean."

"Be off!" exclaimed Thorgils; "I care little for your threats. I commit my course to Him whom the winds and the sea obey."

"Follow me!" said Thor, and his voice was loud and wrathful like the mutter of thunder among the icefields of Eyjafjalla.

In his dream the bonder fancied that he was led by the red-bearded one to the summit of a bluff overhanging the sea.

A brown ragged rack was creeping up the heavens; puffs of wind made cat's-paws on the palpitating bosom of the deep, and then sighed through the stunted grass of the headland.

The god raised his hand, and with a howl the gale descended. Billows heaped themselves up, and thundered against the crags, shivering into white eddies of foam and drifts of brine.

As far as the eye could stretch was a wild dancing ocean, working like yeast, the wave tops cut off by the squall and sent flying in scuds of bitter spray. Gulls wheeled and plunged amongst the foam, Mother Cary's chickens fleeted through the hissing tumbling surf.

For one moment the sun shone forth above the northern horizon, sending a streak of flame over the god's stern countenance, and then it went down beyond the seething deep, leaving wind and sea to their wild strife, in the gloom of night.

"Such storms await you!" said Thor; "be advised in time, and sacrifice to me."

"Never, never!" cried the dreamer; "away with you, foul tempter! He who ransomed earth by His precious blood, shall guard and keep me ever!"

Then the vision faded from his eyes, and Thorgils awoke. His wife was softly breathing in sleep at his side. He roused her, and related to her his strange dream.

"This bodes ill!" she said; "were I to see such a vision, I should hardly venture on the deep."

"It is too late to change plans," quoth Thorgils; "now we must make the best of a bad business, and keep the dream from Jósteinn and the rest."

The sun came out right gloriously on the morning of departure. A breeze sprang up and filled the white sail; the gallant vessel stood out to sea; the snowy heads of the Jökulls lessened in the wake: all promised fair for a prosperous voyage.

Alas! this fair beginning was speedily cut short by storms succeeding each other in rapid succession, so that the ship was beaten about day after day, and week after week, with scarcely any intermission.

The men were worn out with the continued toil of bailing, and the exhaustion of a sickness which had carried off some of the thralls, and left others in a state of great debility. Food ran scarce, and most of the cattle on board perished. The water in the vats was expended, and thirst was quenched by rain and snow.

After the voyage had lasted three months, Thorgils became aware that Jósteinn's men were meditating a sacrifice, to propitiate the offended god of thunder and tempest.

He at once grasped an axe, and walking aft, said to the murmurers, "I understand that you propose offering to Thor! Now mark my words. I shall split the head of the first man who dares do such a thing, and fling him overboard to the fishes."

The bonder spoke with determination; and as he was known to be a man of his word, no one ventured to disobey his wishes.

That night Thor appeared to him again. "Have not my threats proved true?" he asked. "The equinox is at hand, and then these gales will redouble their fury. Return to your old faith, and in seven days you shall be in port."

"Though by rejecting you, I should ensure my never seeing land again, I would renounce you utterly and for ever!"

The god looked at Thorgils with a grave expression; then he said, deliberately, "Though you cut yourself off from me completely, yet pay me my due!"

Thorgils was sorely puzzled at the god's last words, and it was not till morning had broken that their signification flashed across him.

He then remembered having vowed a calf to Thor in his heathen days; this calf had grown into an old cow, and was at that time on board ship.

"If the demon wants the creature he shall have her; one must abide by a promise!" said the bonder: so he cast the beast overboard, notwithstanding his wife's urgent expostulations and representations of the destitution to which they were approaching. "A promise is a promise, even though made to a devil!" quoth Thorgils.

Shortly after this, a bank of snowy Jökulls rose above the horizon, and on the morrow the ship's head was turned towards a bay, girt in between ice mountains, and with a pleasant pebbly beach to the west, up which the surf hissed and tumbled.

The crew cheered, the women came on deck to enjoy the scene, Thorgils shook hands with Jósteinn, and then took the helm himself. The sail was furled, the boats run out, but misfortune had not deserted the settlers.

A shock, a crash—all were thrown on their faces; the vessel was fast on a sunken reef, and was exposed to the violence of the swell, which rolled over the stern and swept the deck. Providentially, all hands were saved, and everything of value was conveyed ashore before the ship went to pieces.

The scene was strangely solemn. Walls of basaltic rock, crowned with green ice, soared up out of the water to the north and south, whilst a mighty mountain of snow rose in terraces on the west, leaving a narrow plain covered with grass and low brushwood between its roots and the beach. Puffins, auks, and guillemots stood in rows along the ledges of rock above the sea. Scoters dived among the waves, gulls and gannets circled around, descending on their prey with a dash; whilst the saucy skuas hovered near, waiting to make the gulls disgorge for their own peculiar benefit.

Thorgils and his party were not idle, but laboured hard to erect a house with the fragments of the ship and the driftwood which littered the shore. The store of meal brought from Iceland lasted whilst they were engaged in building, after which the men fished for their subsistence.

Winter crept on. The birch shed its glossy leaves, and the flowers which had gladdened the brief summer in this solitude, shed their petals and died. The migratory birds disappeared, the nights became long, ice formed in the creeks, and the streams from the Jökulls were congealed.

Yule approached, and Thorgils forbade his men going out after dusk, or indulging in any heathen rites; for Yule is the season when the Œsir ride forth on the blast, and gather up the offerings laid for them by their votaries.

Dreary cries from the snow wastes, and strange sights alarmed Jósteinn's men, warning them that evil drew nigh. Disease broke out among the settlers, and one sank after another: the survivors buried the dead in the sands by the shore, within sound of the ocean roar, which a Norseman loves so well. Jósteinn and his wife died. Thorey was confined to her bed, having given birth to a son, whom Thorgils baptized under the name of Thorfin.

As spring advanced, and the food of the poor settlers improved, the sickness which had made such ravages, ceased altogether, but not till it had carried off nine victims.

Thorey never rallied after her confinement, but remained pale and delicate throughout the summer.

She longed for the time when this bay with all its melancholy associations and privations might be left for ever; but it was found impossible to remove thence during that year, as the boat which Thorgils and his men were constructing, was not yet seaworthy, and, moreover, the ice had not broken up on the sea.

During the second winter died Guthruna, the sister of Kolr and Starkathr, so that Thorey was the only woman of the party remaining. Her husband noticed her increasing pallor with anxiety, but he buoyed himself up with the hope that he should be able to get to sea in the summer, and reach inhabited shores, where she might be provided with proper nourishment.

"Husband," said she one day, taking his hand, and drawing him to the side of her bed, "I have had a dream of bright omen. Methought that I saw a goodly country full of trees and flowers; men in bright clothing came to me, knee-deep in dewy grass, and, extending their hands to me, bid me rejoice, for my sorrows were over. Surely the dream promises that we shall escape from this miserable bay!"

Thorgils stooped over his wife and kissed her.

"Your dream is fair, but it points to another country than any we shall find below. You shall inherit a sunny land, and be holpen by bright angels, for you have led a good life, and passed through bitter sufferings."

Thorey begged her husband to leave the place, and he promised to do so as soon as possible.

One bright day Thorgils proposed an ascent of the Jökull,

that he might look out seaward, and discover whether the ice were breaking up, and blue water visible. The bay itself had been clear for some little while, but great hummocks of ice had encumbered the opening.

Thorey begged her husband not to leave her alone with the thralls, but he laughed at her fears. His son Thorleif, with Kolr and Starkathr, offered to accompany him; and he objected at first, saying that his wife wished some one to remain with her; yet, when they persisted, he gave way, and all three climbed the glacier mountain together. They held on, ascending slowly, till Nones, when a fog came over the mountain, and they were compelled to descend. However, they had seen enough to satisfy them that there was clear water along the coast, as far as the eye could reach; so that, if they were unable to put out to sea, they might, at all events, follow the shore. At the same time, dark blue vapour on the horizon showed that the ice-fields were local, and that the ocean was free at no very great distance, and would soon break up the remaining ice.

The descent was accomplished with difficulty, as the mist was dense, and it was evening when the three reached the shore.

The first circumstance which caused them any alarm was the absence of the new-built boat. Thorgils ran to the house in trepidation. The doors were gone and the hall was perfectly dark within. He stood and listened. The snuffle of a baby was the only sound which broke the stillness.

"Thorey! Thorey!" he cried. There was no answer.

Trembling with the agony of despair, the poor man leaped to his wife's bed. He found her lying cold and dead, on a mattress soaked in blood which had flowed from a stab in her side, made by some small instrument. The baby lay at her breast, encircled by one lifeless arm. When the first burst of misery was past, the other three men examined the house.

Almost every particle of food, the bedding, all the warm clothes, and one of the tents, had been carried off; the tool-chest had been broken open and ransacked, and all the thralls

had vanished with the boat. About the only things left were
a small kettle which had belonged to Thorey, a tent, and
Thorgils' sword and axe, which he had taken with him when
he ascended the Jökull.

That was a sad night for all, but Thorleif did his utmost
to cheer his father.

The whimpering of the poor baby, as it cried for its
mother's breasts, first roused Thorgils from his stupor. He
went to the bed where the little thing lay, lifted it out, and
folding it to his heart, walked up and down the hall, rocking
it to sleep. Every time that he came to the light he looked
down at the tiny tear-blubbered face, so like that of his dead
wife ; and Thorleif saw, by the workings of his father's countenance, that his heart was nearly broken.

But the baby could not be stilled ; it was hungry, and its
little eager fingers plucked and felt at Thorgils' bosom. The
poor father laid the little thing beside the fire, and knelt
before it, watching it thoughtfully and sadly as it sobbed and
held out its dimpled hands.

" Thorleif, the child must not die ! " groaned he ; then
suddenly he exclaimed, " Give me my sword ! "

The young man handed it to him.

Thorgils bared his breast, and with the edge of the blade
slit the nipples. At first there came out blood, but presently
blood and water, and finally milk. So he took the child to
his bosom, and nourished it like a mother.[*]

During the spring and summer the four men toiled at boat
building, but as they were without the requisite tools, their
labour was much increased. Moreover, victuals were not
abundant, so that their time had to be divided between fishing
and building.

One morning Thorgils found a couple of Esquimaux
women on the beach, carrying between them their winter
supply of provisions tied up into a large bundle. Believing
them to be Trolls, whom it was the duty of all good Christians

---

* See a notice of a similar occurrence in Dr. Livingstone's *Africa*, p. 127.

to exterminate, the worthy man grasped his sword and rushed down upon them. The poor creatures took to flight, but did not relinquish their stock of food till Thorgils had hewn off one of their hands.

This stock of victuals was a great catch, and so long as it lasted the work of boat-building progressed rapidly.

The baby was not yet weaned : it seemed to be tolerably healthy, and engrossed its father's whole affection. The bonder washed and dressed his child every day, and brought it to the beach when he went to work on the new smack, that the little one might be near him, and might amuse itself with shells and pebbles.

Winter passed, and with summer the settlers put to sea in their vessel. They kept the Jökulls on one side of them and the ice-fields, which covered the ocean, on the other, steering due south.

The coast of Greenland presents a succession of scenes of savage beauty. Bare crags of black trap, 2,000 feet high, capped by beds of ice, rise abruptly from the water, or terraces of alternate snow and rock soar into the clouds. Stony needles start out of the seas, and their bases in summer are white with foam. Deep inlets between mural precipices crowned by ice, wind for many miles inland, and terminate in glaciers of prodigious magnitude. These fjords, now twinkling in the sun, now shaded in gloom, are the resort of countless wild-fowl, relieving the desolation of the landscape. Icy platforms dip to the sea from the high table land of the interior, and their tall blue cliffs, continually undermined by the surge, fall into the waves with a roar like thunder, and stud the sea with icebergs. At rare intervals, a sheltered bay is green with meadows and bushes of service tree, willow, and birch, but the prevailing hues of the shore are black and white.

The settlers wintered in a little island, the caves of which were thronged with seals ; and when summer came they put to sea again.

Among the rocks they found the eggs of a black-backed

gull ; and the poor fellows boiled them, and gave one to the child, who eat half of it, but declined taking any more. On their asking him the reason, he replied that all around him stinted themselves in food, so he would stint himself also.

A broken oar washed up on the shingle gave the men hopes that they were nearing inhabited country. A few days after this they reached a craggy frith, and ran the boat ashore.

Thorleif and his foster-brothers brought out the tent and pitched it on a grassy spot, whilst his father went in search of eggs and shell-fish.

The kettle was brought on land, and drift-wood was collected for a fire, as the weather was still piercingly cold.

After a scanty supper all went to sleep, intending to spend a few days in the fjord fishing, as their provisions had run short, and then to continue their course.

Kolr was the first to wake next morning, and he went down to the shore. The boat was nowhere to be seen. The young man returned to the tent, and without saying a word lay down.

Thorleif rose next, and he descended to the sands. His heart sank when he found that the little smack was gone ; and he also lay down again, without having the courage to tell his father of their loss.

Thorgils woke now, and went out of the tent, followed by Kolr and Thorleif, both silent and crushed by the disaster. When the truth flashed upon him, he saw that all hopes of life were gone.

Jökulls flanked the fjord, the beach was too diminutive to allow for much lodgment of drift-wood, so that there was no chance of the men being able to build house or boat, and the situation, exposed to the north, would be insupportable through the winter.

"The baby!" moaned Thorgils, and his eye flashed wildly with despair; take the poor little thing, Thorleif! and put it to death, that it may not languish with starvation before my eyes."

"Father!" exclaimed Thorleif, "I cannot do what you bid me."

"Kolr! then you take it. There are no hopes of our ever being able to leave this spot." And when the men hesitated, Thorgils started up, with all the savageness of frenzy, and dared them to disobey him.

The two young men then went to the tent, lifted the child, and walked away with it.

"You must slay it," said Kolr; "I have not the heart to do it myself."

"Never!" answered Thorleif; "it would ill become me to kill my little brother. Besides, I know full well that if the child dies, my father will pine away. We had better leave the little fellow somewhere out of sight, till the first agony of despair is past."

Kolr gladly acquiesced, and little Thorfin was placed in a sunny nook, and left to play with the buds and mosses around him.

On the return of the youths to the tent, Thorgils asked after the child, and when he learned that they had not killed it, he grasped their hands, and thanked them fervently for having spared the little life so dear to him. Then he ran to the spot where his baby lay, and hugged it passionately to his breast. He would not part with the poor thing now, but kept it locked in his arms throughout the night.

Next morning he woke Thorleif to tell him his dream. "I fancied myself back at Tratharholt," said he; "and I saw a swan come in at the door. It shunned me at first, but presently I caught the white bird, and stroked its soft feathers. Then it laid its neck on my shoulder, and caressed me with its bill."

"This dream signifies, that you shall marry a young wife, father! and that at first she will care little for you, but after a while, she will love you very tenderly."

"I dreamed again," continued Thorgils; "and methought I sat in the garden of Tratharholt. Seven leeks grew around me and waxed great, so that they spread till they filled the garden. But one of them shot higher than all the rest, and its head resembled a cluster of golden flowers."

"Father! that signifies that you shall have seven children, and that their descendants will spread through Iceland; of them shall spring one who is to excel all the others."

"I dreamed once more," said Thorgils, "and saw my guardian spirits come to me and bid me cheer up, for that they had brought my boat again to shore." Then he started up, girded on his sword, and hurried down to the strand.

The little smack was soon found, washed up in a creek; and, near at hand, was something which was calculated equally to rejoice the poor man's heart. This was a bear with one of its fore arms broken, and nearly dead with cold, struggling in a crack of the ice, trying to scramble up on the hummock from which it had slipped.

Thorgils drew his sword, leaped towards the beast, and smote it to the heart. Then catching at its ears, to prevent Bruin from sinking, he called for Thorleif to assist him in drawing it out of the water. When the brute was brought to land, it was found to have been frost-bitten in its foot; "hence," says the Saga writer, "you may judge how Thorgils must have suffered from the severity of the cold and frosts."

This godsend was soon skinned and cut up. Thorgils portioned off each man's share, and put aside that which was to be reserved for another time. The poor starving fellows were wofully disappointed, to see how little he set before each.

"You give scanty rations, father!" said Thorleif.

"My boy!" answered the bonder, "we do not need food now in the same degree as we shall when we are hard at work."

Soon after this, Thorgils and his party rowed off, and kept farther out to sea than had been hitherto possible.

Their advance was very slow. Sometimes they had to lift their boat out of the water, and carry it across a sheet of ice; at other times, they pushed it through rifts in the ice-floes, which were too narrow to admit of their rowing.

They suffered severely from thirst. On one occasion, Starkathr asked leave of Thorgils to wring into a quiver the shirts drenched with perspiration, from the toil of continual rowing. The bonder would neither forbid nor give consent, and the poor fellows succeeded in extracting, by this means, a scanty supply. But before tasting it, they offered the few drops to Thorgils.

He took the quiver from their hands and rose up in the boat. The men watched him, expecting that he would propose a toast, as he held the arrow-case before him, looking into it thoughtfully. Then, stepping to the bows, he poured it out, exclaiming—

"Foul fiend! thou hast plagued us too long! Now, in the name of God, I bid thee depart, and molest us no more!"

As he said the words, there was a rustling in the bows, and a large bird, strange and hideous, something like an auk, rose from among the tackle, and, spreading its wings, flew north, hoarsely croaking.

"Thanks be to God!" said Thorgils. "We are delivered from the hands of our enemy; our privations are over and we shall now meet with success."

A glistening heap of ice was at no great distance from the boat; the men pulled towards it, and found that the hollows were full of the purest water, which had trickled from the sides of the iceberg, as the sun melted its glittering spires.

Shortly after this they made for land, and ran the boat up on the shore of a small island. Here they found a tent, and, on entering it, discovered one of the runaway thralls in the last stage of sickness. He told them that he had been forced by the others to join them; that he had secured nothing of the stolen property, except the tent; and that he was guiltless of the blood of Thorey, who had been murdered by the thrall Snœkollr.

Before Thorgils and the others left the island, the unhappy man died, and they buried him in the sands by the shore.

Autumn approached, and Thorgils looked out for a
sheltered spot where they might winter. Finding a bay
facing the south, he ran the boat in, and his heart danced
at the sight of a boat-house near the water's edge, and a
line of blue smoke curling from a birch-grown dell, hard by.

You may well imagine that the poor battered smack was
soon run ashore, and that the travel-worn men hastened in
the direction of the smoke, with all the speed that hope
could give.

They found a snug farm pitched on a grassy slope, with
the cows gathered around the door for milking. A red-faced,
good-tempered man came forward and greeted the strangers.
On hearing their story, he begged them to winter with him,
and they were only too thankful to accept his offer.

Thorfin was given into women's keeping to be properly
weaned.

A pleasant winter slipped by, and in spring, Rolf, as the
farmer was called, offered his guests a ship if they wished to
depart. Thorgils accepted the vessel, but before leaving
Greenland, he visited Eric the Red, at whose invitation he
had come. Eric received him very coldly, as he had calculated
on Thorgils arriving in a well-laden ship, and was not prepared
to receive him when he was destitute of everything.

Thorgils accordingly returned to Rolf. From him he
learned that Snœkoll and the other thralls were in the country;
so he went after them, captured them, recovered his stolen
property, and then sold the wretches into bondage in Greenland. Thorgils, his sons, Starkathr and Kolr, then left
Greenland on their return to Iceland.

They again encountered storms, and, on nearing their
native isle, were beaten about for twelve days, after which they
sighted Iceland. The wind now changed, and rolled up from
the south laden with rain; it blew a gale for two days, so
that they could not venture to run the vessel ashore. For
these two days, Thorgils had been labouring almost incessantly
in baling, up to his middle in water, as eight enormous waves
had rolled over the ship.

Starkathr came up, and begged him to leave the hold, and let him take his place. Thorgils did so, and seated himself near the opening to the hold, with Thorfin on his knee. At that moment, a huge green billow rolled over the vessel, threw Thorgils from his seat, and washed the little boy overboard.

Then Thorgils exclaimed : " Such a surge has swept over us that baling avails us no more ! "

At the recoil of the wave, the child was brought back into the ship alive. The little fellow cried out—

" That is well over, papa ! "

Up sprang Thorgils, shouting, " Bale he who can now ! " The men worked with might and main, and cleared the ship of water.

Thorgils took the boy to bed as he had been completely drenched in the brine. He spat blood that evening, and, after lingering two days, died on a golden morning as the vessel sighted Hjörleif's Head.

The vessel ran into harbour and dropped anchor at Arnar-bœli, the " Eagles' haunt." The men wished to remove the body and bury it, but Thorgils would not suffer it to be taken from his lap. " We have been constant companions in hard-ship, night and day," said he; " and now we shall not be parted."

His friends consulted together what should be done, and at last hit upon a plan.

They went ashore and picked up a quarrel with a farmer named Sigmundr. Kolr then hastened to the vessel, and told Thorgils that there was a fight ashore, and that his son, Thorleif, wanted help.

The bonder started up, girded on his sword, laid the dead child gently on a bed, slung himself over the ship's side, and hastened to the scene of conflict.

He soon succeeded in patching up the quarrel, which was only a fictitious one ; and then he returned to the vessel.

In the meantime Kolr had taken the corpse to a church and buried it. Thorgils was furious at what had been done, and was hardly restrained from slaying the faithful Kolr on the spot. When, however, the first burst of passion was over, the poor father regretted his violence, and going up to Kolr, he shook hands with him. For four days and nights he lay without eating or sleeping, and said that he could not blame women for so dearly loving the bairns which they have suckled themselves.

Thorgils then went home to Tratharholt, and wondered how he could ever have left it, so rich and fertile did the farm look after the icy terraces of Greenland.

He married a young wife soon after his return, and before his death saw seven children growing up around his knees. From him is descended the blessed Thorlack—Iceland's greatest saint. This is what was signified by the golden flower in the bonder's dream.

I may inform those who are curious about the discovery of Greenland and America by the Icelanders, that there is a very accessible account of it in Mr. Blackwell's edition of Mallet's *Northern Antiquities*, published by Bohn, price five shillings.

Mallet's book is valueless, but the additions and notes of the editor are excellent.

For those who understand Danish, there is the work, *Grönlands Historiske Mindesmærker*, 3 vols. 8vo. 1838-45, containing extracts of the Sagas relating to Greenland, and consisting of 2,588 pages.

CHAPTER XXIII.

CONCLUSION.

Leave Geysir—Last View of Heckla—The rumoured Eruption of Skapta—
Return to Thingvöllum—Latin Conversation—Seljadalr—The Plague of
Flies—Halt at a Farm—A fair Haymaker—The Spell broken—Return
to Reykjavík—Sale of Horses—Icelandic Ponies—Their strong and
weak Points—Leave Iceland—The Captain's Joke—Reach England—
Advice to Travellers.

WE were sorry enough to leave Geysir where we had spent
some joyous days, but the steamer waits for no man, and we
were obliged to be back in Reykjavík some days before she
sailed, so as to dispose of our horses by a public auction.

Farewell, Geysir! We took one last look into the calm
steaming basin, tossed one final load of turf into Strokr and
galloped off. We stopped at Uthlith to shake hands with the
farmer and his wife. Really, their clean cheerful faces did
one good!

They seemed to be quite pleased to see us again, and
offered us bowls of milk which we emptied thirstily. In Iceland one learns to live and fatten upon milk. We took a
parting glance over the tún wall at the glorious panorama of
snow peaks beyond the plain of green morass. Heckla was
snow-clad still, its ridge starting into three teeth, one of which
is perfectly black. Far away to the south were the twin peaks
of Tindfjalla, the tops sunlit, and the bases lost in swimming
blue. More distant still rose Eyjafjalla, like a golden cloud
on the horizon. Heckla is distant from Uthlith, as the crow

25

flies, about thirty-six miles; Tindfjalla, forty-five; and Eyjafjalla, fifty-six miles.

My intention had been to have gone on to the Skaptár
Jökull, but I found now that the rumours of its being in
eruption were without foundation. At Haukadalr, a farm near
Geysir, we were told that the eruption was supposed to be
taking place at Krafla, near Myvatn, whence we had come!
On further inquiry it proved that only one man pretended to
have seen any signs of it, and he had come from the
Lómargnupr, in the south. He did not, however, assert that
he had noticed anything except rising columns like smoke—in
fact, the sand clouds which we had observed. At Thingvöllum we hoped to obtain further information, but we were
disappointed. The people said that there might be an eruption
somewhere, as there was so much sand in the air, but they
could give us no account of the outbreak. One thing they
all were agreed in, that Skapta was quiescent; so that there
was no advantage in my journeying thither.

On my arrival at Reykjavik I made further inquiries, and
learned that the postman from Eyrarbakki declared that he
had seen flames in the direction of Trölladyngja. This
mountain is just 140 miles from Eyrarbakki, as the crow flies,
and the lofty Tungnafells Jökull intervenes. I think, therefore, that the statement of the postman is questionable. It
is possible that Trölladyngja may have erupted, but, if so, the
outburst must have been very slight, or we should have heard
some account of it from the Möthrudalr men, whom we met
at Myvatn.

The Jökulsá has two sources; one of these is at the foot
of Trölladyngja, and if the volcano had been active it would
have melted the snows on its head, and the river would have
been very full and discoloured. This was not the case; I
found the Jökulsá lower than it is in general, on account
of the coldness of the summer; and the water was milky,
with partially dissolved snow, like all rivers rising among
Jökulls.

At Thingvalla parsonage we again pitched our tents near

the church, and retired late to bed, after having paid a farewell visit to the Logberg, or Hill of Laws.

Next morning I was roused by the sound of voices outside
the tent; and on putting my head out I observed Mr. Briggs
and the pastor of Thingvalla engaged in an animated Latin
conversation.

" Salve, Domine, Dii tibi benefaciant!" began my fat
friend. "Diluculo surgere saluberrimum est! Haud nitet sol,
hodie, serenus!"

" Nunc quidem adstitit imber, non tamen longa est mora
quin nubes fugentur," replied the parson.

To which Mr. Briggs replied, with promptitude, " Quo-
cunque aspicias, nihil est nisi pontus et aër, nubibus hic
tumidus, fluctibus ille minax. An voles gustare Brandæum
nostrum Gallicum?"

" Paululum sine me bibere."

Mr. Briggs then poured out a bumper, and handed it to
the priest, who took a draught, but did not finish the tumbler.
He put it down, shaking his head, and saying that the brandy
was too hot for him.

" Æstuosus nonne?" quoth my portly friend. " Aqua
Brandæum emollit, nec sinit esse ferox. Dignissime pastor!
Nunc est bibendum, nunc pede libero pulsanda tellus."

Then offering brandy with one hand and whiskey with the
other, he said, " Utrum horum mavis accipe!"

" Gratias tibi, jam satis est potatum."

" Jam satis!" exclaimed Mr. Briggs. " An placeat cum
nobis illud modicum prandium, Anglicè breakfastum, sumere?
Sis exorabilis, Domine! habeo quæ tibi offeram: frigidam
bulletam, carnem vervecinam, caseum Stiltonæum, Fortnumi
Masonique jusculum ex caudis boum extractum, succi plenum.
Siste, reverendissime pater! sume, gusta nostra vegetabilia,
pastinacas fabas, potatosque, in parva stannea arcula compressa. Hæc omnia in lebete decocta cupedias faciunt in
Magna Britannia hodgepodge dictas, sed Græce lepado-
temacho-selacho-galeo-kranio-leipsano, et quæ præterea sciunt
Diabolus, Liddellæus Scottusque."

25—2

" Gratias plurimas, jam pransus sum," replied the pastor, with a tremulous curl of the lip.

" Quæso, Domine ! " continued Mr. Briggs, nothing abashed ; " intra tentorium, conspice nostram tam exiguam, tam fragilem domum ! intra, precor ! "

Then, finding us all in bed as he drew back the door-flap, Mr. Briggs pointed at us derisively, and said—" Implentur veteris Bacchi, pinguisque ferinæ ! "

The pastor was persistent in his refusal of the proffered breakfast ; I think that the enumeration of victuals had somewhat alarmed him, and possibly he fancied that the extraordinary compounds which suited an English stomach might disagree with that of an Icelander.

Whilst I was dressing I heard snatches of conversation between my friend and the parson, and was highly amused at the fragmentary character of Mr. Briggs's observations, ranging from statements of facts in natural history, such as " Hippopotamus bellua in Nilo habitat," and " In summis montibus tantum est frigus ut nix ibi nunquam liquescat," to matters of theological importance, such as, " Pii orant taciti," and "Vilius argentum est auro, virtutibus aurum, i. e. quam aurum, quam virtutes sunt."

At last Mr. Briggs having exhausted every other topic, plunged into an animated defence of Balbus. The priest, apologizing for his ignorance, asked who that worthy was.

" Nescis Balbum ! " exclaimed Mr. Briggs, in amazement ; "nescis Balbum, qui Caium accusavit, qui murum ædificabat ! Balbum, qui nec domi nec militiæ mecum fuit ; qui quum manus in aquam immersisset, abiit ; qui omnem occasionem exercendæ virtutis arripiebat, et barbaram consuetudinem immolandorum hominum retinuit, et patriam auro vendidit ; qui cum gallinis cubitum it vesperi ! Balbum qui prœlio interfuit, qui viribus fretus vincula carceris rupit, et oves totondit non deglupsit ! Si Balbum nescis, omnia nescis ! "

The Yankee was much astonished at this display of learning on the part of Mr. Briggs. I was malicious enough to tell my portly friend that his observation had been familiar to me for many years.

"That may be," answered he, laughing; "but, as neither the *Eton Latin Grammar*, nor *Henry's First Latin Book*—the only classic works with which I have the least acquaintance—are likely to have found their way into Iceland, I have not the smallest doubt that all my remarks appeared to the priest to be fraught with singular originality."

As a change we returned to Reykjavík by Seljadalr instead of by Mósfell. The landscape was dressed now in very different colours from those in which it had appeared on our ride to Thingvöllum. Then, all was dull under a clouded night sky; now the heavens were bright and blue; the lake twinkled in the sun; the mountains were bathed in light; and all was fair and goodly except the eastern sky, over which hung a lurid red cloud rising far up into the blue vault, and creeping stealthily over the face of the country, blotting out the landscape. This proceeded from sand raised in whirlwinds, as we had seen from Ók. These sand columns had risen daily whilst we were at Geysir, and had caused us considerable annoyance, for the air was so thickly impregnated with dust that the particles were for ever entering and clogging our eyes, ears, and mouths. The sand was carried as far as to Reykjavik, a distance of over fifty miles, and was there collected, and pronounced to be volcanic ash erupted from Skapta or Trölladyngja.

That such was not the case I am positive, as the sky to the east and south-east of the district whence rose these sand columns was perfectly clear.

We baited at Seljadalr, a small vale full of humps of coarse grass, and dwarf willow. The flies were intolerable. Multitudes of black loathsome hunch-backed little fellows swarmed around us, irritating the horses, and driving the riders into a furious state of temper with themselves and each other.

I was provided with a green butterfly-net, into which I forthwith thrust my head: this afforded but a momentary

relief, as the horrible creatures, after buzzing round it and
taking cognizance of its weak points, perched on my coat,
crept in under the folds of the gauze, and—I had a dozen
buz—buz—buzzing in the bag along with my head.   Another
lot settled on my boots, and proceeded in exploring parties up
my legs, whilst a third set promenaded along my arms tickling
me horribly with their proboscises.

It gave us inexpressible relief when we emerged from
Seljadalr, and the fresh wind freed us from our tormentors.
Iceland is cursed with the plague of flies; in summer the
banks of rivers, the shores of lakes, the meadows and
morasses are alive with countless swarms, gathering so thick
that, without exaggeration, it is hardly possible for one to see
the rest of one's caravan a few yards off.   At Myvatn the
men wear a peculiar cap shaped like a bassinet, to protect
their heads and necks from them.   The cap is made of black
cloth, it covers the shoulders and back, is tight round the
throat, tight round every part of the head except immediately
over the eyes, nose, and mouth, where a lappet is cut which
can either be turned down to cover the face completely, or be
erected, very much like a vizor.   On a cloudy or windy day
the flies do not appear, they lie under the blaeberry leaves, or
cluster in lava fissures; but the moment the sun comes out,
the air is black with them.

Seljadalr is an interesting spot to the geologist, as the
stream which flows through it has torn itself a way through
the rock, and laid bare a section of palagonite tuff resting on
a bed of volcanic cinder, through which a mass of basaltic
lava has been protruded, partially fusing the ash in contact
with it.   The depth of the cutting is from fifty to seventy feet,
and the stream at the bottom prattles over fragments of tuff
and basalt confusedly mingled together.   Three or four varieties
of willow grow in Seljadalr, the most common of which is the
round-leaved Salix caprea.

We passed a lone tarn girt in by bare hill-sides, with a
few reeds growing around the marge, but did not draw rein
till we reached a byre, where haymaking was going on.

Here we rested for a few minutes and drank milk. A bowl of the size of an ordinary wash-hand basin was brought out full of milk. Mr. Briggs very nearly emptied it, and it was twice replenished for the rest of us.

During our milk-drinking, rapturous bursts of "A houri! a tinted Venus! a valkyrie!" from my fat friend made me look round.

I saw that he was lounging over the tún enclosure with his opera glasses directed towards the haymakers. One of those so engaged had attracted his attention; she was a pretty girl with golden hair flying loose in the wind from beneath her berretta-like cap, which was put jauntily on one side, with the long silk tassel passed through a silver ring, and dangling against her cheek.

"Would that I were that tassel," murmured Mr. Briggs. Like most Icelandic maidens, she was tall, and slim as a willow wand; her face was fresh and bright with colour; her moon-like eyes blue as the neighbouring tarn. She was dressed in a black skirt, her jacket was off, and she wore a white bodice; her neck and arms were bare, the former with only a little dark green handkerchief knotted round it. When her sunny hair fell over her face, she tossed it aside, swept the yellow strands off her forehead, and brushed it behind her ears with her fingers in a most bewitching manner. Impassioned glances were cast through the opera glasses by my stout companion, but they either missed their mark, or else the damsel was invulnerable. Mr. Briggs began to consider whether he had not got some treasures of handkerchiefs at the bottom of his box, which he might draw forth and present to the fair maid, but Gúthmundr protested that it would take half an hour at the least to unpack the box, and that we ought to be on the move, if we hoped to reach Reykjavík before midnight. I very much doubt whether we should have got Mr. Briggs away had not the houri of her own accord dispelled the glamour flung over him. Her mouth opened. A thrill ran through the frame of Mr. Briggs as he saw her ranges of teeth white as the crests of Eyjafjalla.

The girl felt for something in her pocket, and drew forth a tough leathern piece of stockfish. The moonlike eyes rested on it lighted up with the ardour of love; then, suddenly, with a crack, snap, her beautiful teeth fastened on it. She held the tail by both hands and clenched the body between her jaws; she bit, she dragged, she tore; she became purple in the face, she sparkled at the eyes, she rived off great shreds of skin from the fish with her pearly front teeth, she ground the bones to impalpable dust between her snowy molars with a horrible crunch, crunch, crunch—frightful to conceive, impossible to describe.

A moan burst from Mr. Briggs's pallid lips; with a bound he reached his saddle, he drew his sou'-wester over his eyes, and fled the scene with precipitation.

It was not till we reached Reykjavík that he discovered that he had left his opera glasses behind him on the tún-wall.

We passed a small river very productive in salmon, but preserved by M. Thomson, a Danish merchant, who has built himself a pretty farm in a very superior style by its side. This stream flows from Elithavatn, a circular lake enclosed within tuff heights, and bearing every sign of having once been an active volcanic vent. In another hour we were cantering into Reykjavík, when a wild, unkempt Icelander started out of a turf-burrow by the roadside, and, with a face of awe, informed us that the annual meeting of Althing had taken place, and that, after several days of anxious deliberation, the great national council had promulgated a decree that thenceforth no more fast riding should be permitted in either of the streets of the metropolis.

We hurried at once to the agent for the steamer to inquire after letters and newspapers, both which we devoured with intense eagerness.

A couple of days later, our horses were sold, the town having been summoned to the auction by roll of drum. The sheriff acted as auctioneer, and held the sale in a yard before the door of the inn. Amongst the horses was one which Grímr very kindly let me have in exchange for a noble fellow who had fallen

lame *en route*, and had been left behind in the Vale of Smoke.
The sale was tolerably satisfactory; in that I got about a
quarter of what the ponies cost me, when I take off the
sheriff's fee and the discount for ready money, together with
some other small items which appeared in the bill, but which,
on the principle of "many mickles make a muckle," reduced
the receipts considerably. Little Bottlebrush I did not sell.
I could not find it in my heart to part with the brave fellow
who had carried me so well over the island. I accordingly
booked his place in the *Arcturus*, that he might come with me
to England and enjoy our luxuriant grass.

I should not advise a traveller to follow my example in
this respect, as Icelandic ponies are not suited to England;
they become sleepy and lazy, lose their pluck, and become
overwhelmingly fat. An Icelandic horse is a most remarkable
object; it stands about fourteen hands high, is strongly built,
with short sturdy legs, large solemn head, with a sort of beard
under the chin, the nose rounded off, the neck short, and
crested with a thick upright mane, like those of horses in
antique sculpture.

The pony will go six hours a day without food, and
requires only a bait of an hour to set him in condition to
go another six without stopping. He runs like a cow, both
legs on the same side of the body moving at once, and does
not trot. As he spins along, he holds his head towards the
ground, observing it intently, so that he seldom trips, and
when he sees a crack or hole in the lava, he swerves rapidly
and avoids it.

An English judge of horses is somewhat at fault in Iceland;
the legs of a pony are always in good condition, the hoofs
generally so; the weak point is the back. The ponies suffer
from the rot—a horrible complaint, which they get whilst
travelling. The turf sods which are laid against their sides
to keep the boxes from bearing on them, become wet with
rain, and rub the skin till a sore is made which becomes an
ulcer. Rot marks are carefully concealed by horse-vendors, by
rubbing a little peaty soil over the place; they may, however,

be detected by passing the hand along the back from the shoulder. If a slight knot is felt the animal must be rejected, as the wound is sure to reopen after a day or two. In choosing a pack-horse, the tail should be examined, as the poor beasts often suffer from the wear of the crupper, and one which has lost skin at the root of the tail will not be fit for work as a sumpter animal. The Icelanders have a prejudice against mares, which they say are not up to doing hard continuous work.

The horse should be provided with shoes having six nails; one often meets with them having only three; they are then very likely to tear their hoofs to pieces. A pony sometimes has his shoe ripped completely off in crossing a lava field; of this there is less chance if the nails be small and numerous. The hoof of an Icelandic horse is remarkably strong; on account of the cold water and snow in which the poor animal stands. An Icelander thinks nothing of riding his horse for some days helter-skelter over rocks and through rivers without its being shod, and I was compelled to ride one of mine from Vithimyri to Akureyri without shoes. This may, however, be done once too often, as was the case with a poor brute I saw in the North, standing in a pool of blood with its hoof split.

I was sorry to leave Iceland, for I had spent many happy days in it, and had learned to feel a very strong attachment to the wildly beautiful island. The parting would have been more trying but for the confidence I had of revisiting the island, if all went well, on some not very distant occasion.

On the 2nd August we sailed, after having bidden farewell to all the friends I had made in the little town. The morning was most glorious, and as we steamed out of the noble Faxafjord, sixty-five miles across, we saw the tall dome of Snœfels Jökull standing on our right out of the sea to the height of 4,400 feet, covered with snow.

We had a fair voyage, with only a little rough weather during a night and a day. One of our party had brought some Arctic foxes with him, another had an eagle on board;

when the sea was rolling heavily the cage broke loose and the bird was nearly carried overboard; as for the foxes, they were washed by every wave which went over the deck, which was satisfactory, for the creatures smelt most horribly.

One morning the captain came to the side of my berth with a grave face.

"Is your pony the little piebald?"

"Yes," I answered, starting up; "has anything happened to him?"

"Well, he has had his tail jammed between two beams, and it has been all but torn off, it now hangs on by a single thread."

I was much concerned, for the beauty of the animal lay in the tail.

"What is to be done?" asked Captain Anderson; "shall we amputate the tail? That will be a sad pity, too, for you can never ask a lady to ride in Rotten Row on a tailless pony. I think the poor thing had better be thrown overboard."

Whilst the captain was speaking I undid my huswife and threaded a darning needle.

"Captain!" said I, "you will do anything to oblige me, I know."

"To be sure I will, what shall I do?"

"Why, just stitch the tail on again, there is a good fellow; and—look here—I daresay you have no notion of making a knot, so observe—when the sewing is done, make a loop with the string, run the needle through, and then bite the thread off close with your teeth. I'm sorry I have no scissors to lend you."

The captain was fond of a joke—he was; and I suspected from the first that he was making fun of me; still it was a great relief to me to find, when I did descend into the hold to look at poor Bottlebrush, that his tail was on without a scar.

On the 9th August we were in Liverpool, after an absence from England of just two months.

And now for a few hints to the traveller:—

1. Take plenty of money in small change, and English gold. Your expenses will amount to about a guinea-and-a-half per diem. The money used in Iceland is Danish.

					s.	d.
16 skillings = 1 mark	.	.	.	.	. = 0	4½
6 marks = 1 rigs dollar	.	.	.	. = 2	3
2 rigs dollars = 1 specie dollar	.	.	.	. = 4	6

English money can be changed at Reykjavík by any of the Danish merchants.

2. The steamer *Arcturus* sails once a month during the summer, starting from Grangemouth, near Falkirk. The vessel is manned and commanded by Danes. Information respecting the times of departure can be obtained from Messrs. Robertson, at Grangemouth.

3. The names of the guides at Reykjavík are, Olavur Steingrímson, Oddr Gislason, Bjarni, Arni Sigurthsson, Gúthmundr Jónsson, and Magnús. Next time that I go to Iceland, I shall take an English servant with me, and hire an Icelander just to look after the horses. You must be provided with a compass and Gunnlangson's map, then you can find your way as well without as with a guide.

The map can be obtained at Reykjavik for 16s.; from Messrs. Williams and Norgate, Henrietta Street, Covent Garden, for 30s.; or from Mr. Stanford, Charing Cross, for 2l. 12s. You had better provide yourself with a pocket Danish-English and English-Danish dictionary, price 4s., which can be procured from Messrs. Williams and Norgate. It will be found very handy.

4. Take the least possible amount of luggage with you; say, two flannel shirts, a comb and brush, two towels and soap, an oil-skin pilot coat, which can be procured at Grangemouth for 6s. 6d., a sou'-wester, and fishermen's boots. I found my india-rubber contrivances very useless, as the buttons were continually coming off and the material tearing.

Fishing stockings are, however, serviceable, but they must be taken great care of, a pair of Icelandic stockings must be drawn over them, and wading shoes over these again. Be provided with rope, hammer, and nails. Horse-shoes must be purchased at Reykjavik. Take also biscuits, portable soup, salt, and a spirit-lamp for boiling coffee or tea. I should advise every tourist to be provided with a veil, as the flies are intolerable near lakes and standing water. Let all the goods be packed in a couple of strong wooden boxes, 15 inches high, 10 wide, and 22 long; the wood of which they are made must be ¾ inch thick, and the sides must be morticed into each other. The lid should be arched to let off the rain. These boxes can be procured from Messrs. Day and Son, 353 and 378, Strand. Let the traveller ask for them made after Mr. Shepherd's improved plan. When they are packed take them to the top of the house and roll them down stairs; if they stand the test they will do for Iceland.

I recommend that pack-saddles should be brought from England; we found considerable difficulty in getting good ones at Reykjavik.

Take also with you a light saddle without a tree, commonly called a pilch.

If you are not prepared to undergo the discomforts of lodging in an Icelandic farm, you must take with you a tent, and rugs for a bed.

You will then require eight horses; if you do without a tent, six will suffice.

5. Pay a good price for horses; no animals which cost less than 2*l.* 10*s.* are fit for anything. For riding ponies you will have to pay about 5*l.* Do not bring an Icelandic horse home with you, the climate of England does not suit it, and it becomes fat and desperately lazy.

6. If you go in quest of birds, take a water-dog with you, or an india-rubber boat; the latter would necessitate the purchase of two more horses.

7. Extraordinary precautions must be taken to preserve

thermometers from being shivered to atoms.   Of ten which a
friend of mine took to the Geysir wrapped in wool, seven were
broken in two days.

" Claudite jam rivos, pueri, sat prata biberunt."
                                                    Virgil.

APPENDICES.

APPENDIX A.

NOTES ON THE ORNITHOLOGY OF ICELAND,

By Alfred Newton, M.A.,

Late Fellow of Magdalene College, Cambridge;
Foreign Member of the Icelandic Literary Society (Reykjavik Branch);
F.L.S., F.Z.S., &c.

Though several British naturalists of no mean repute have visited Iceland, I believe that hitherto no connected account of its Ornithology has ever been published in the English language. What is chiefly known of the subject in this country has either been derived from foreign works, or from the communications made to our standard authors by British travellers. Thus, in 1833, Mr. G. C. Atkinson, accompanied by Mr. Cookson and Mr. William Proctor, the present curator of the Durham University Museum, undertook a voyage thither, and, on his return, supplied his friend Mr. Hewitson with a series of valuable remarks, which are to be found in the various editions of that gentleman's work on the *Eggs of British Birds.* Four years later, also, Mr. Proctor again visited the island, and, besides contributing a few notes on his tour to Mr. Neville Wood's magazine, *The Naturalist,* for 1838, furnished the late Mr. Yarrell with a good many specimens and some further observations on the birds of Iceland, the latter of which are embodied in his well-known *History of British Birds.* But neither Mr. Atkinson nor Mr. Proctor penetrated far into the interior of the island. In 1833, their researches were confined to the neighbourhood of Reykjavik, or, at most, to the south-western portion of Iceland, and in 1837, Mr. Proctor visited only the district of Myvatn, and the interesting, though inhospitable, islet of Grimsey.* In 1846,

---

* Nearly two years ago Mr. Proctor was kind enough to place in my hands, with liberty to use them as I thought proper, the original journals which he kept on each of his voyages, and they have been of no small service to me.

Mr. Henry Milner, attended by Mr. David Graham, of York, traversed the whole island from Eyafjörðr, by way of Arnavatnsheiði, to Reykjavík, and formed a collection of Icelandic birds and their eggs, but he has never published any account of his experiences. Others have, I believe, done the like, and last year (1862) my friend Mr. G. G. Fowler, in company with Mr. Shepherd, spent the summer in Iceland, and from him, as also from Mr. Milner, I have privately derived considerable information respecting its birds. Of my own visit I need say little. In 1858, I passed more than three months in the country, accompanied by that unwearied explorer and talented naturalist, the late Mr. John Wolley, but with the special object we had in view (that of solving, if possible, the moot point of the Gare-Fowl's present existence), we had few opportunities of learning much, from our own general observation, respecting the ornithology of Iceland.

Abroad the case has been different. Numerous have been the works published which relate to the zoology of Iceland. From the middle of last century, when Anderson's posthumous *Nachrichten von Island* called forth Horrebow's celebrated *Tilforladelige Efterretninger von Island*, through later years, in which appeared Brünnich's *Ornithologia Borealis*, Olafsen and Povelsen's *Reise igiennem Island*, Olavius' *Oeconomisk Reyse*, and Mohr's *Forsög til en Islandsk Naturhistorie*, the stream has been flowing almost uninterruptedly till the present time. But I must especially mention the various works of Friedrich Faber—known almost everywhere in Iceland, even now, as "Fugl Faber"—and particularly his *Prodromus der isländischen Ornithologie*, which contains the result of a year and a half's close and careful research. Though the progress of knowledge in the last forty years has of necessity invalidated some of the author's remarks, the extreme value of this little book is not to be questioned for a moment. I may say that, on two or three almost trivial points, I can myself bear witness to Faber's minute truthfulness. In the German ornithological magazine, *Naumannia*, for 1857, there is contained an excellent series of papers by a very trustworthy traveller, Dr. Theobald Krüper, recording the observations on the birds of Myvatn and its neighbourhood, made by him during a visit thither in the preceding summer. These are all the more interesting, though, perhaps, to us English the less instructive, as the district was one of the chief scenes of Mr. Proctor's labours, already mentioned. Lastly, I must mention that, in 1860, M. G. Benguerel, a Swiss gentleman, made a prolonged tour in Iceland, and, as he was good enough to tell me, formed a considerable ornithological collection, a brief notice of which he subsequently communicated to the Society of Natural Sciences at Neuchatel, in whose *Bulletins* it will be found; while, simultaneously, Herr William Preyer, accompanied by Dr. F. Zirkel, was performing a similar expedition, the account of which they last year published. To this, their *Reise nach Island*, the first-named of these gentlemen appended a systematic review of Icelandic vertebrate animals. It undoubtedly contains by far the most complete notice of the birds that has been published since Faber's time, but I am bound to express my opinion that the writer has not shown sufficient discrimination in its compilation.

The only work of an Icelander on the ornithology of his own country, that

I am aware of, is a short and unfinished treatise by the late Jonas Hall-grimsson, who died in the prime of life about twenty years ago. It appeared in a periodical bearing the name of *Fjölnir*, published at Copenhagen in 1847, edited by my friend Prof. Halldór Kr. Friŝriksson, and my introduction to it I owe to my friend Cand-Theolog. Eiríkur Magnússon. It contains a list of forty species of Icelandic birds, with brief but useful notes upon them. However, it extends no further than the end of the order *Grallæ*, and I can, therefore, only express my hope that before long some native of the island may arise to complete the work, or, better still, be patriotic enough to give his fellow-countrymen a yet fuller account of the birds which are met with around them.

The following is a list of authors and their writings, which may be usefully consulted by any one wishing to make himself acquainted with the Ornithology of Iceland, so far as it is known. It has no pretensions to being perfect, and it must be remembered that much additional information on the subject is to be gathered from works, the scope of which is general instead of special, as are these here enumerated :—

JOHANN ANDERSON : *Nachrichten von Island, Groenland und der Strasse Davis,* &c. Frankfurt und Leipzig, 1747. 8vo.

G. C. ATKINSON : Various observations on the birds of Iceland contributed to Mr. Hewitson's *Illustrations of the Eggs of British Birds.*

G. BENOUZREL : "Voyage en Islande," *Bulletins de la Société des Sciences Naturelles de Neuchatel,* V. p. 445. Neuchatel, 1861. 8vo.

M. TH. BRUENNICH : *Ornithologia Borealis,* &c. Hafniæ, 1764. 8vo.

EDWARD CHARLTON : "On the Great Auk," *Transactions of the Tyneside Naturalists' Field Club,* IV. part ii. p. 111. Newcastle-upon-Tyne, 1859. 8vo.

J. W. CLARK : "Travels in Iceland," *Vacation Tourists in* 1860. London, 1861. 8vo.

FRIEDRICH FABER : *Prodromus der islaendischen Ornithologie.* Kopenhagen, 1822. 8vo.
Supplement to the above. *Isis,* 1824, p. 792.* Jena. 4to.
Ueber das Leben der hochnordischen Voegel. Leipzig, 1826. 8vo.
"Beitraege zur arctischen Zoologie." *Isis,* 1824, st. 4, 7, 9 ; 1826, st. 7, 8, 9, 11 ; 1827, st. 1, 8. Jena. 4to.
"Mohr's Beschreibung der islaendischen Voegel." *Ornis,* I. II. III.* Jena. 8vo.

E. FAIRMAIRE : "Liste des espèces d'oiseaux recueillies par l'expédition du Prince Napoléon," &c., *Edinburgh New Philosophical Journal,* VI. p. 191.* Edinburgh, 1857. 8vo.

---

* These I myself have not seen.

26

J. C. H. Fischer: "Der faroeische Zaunkoenig," *Journal fuer Ornithologie*, IX. p. 14. Cassel, 1861. 8vo.

Ph. Gliemann: *Geographische Beschreibung von Island.** Altona, 1824. 8vo.

Jonas Hallgrimsson: "Yfirlit yfir Fuglana á Íslandi," *Fjoelnir*, IX. Ár. p. 58. Kaupmannahoefn, 1847. 8vo.

John Hancock: "Remarks on the Greenland and Iceland Falcons," *Annals of Natural History*, II. p. 241. London, 1838. 8vo.
"Note on the Greenland and Iceland Falcons," *Annals and Magazine of Natural History*, 2nd ser. XIII. p. 110. London, 1854. 8vo.

W. J. Hooker: *Journal of a Tour in Iceland, &c.* 2nd ed. London, 1813. 2 vols. 8vo.

Niels Horrebow: *Tilforladelige Efterretninger om Island, &c.* Kjoebenhavn, 1752. 8vo.

Theobald Kruefer: "Der Myvatn und seine Umgebung," *Naumannia*, VII. Heft i. p. 33, Heft ii. p. 1; "Die Inseln des Myvatn," *idem*, Heft ii. p. 33; "Ornithologische Miscellen," *idem*, Heft. ii. p. 436. Leipzig, 1857. 8vo.

Frederick Metcalfe: *The Oxonian in Iceland, &c.* London, 1861. 8vo.

N. Mohr: *Forsoeg til en Islandsk Naturhistorie, &c.* Kjoebenhavn, 1786. 8vo.

Alfred Newton: "Abstract of Mr. J. Wolley's Researches in Iceland respecting the Gare-Fowl," *Ibis*, III. p. 374. London, 1861. 8vo.
On the Zoology of Ancient Europe, &c. London and Cambridge, 1862. 8vo.

Eggert Olafsen: *Reise igiennem Island, &c.* Soroe, 1772. 2 vols. 4to.

Olaus Olavius: *Oeconomisk Reyse, &c.* Kjoebenhavn, 1780. 2 vols. 4to.

Thomas Pennant: *Arctic Zoology, &c.* London, 1784. 3 vols. 4to.

William Preyer: *Reise nach Island im Sommer 1860.* Leipzig, 1862. 8vo.
"Ueber Plautus impennis," *Journal fuer Ornithologie*, X. pp. 110 and 339. Cassel, 1862. 8vo.

William Proctor: "Notes on an Ornithological Tour in Iceland," *Naturalist*, III. p. 410. London, 1838. 8vo.
"Clangula Barrovii a Native of Iceland," *Annals of Natural History*, IV. p. 140. London, 1840. 8vo.
Various observations on the birds of Iceland, contributed to Mr. Yarrell's *History of British Birds.*

Johannes Reinhardt: "Om den islandske Svane," *Naturhistorisk Tidsskrift*, II. p. 527; * "Om Gejerfuglens Forekomst paa Island," *idem*, p. 533. Kjoebenhavn, 1839. 8vo.

---

* These I myself have not seen.

JAPETUS STEENSTRUP: "Et Bidrag til Geierfuglens Naturhistorie," &c., *Videnskabelige Meddelelser for Aaret* 1855, p. 33. Kjoebenhavn, 1856–1857. 8vo.

CHARLES TEILMANN: *Forsoeg til en Beskrivelse af Danmarks og Islands Fugle,* &c.* Ribe, 1823. 8vo.

F. A. L. THIENEMANN: *Naturhistorische Bemerkungen,* &c.* Leipzig, 1824 (?). 8vo.

WALTER CALVERLEY TREVELYAN (?): *Historical and Descriptive Account of Iceland,* &c. 4th ed. Edinburgh, 1840.

J. C. O. WALTER: *Nordisk Ornithologie,* &c.* Copenhagen, 1842. Fol.

JOHN WOLLEY: "On the Birds of the Faroe Islands," &c., *Contributions to Ornithology,* III. Edinburgh, 1850. 8vo.

OLAUS WORM: *Museum Wormianum,* &c. Lugduni-Batavorum, 1655. Fol.

From a consideration of the above-mentioned works, or at least of such of them as I have examined, coupled with my own personal experience of the country, I am inclined to believe that Iceland offers a field of considerable promise to the ornithologist; and though it is not to be at all expected that any previously undescribed species of birds will reveal themselves, yet many possessing great interest commonly frequent both the coast and the interior. Besides which, it is not beyond the bounds of probability that one or two of those whose places of retreat during the nesting season, if not altogether unknown, are still shrouded in much mystery, may be found breeding on some lonely Icelandic "heiði." Of these I might mention the Knot, and the Sanderling, and perhaps even the Grey Plover (*Squatarola helvetica*) —though this latter bird, of almost ubiquitous occurrence, does not seem hitherto to have been met with in the island—as likely to reward the search of some future investigator. The character of the Avi-fauna of the country, as might have been expected from its geographical position, is essentially European; just as that of Greenland has American tendencies. Indeed, dismissing from our consideration the species of purely Polar type, which are common to the whole Arctic region, there are, as far as my knowledge extends, only four or five which make Iceland their home without inhabiting some other part of continental Europe. These are the Iceland Falcon, the Northern Wren (which, however, does occur as a resident in the Faroes), the Iceland Ptarmigan, the Iceland Golden-eye, and the Harlequin Duck. The first is by most ornithologists of the present day recognized as distinct from the true Gyr-Falcon, and though the differences between them are but slight, I believe no one has ever observed the characteristics of the Scandinavian form in an Icelandic specimen. The second has been but lately separated from our own Common Wren, which is a bird as well known throughout the greater part of the Continent as in this country, but I believe the separation is deserved. The third, the Ptarmigan, certainly differs in some respects very considerably from the bird which occurs in Scotland and Norway, and much more nearly

---

* These I myself have not seen.

resembles the form found in Greenland. The fourth and fifth are most unquestionably distinct species; and both are found breeding over a good part of the Arctic portion of the New World, while neither occurs in the rest of Europe, except accidentally. I am only aware of one species which does not properly belong to Europe, and which yet occurs frequently in Iceland without breeding there—this is the Greenland Falcon.*

Before proceeding to a detailed and technical list of the birds of Iceland, the reader of this work might perhaps wish to have placed before him a sort of general summary of the ornithology of the country. For it always happens that many of the species which swell the bulk of a local catalogue make but little show in the eyes of a traveller, and are entirely wanting in the pictures which memory recalls to his mind. To begin then with the Falcons, which (for so many centuries more highly prized than any others by all the nations of Europe) are yet to be found in greater plenty in Iceland than elsewhere, and are as much sought for by collectors now as formerly by kings or emperors. Almost exterminated in the British Islands, in Iceland the White-tailed Eagle is still constantly seen perched in solitary grandeur on the rocky shore, while the courageous little Merlin glides over the hill-tops, striking fear and silence into the Titlark and the Wheatear, the White Wagtail, and the Snow Bunting. Wherever the birch or the willow attains the height of a man, there may the monotonous twitter of the Redwing, followed by a low inward warble, be heard by the traveller. Companies of Ravens throng every fishing settlement, and obtain a plentiful subsistence from the offal by which it is surrounded. The Ptarmigan, as I have above said, in plumage if not in species distinct from that which haunts the mountains of Scotland, the fjelds of Norway, or the Alps and Pyrenees of Southern Europe, utters its strange guttural croak among the contorted slabs of the lava streams. Where the turf is softest and greenest, the Golden Plover, by its tameness, provokes the passer-by to unsling his gun—unless, indeed, his hunger being satisfied (not an every-day event in Iceland), he is disposed to take a more merciful view of its familiarity. Along the shore, flocks of wheeling Turnstones, Ring-Dotterels, Dunlins, and other less common kinds of Sandpipers, attract the attention of even the most unobservant. The merry whistle of the Redshank contrasts with the discontented wail of the wary Whimbrel, as keeping well out of shot, he rises lightly from the barren moor. While from the marsh the "zick zack, zick zack" of the Snipe sounds cheerily, and suggests to the sportsman recollections of former, or visions of future, visits to some well-remembered bog or fen, far away across the south-eastern sea. As he strives to ascertain the source of some secluded hot-spring, which in more accessible districts would outrival Buxton or Aix, he may perhaps catch a glimpse of a Water-Rail creeping stealthily through the luxuriant herbage. At almost any of the numerous pools throughout the country, the Red-necked Phalarope is to be seen busily seeking its food round the margin, or, like a graceful naiad, reposing quietly on the smooth surface in the

* A pretty full recapitulation of all that has been written on the question of the Great Northern Falcons (*Falco gyrfalco, Falco islandicus,* and *Falco candicans*) will be found in the *Ibis* for January, 1862, pp. 43-53 inclusive.

softened glow of the Northern midnight. Here, on a patch of semi-natant
bog-bean, the weird-looking Horned Grebe piles a mass of water-weeds,
dragged from the muddy bottom, whereon to deposit its chalky eggs; while
overhead a swarm of Arctic Terns assail the wayfarer's ears with their shrill
shrieks, and should he stop to examine one of the mottled living powder-
puffs he finds crouching in the grass, almost threaten his eyes with their
sharp beaks.

On some wide lake or far-receding fjord, a single Northern Diver may
be descried, spirit-like, disappearing and reappearing almost without causing
a ripple, until, having finished his fishing, he flaps heavily along the
surface, leaving a wake like that of a steamer, and then mounting to a vast
height finally vanishes in quest of his mate, whom he left brooding her dingy
eggs far beyond the rocky ridge, to cross which would cost us two hours' hard
work; while long after he is out of sight, his wild scream, like a cry of human
agony, reaches us, and jars our feelings by its discord with the placid scene
so lately before us.

Where the stream rushes fastest and foams most furiously over its stony
bed there is the home of the quaintly-marked and yet beautiful Harlequin
Duck, so rare a visitant to other parts of Europe. To the upland tarn resorts
the Long-tailed Duck, while the Teal, the Widgeon, and the Pintail frequent
the less bleak lakes, on the islands of which, secure from the ravages of the
Arctic Fox (the only beast of prey in the country), they rear their young, in
company with the Scoter, the Scaup, and the Icelandic Golden-eye. On yonder
grassy plains, intersected by the innumerable rivers that spring from those
distant jökuls of eternal ice, there yet assemble (but not, alas! in numbers
as of yore) a goodly company of wild Swans. There they make their domestic
arrangements, proclaiming their completion with the glorious sound of the
trumpet.   There they lay their elephantine eggs ; and there—O joyful
moment!—they lead forth their infant train, too often, indeed, only to suffer
capture and death at the hands of the neighbouring peasants, or, if they survive
these casualties, to fall the victims of Southern gunners.

The Eider, throughout Northern Europe the chief friend of man amongst
birds, inhabits islets, either natural or artificially constructed, which are
guarded from intrusion by the lords of the soil as jealously as any Norfolk
game-preserve or Oriental harem, and testifies, by its familiarity, to the
effectual means taken for its security. It is, indeed, in Iceland, as in other
lands over which it ranges, almost a domesticated fowl, and readily occupies
the nesting places prepared for its accommodation, paying valuable tithe and
toll in down and eggs for the protection it enjoys.

Seawards small parties of Gannets may be seen circling over the same
spot, heavily plunging one after another beneath the surface, and each, as it
emerges with its prey, shaking the water from its wings and joining its
brethren aloft to repeat the same process. That ridge of seaweed-covered
rocks, left bare by the falling tide, is surmounted by a cluster of Cormorants,
some slumbering in the sunshine, while others are intent on preening their
feathers. Near the mainland, the Great Black-backed Gull soars in dignified
majesty around the intruder, expressing his anger in notes of the deepest

bass, until the alarm being spread abroad, a cloud of Kittiwakes, obedient to his summons, hurry from the neighbouring shallows and awaken the echoes with their petulant exclamations; which are redoubled, should a Skua, that Viking among birds, make his appearance. Still and ghost-like in the distance, buoyant Fulmars wing their way, wheeling round with scarce a beat of their wide pinions.

The insulated stack or precipitous cliff affords a footing, and where a footing, a nesting-place, to countless Razorbills and Guillemots, which crowd there in numbers even more confusing than may be seen by a London excursionist to Flamborough Head or the Isle of Wight. But of all the birds of Iceland the greatest interest centres in the Gare-Fowl, or Great Auk (*Alca impennis*, Lin.), the only wingless, or rather flightless, species of the Northern hemisphere. There is no doubt that in former days it was plentiful enough, at certain seasons, in certain localities to which it resorted to breed. From such of these as were easy of access to the inhabitants of the nearest villages, it has been thoroughly exterminated; but until the farthest rock of the group called the "Fuglasker," lying off Cape Reykjanes, be examined, I think the question of its utter extinction must not be considered settled. Indeed it is probable, from the interest that has now been excited on the subject, that no great period will elapse before this rock, the Geirfugladrångr, is visited. But the expedition is one of no small danger, and during the last five years the weather has not admitted of its being undertaken. I can only express my sincere wish that whenever this rock is reached, the bold adventurer may reap a fitting reward; but at the same time I must declare my hope that in such case, he will be careful to see that the best possible use is made of the spoil. The mere addition to the already considerable number of stuffed skins and blown egg-shells [a] of the species which are dispersed in various collections will be no addition whatever to science. If the bird is doomed to extinction, and such, I fear, is its fate, all who are concerned in bringing about the catastrophe are bound to see that the most is made of whatever chance may throw in their way. It should accordingly be their object if possible to capture the examples alive, and transmit them as speedily and as carefully as possible to the gardens of the Zoological Society of London, where they are sure of receiving every attention; where their gestures may be studied, and their attitudes transferred to the painter's canvas.[†] Or should circumstances hinder the birds from being taken alive, the whole bodies should be preserved in spirit or brine, or by the application of pyro-ligneous acid, and thus rendered serviceable for the anatomist's scalpel. The same may be said of the eggs; their contents should on no account be thrown away, but taken care of in the same manner as the birds, for

[a] Of the former I can now enumerate no less than sixty-one, and of the latter fifty-nine, specimens. There must be several besides, of which I have as yet no knowledge.

[†] There are two instances on record of Gare-Fowls being kept some time in confinement. One is mentioned by Olaf Worm, in his work quoted in my list (p. 300), the other by Dr. Fleming (*Hist. Brit. Anim.*, p. 128, and *Edin. Phil. Journal*, vol. x, p. 97). Worm's bird survived the voyage from the Faerö Islands to Copenhagen, and lived in his possession several months afterwards; and this was *more than two centuries ago!*

it is difficult to exaggerate the value of embryology in the present state of scientific research. Professed naturalists are of course aware of all this, but these words may haply be read by some who would otherwise think that by neatly preparing the skins, and according to the most approved method blowing the eggs of the very last of the Gare-Fowls, they were advancing the study of natural history.

Having thus indicated the most prominent features in the ornithology of Iceland, I will conclude by giving a list of the birds, which, as nearly as I can ascertain, have been hitherto met with in that country; in drawing it up my object has been to exclude all about the occurrence of which any reasonable doubt may be said to exist, though reference is made to most of them in the notes. This list is rather intended for the scientific than the general public. It will be found that I have quoted especially from Faber, whom I consider by far the best authority on the subject, and I only hope I have not been unnecessarily critical, especially when speaking of the labours of Herr Preyer, from whose opinion I am unfortunately so often compelled to dissent.

WHITE-TAILED EAGLE. *Haliaetus albicilla* (Linn.)—Œrn. Assn.
Generally distributed throughout the island in the vicinity of water, but nowhere very abundant. Breeds and, according to Faber, remains during the winter.

ICELAND FALCON. *Falco islandicus* Gmel.—Fálki. Veitifálki. Valur.
Probably of universal occurrence in Iceland, but certainly more common near Myvatn than anywhere else in the island, owing perhaps to the great facilities for breeding afforded them by the inaccessible precipices in the neighbourhood, and to the abundance of food in the immediate neighbourhood.

GREENLAND FALCON. *Falco candicans* Gmel.—Hvitifálki.
Much less abundant than the preceding, with which it was so long confounded, but of regular occurrence in the winter. Has not been known to breed in Iceland, in fact the only instance I am acquainted with of its being met with there in summer is the one mentioned by Herr Preyer, who states that he saw one in a lonely valley between Háfnafjörðr and Krísuvík, apparently towards the end of June. The pale colour of the beak and claws, especially observable in freshly killed specimens, always serve as a ready means of distinguishing this from the true Iceland Falcon.

> Faber mentions that he shot an example of *Falco lanarius* (Linn.) at Akureyri, Sept. 18, 1819, but from what we know now of the geographical distribution of the Lanner, I think it highly probable that he was mistaken, and the more so that it was not until nearly twenty-five years afterwards that the confused nomenclature of the Falcons began to be cleared up. I quite agree with Dr. Krüper's suggestion as to Faber's bird having been a Peregrine Falcon (*Falco peregrinus*, Gmel.), but I do not know that any subsequent observer has noticed this species in Iceland.

MERLIN. *Falco æsalon* Linn.—Smirill. Dvergfálki. Smiri.
Arrives, according to Faber, at the end of March, and goes away at the beginning of October. Very common everywhere, and breeds on the moors.

SNOWY OWL. *Surnia nyctea* (Linn.)—Snjó-ugla. Náttugla.
Not unfrequently observed in winter, but rarely seen in summer. It may
very possibly breed in some parts of the interior, but probably most of the
examples met with come from Greenland, though not necessarily in a ship, as
Horrebow supposes.

SHORT-EARED OWL. *Otus brachyotus* (Gmel.)—Trjáugla. (?) Brandugla. (?)
Olafsen gives (tab. xlvi.) the figure of a bird, which to my mind can only be
intended to represent the above-named species, though Jonas Hallgrimsson
says he can make nothing of it. Unfortunately the reference on the plate is
to a page of the work where no mention is made of an Owl. But at page
705, speaking of the birds of the north of Iceland, and of what is no doubt
the preceding species, he mentions another smaller Owl, which may be seen
in the south and west, though but seldom. This, he says, is of a yellowish
brown, speckled with white and black all over, and these particulars are not
irreconcilable with Dr. Krüper's theory that *Otus brachyotus* is meant.
Faber seems to have thought that the second species of Owl found in Iceland
was the *Strix aluco* of Linnæus, but it is hardly likely to occur in such a
country.

HOUSE MARTIN. *Chelidon urbica* (Linn.)—Bæjarsvala.

SWALLOW. *Hirundo rustica* Linn.—Landsvala.
Both the above-mentioned species seem to occur annually, but do not remain
long in one place.

NORTHERN WREN. *Troglodytes borealis* Fischer, *Journ. f. Orn.* IX. p. 14.
          —Músarbrótir. Músarrindill.
In 1861, as above quoted, my friend Herr J. C. H. Fischer, of Copenhagen,
pointed out the distinctions observable between Færoese and Continental
examples of the hitherto undivivided *Troglodytes parvulus* Koch. To the
kindness of Mr. John Hancock I am indebted for the opportunity of examining a specimen of the Wren of Iceland. This I find to correspond with Herr
Fischer's description of *T. borealis*, and accordingly I here adopt that name,
believing that a sufficiently good species is thereby indicated. It is not
unworthy of remark that the Icelandic example now before me is to some
degree midway in appearance between our British form and the *T. aedon* of
North America. The much larger bill and feet of the Northern Wren are
characters which enable it to be easily distinguished from our own familiar
bird.

BLACKBIRD. *Turdus merula* Linn.
Seems to have occurred on two occasions in Iceland. The first, in 1823, is
mentioned by Herr Preyer (p. 428), on the authority of Gliemann ; the
second, in March, 1860, by Mr. Metcalfe (pp. 191, 192). But even if there
was no mistake in either case, it must be regarded as a very exceptional
visitor.

REDWING. *Turdus iliacus* Linn.—Skogar-thrœtur.
An annual migrant, and found in suitable localities throughout the island.
Breeds early, beginning its nest before the snow is well off the ground. Mr.

Fowler found that by the 26th May most of the young had hatched, but it must also be double-brooded, as Mr. Proctor found a nest on the 7th August only just left by the young.

BLACK REDSTART. *Ruticilla tithys* (Scop.)
First noticed in Iceland by Herr Preyer, who says that he saw it at Viðey, on the 17th June, 1860, and thought that it had a nest in a hole in the wall of the little chapel there. This may have been the species mentioned by Olafsen (p. 586, § 679, b), as seen by him in September, 1763.

WHEATEAR. *Saxicola œnanthe* (Linn.)—Steindepill. Steinklappa. Grá-dílóttur.
Rather plentiful over the whole island, but, of course, only a summer visitant.

WHITE WAGTAIL. *Motacilla alba* Linn.—Máriu-erla. Máriátitlingur. Máriatla.
Not quite so common as the last, but, from its more familiar habits, more frequently observed. Arrives at the end of April, leaves in September.

MEADOW PIPIT. *Anthus pratensis* (Linn.)—Grátitlingur. Thúfutitlingur.
Common on low grounds over the whole country. Migratory as the last.

LAPLAND BUNTING. *Emberiza lapponica* Linn.—Sportitlingur.
Very seldom observed in Iceland. Faber saw a single example in the south in the spring of 1821. I do not know of any other unquestionable instance of its occurrence. Herr Preyer supposes that the bird mentioned by Olafsen (p. 586, § 679, a) may have been this species.

SNOW BUNTING. *Emberiza nivalis* Linn.—Snjótitlingur. Sólskríkja (æstate). Titlinga-blíke (mas).
Perhaps the commonest of Icelandic small birds. Most of them pass the winter in the country, according to Faber. The nests are pretty easy to find with a little patient watching, but difficult, and sometimes impossible to get at, from being situated so far in crevices of the rock.

MEALY REDPOLL. *Linota linaria* (Linn.) (*sed non auctt. britt.*).—Auðnu-titlingur.
Rare in Iceland. Faber found a nest on the 13th July, 1820, in the north, but the young had flown from it. It occurred in small flocks at Akureyri in the following winter, and Mr. Proctor met with a pair near that place, August 10, 1837. Dr. Krüper was more fortunate, obtaining several nests with eggs at Myvatn. Olafsen is thus entirely justified in his supposition (p. 586) that it bred in Iceland. He also mentions its appearance on the islands of the Breitifjörðr. I think Dr. Krüper is right in suggesting that the bird which Faber says (p. 14) he killed at Húsavík, Sept. 12, 1819, and which he identifies with the *Loxia serinus* of Scopoli, must have been only the young of this species. (*Naumannia*, VII. i. pp. 63, 64, note.)

RAVEN. *Corvus corax* Linn.—Hrafn. Krummi.
Very abundant, and resident all the year. The pied varieties, which, under the impression that they were peculiar to the Færoes, Vieillot, and others after him, have considered entitled to specific recognition, and named *Corvus*

leucophæus, have, since Faber's time, been also observed in Iceland. Herr Preyer says that he saw one at Fremriket in the north, and another at Hruni in the south. Mr. Baring-Gould observed one on Oxnardalsheiði. I have also known two pied Ravens in England. Mr. Wolley ascertained, on his first visit to the Færoes, of which an account will be found in Sir W. Jardine's *Contributions to Ornithology*, that it was no unusual thing for a mottled individual to be hatched from the same nest as entirely black ones. It appears to me that Herr Preyer has entirely misapprehended the statement of M. Temminck's, which he quotes (p. 389).

HOODED CROW.　　*Corvus cornix* Linn.—Kráka.

Does not inhabit Iceland, but occasionally pays a visit thither. Faber saw some in the north in July and August, 1819.

　　Certain black Crows are said to have occasionally made their appearance in the south-west of Iceland, and Faber supposes them to have been the *Corvus corone* of Linnæus, but Jonas Hallgrimsson (p. 65) especially says that they have been bald at the base of the bill, and hence I should infer that they were certainly Rooks (*Corvus frugilegus*, Linn.) The more so as the black Carrion Crow does not occur even in the Orkneys, Shetlands, or Færoes, nor is it found in Norway. Jonas Hallgrimsson further mentions that on one of these birds which was taken at Viðey, Olafsen composed a poem, and applies to it the name of "Færeyja-hrafn."

ICELAND PTARMIGAN.　　*Lagopus islandorum* (Faber).—Rjúpa.　Karri (*mas.*)
Rjúpkarri (*idem*).

Faber was the first to show the distinction between the Ptarmigans of Iceland and the rest of Europe, but it is by no means certain to me whether the former is not identical with that of Greenland (*Lagopus reinhardti*, Brehm), and this perhaps again with the *Tetrao rupestris* of older authors. I have a considerable series of skins of the Iceland bird in my possession, and they appear not only to differ constantly from Scotch and Norwegian specimens, but to differ much more from them than the latter do from one another. In Iceland the birds are pretty numerous, and not confined to the mere mountain tops as are their brethren in Scotland and other parts of Europe, but may be found almost in all places where berries grow.

WATER RAIL.　　*Rallus aquaticus* (Linn.)—Keldu-svín.

Rare in Iceland, though apparently a resident there. Faber obtained one in the north on the 23rd December, 1819. Dr. Krüper says he saw two of its eggs in a collection at Reykjavík in 1856, and these were probably the specimens which were obtained two years after by Mr. Wolley and myself, and are now in my possession. M. Benguerel also seems to have met with the bird, concerning which wonderful stories are told by the Icelanders.

　　I may here take the opportunity of remarking that Olafsen speaks of this bird (p. 227) as a *Tringa*, which, I think, rather diminishes the value of his statement about his puzzling "Lækiaduðra," also described by him as belonging to the same genus. I am myself

unable to suggest what this last species may be, but I think that
had Herr Preyer read the passage (pp. 985-6) carefully, he would
never have supposed it could refer to *Totanus ochropus*.

Coot. *Fulica atra* Linn.—Vatnshæna.
Faber mentions a pair killed at Reykjavík late in the year 1819, and one
caught in the sea off Grindavík, in April, 1821. In 1858, one was killed near
Utakála, the skin of which is in my possession. I am not aware that it has
occurred in Iceland except in the south-west. It is worthy of remark that
the Coot, which has been met with occasionally in Greenland, is the corresponding Transatlantic species, *F. americana*, Gmel.

Lapwing. *Vanellus cristatus* (Meyer).—Vepja. Isakráka.
Occasionally wanders to the south-west portion of the island, chiefly in
autumn. Faber mentions its occurrence at Hafnafjörðr in 1818, and on the
Vestmanneyjar in 1820; Jonas Hallgrimsson corroborates this assertion.

Golden Plover. *Charadrius pluvialis* Linn.—Lóa. Heylóa.
Quite the commonest bird in Iceland, and of great use to the traveller, who
by its means often obtains a good meal in the desert. In Greenland its place
is taken by the nearly allied *C. virginicus* Bork., which is easily distinguished
from it by its grey axillaries.

Ringed Plover. *Aegialites hiaticula* (Linn.)—Sandlóa.
Not rare on the sea-coast and on some of the moors in the interior.

Turnstone. *Strepsilas interpres* (Linn.)—Tildra.
Said by Faber to be of commoner occurrence in the south and west than in
the north; yet he found it on Grimsey, in June, 1820. It arrives in Iceland
about the last week in April; and I have little doubt breeds there, for
Mr. Proctor has received its unmistakable eggs from the north. It mostly
leaves again in the autumn, but Faber obtained one at Reykjavík, on the
11th December, 1820. In 1858 it was very common in the south-west
about the end of May.

Oyster-catcher. *Haematopus ostralegus* (Linn.)—Tjaldur.
Like the last, appears to be more common in the south than in the north.
Faber considered it to be resident throughout the year, for it remains in large
flocks during the winter in the south. It is of course most abundant on the
sea-coast, but Herr Preyer met with it on some of the inland waters.

Grey Phalarope. *Phalaropus fulicarius* (Linn.)—Thorahani. Flatnefj-
aðdur-Sundhani. Rauðbrystingur (*partim*).
This bird has been but seldom observed by strangers in Iceland, yet in 1858
I found that it was very well-known to the natives of the district where
Faber had seen it in 1821. On the 21st June in that year he obtained a
pair, which were swimming in a flock of the commoner species, next to be
mentioned. The female contained largely developed eggs. On the following
day he found a single pair at their breeding-place, in the neighbourhood of
the same locality, and searched in vain for their nest. Finally, on the
9th July, he met with a family party some miles to the eastward. In 1858,

I discovered two pairs on a lake in the same district ; but a few days afterwards they had disappeared, and they certainly did not remain to breed there that year. Last summer, a friend of mine sent me four eggs as those of this bird, which had been taken under his special superintendence. Setting aside the excellent authority on which their identification rests, they are so entirely different from those of any other Icelandic bird I know, that I can hardly doubt their genuineness.

RED-NECKED PHALAROPE. *Phalaropus hyperboreus* (Linn.)—Óðinshani. Sundhani.

Very common all over the island on all ponds and lakes. Arrives late in May, and at once begins the duties of nidification. On one occasion the month of June I saw a flock of at least a hundred sitting on the surf, between the breaking waves and the shore.

REDSHANK. *Totanus calidris* (Linn.)—Stelkur.

Very commonly met with throughout the island. Arrives about the middle of April, and, according to Faber, some remain till the end of the year.

BLACK-TAILED GODWIT. *Limosa aegocephala* (Linn.)—Jaðreka.

Arrives the last week in April, according to Faber. I do not know that any naturalist has found the eggs of this bird in Iceland, but have little doubt it breeds there. It seems to be rare, if it occurs at all, in the north. I obtained a fine pair in the flesh at Reykjavík, towards the end of June.

RUFF. *Philomachus pugnax* (Linn.)—Kragi. Áflogakragi.

Faber records a single instance of the occurrence of this bird in Iceland, a female, near Reykjavík, at the beginning of September, 1820.

KNOT. *Tringa canutus* Linn.—Rauðbrystingur (*partim*).

This bird arrives late in May, and Faber considers that it breeds on the uplands, a supposition which I consider very probable. However, Mr. Fowler, whose attention I particularly called to it before he started for Iceland, says:— " I had my eyes very wide open for this bird. I never once got a glimpse of it, and do not believe in its existence in the island at the breeding-time, at any rate inland. Though I questioned the natives very closely, I could hear no tidings of it." On the south-west coast it is very well known as a bird of passage. One morning, at the end of May, 1858, I found the shore at Kyrkjuvogr literally alive with a large flock of Knots, all in their beautiful red plumage. There had been none there the day before. They stayed about a week, their numbers gradually diminishing until at last only two or three were to be seen. This is one of the birds possessing great interest to the oologist, for, I believe, no collector has well-authenticated specimens of its eggs. Notwithstanding Mr. Fowler's evidence, I still conceive it possible that a few pair may remain to breed in the island, though undoubtedly the majority pass on to Greenland, or perhaps to land farther north of which we have no knowledge.

PURPLE SANDPIPER. *Tringa maritima* Gmel.—Selningur. Fjallafala (*æstate*).

Common everywhere in the neighbourhood of the coast, and occasionally to be seen inland, where it also breeds. According to Faber, a resident. Hatches its eggs about the middle of June. Great numbers are shot near Reykjavík in spring, and are sold for the table.

DUNLIN. *Tringa alpina* Linn.—Lóuthræll.

Not so abundant, according to my observation, as the last species, but frequents the same localities. According to Faber, appears only from the middle of April till the end of October. Its well-known habit of flying after the Golden Plover gets it its Icelandic name, as well as its corresponding appellation in some parts of England, of "Plover's Page."

> The *Tringa schinzi* of Herr Preyer (p. 402) is doubtless the small race of Dunlin well known in some parts of Europe, and not the American bird to which that name is often misapplied.

SANDERLING. *Calidris arenaria* (Linn.)—Sanderla.

Possibly more common in Iceland than has been thought. Faber, and after him Mr. Proctor, observed it on Grimsey, where it has been said to breed ; but Dr. Krüper rightly remarks that the eggs fathered upon it by the inhabitants are doubtless those of the Ringed Plover (*Naumannia*, VII. ii., p. 18, note). At the same time, I may remark that I know of no one who has an authenticated egg of this species. In 1858, I saw several in the south-west, and shot a female, with a very backward ovary, on the 21st May, at Bæjasker. Mr. Fowler saw it in 1862 at Akranes.

COMMON SNIPE. *Gallinago media* (Leach).—Hrossagaukur. Mýrispíta. Mýriskitur. Mýrismpa.

Fairly abundant in suitable localities. According to Faber, arrives the last week in April, and leaves about the middle of October, though a few probably remain through the winter, for he saw three at a warm spring in hard frost, on the 3rd February, 1821.

CURLEW. *Numenius arquatus* (Linn.)—Nefboginn-Spói.

Faber records one that was shot at Reykjavík, 6th September, 1819. Dr. Krüper was told that the autumn before he was there, six examples had been killed at the same place.

HUDSONIAN CURLEW. *Numenius hudsonicus* Lath.

This is the *Numenius borealis* of Wilson (but not of Forster), of which Dr Kæjrbölling states (*Naumannia* VI., p. 308) that he had received a specimen from Iceland. In the event of its occurring there again, I may mention that it is easily distinguished from the next species by its rufous axillary plume.

WHIMBREL. *Numenius phaeopus* (Linn.)—Spói.

A very common bird, and one of the most characteristic of Iceland. Arrives at the end of April, breeds on the moors, and departs by the middle of September. The late Prince C. L. Bonaparte considered that the specimens of a *Numenius* obtained on his cousin's expedition to Iceland and Greenland, were distinct from the common *N. phaeopus* (*Comptes Rendus*, Août 2, 1856, XLIII., p. 1021). I agree with Professor Reinhardt (*Ibis*, 1861, p. 10) in doubting the necessity for so doing.

COMMON HERON. *Ardea cinerea* Linn.—Hegri.

According to Faber, stray examples occur every spring and autumn, for the most part in the south ; but he mentions one taken at Grimsey, in September, 1819. It occasionally wanders to Greenland.

WHOOPER. *Cygnus ferus* Leach.—Ālpt. Svanur (*poetice*).

According to all accounts, common in some districts of the country, as testified also by the numerous places which take their names from it. Faber and others say it is resident during the whole year. In the winter congregates in large flocks, which, in the summer, separate first into small parties, the members finally pairing off among themselves. Breeds in many places in the interior. Bewick's Swan (*Cygnus minor*, Pallas) has, by many authors, been said to occur in Iceland; but the late Etatsraad Reinhardt asserts (*Naturhistorisk Tidsskrift* II., p. 532) that this is an error, and I fully believe he does so with justice. The same excellent authority shows also that the Swan inhabiting Iceland is identical with that of the north-west of Continental Europe, and that there is no good foundation for its recognition as a distinct species under the name of *C. islandicus* of Breham.

—GOOSE. *Anser*—(? sp.).—Grágaes. Grágás.

I consider that we are not yet in a position to decide which species of large Goose visits Iceland, or whether more than one species does so. I have never seen any examples killed in the island, though it is no doubt common enough there. I do not believe that specimens of eggs obtained there, some of which are in my collection, afford evidence at all conclusive on the subject. Some writers identify the "Grágaes" with *Anser ferus* (Gmelin), others with *Anser segetum* (Gmelin). The former is easily distinguished from the latter by several characters, among them the *white* "nail" at the end of its bill, which in the Bean Goose is *black*. It seems to me not improbable that the Icelandic bird may be the Pink-footed Goose (*Anser brachyrhynchus*, Baillon). If travellers would but save merely the heads of any Geese they may obtain in Iceland, and submit them to naturalists at home, the question might be easily settled.

WHITE-FRONTED GOOSE. *Anser albifrons* (Linn.)—Helsingi.

Faber only observed this species in the south of Iceland. He says it breeds there, a statement doubted by Dr. Krüper. Faber never states, as asserted by Herr Preyer, that he did not obtain it. On the 11th of May, 1858, I saw several freshly-killed examples at Reykjavík, one of which I purchased, and had it preserved as a specimen. All the Icelanders who saw it recognised it as "Helsingi."

BERNACLE GOOSE. *Bernicla leucopsis* (Temm.)

Faber says this species arrives in Iceland a few days earlier than the "Grágaes," that is to say, about the middle of April, and departs about the middle of October. He found it most abundant in the south-west, though not rare in the north. He hardly believes it to breed in the island; for it is never seen in the summer. I do not know that later travellers have themselves observed it. Olafsen figures (tab. xxxiii.) an Icelandic specimen.

BRENT GOOSE. *Bernicla brenta* (Linn.)

According to Faber, arrives at the same time as the last species, but is rare, being only occasionally found here and there throughout the island. Yet he says a nest found at the end of June, 1819, inland from Eyjarfjörðr, belonged to this species. At Myvatn, the people told Mr. Baring-Gould that this, as well as three other kinds of Geese, bred on the islands in the lake, and that the name of this was "Margás." But I do not think that any recent naturalists can speak of the species from their own observation.

It will be seen from the above remarks that our knowledge of the different Wild Geese frequenting Iceland is very imperfect. Olafsen (pp. 548, 549) speaks of three species. He calls the largest of them "Hraagaasen;" the second, "Helsingen;" and the third and smallest, "Hrota" or "Mar-gies." I doubt not that future investigation will in time thoroughly clear up the subject, at present in such an unsatisfactory condition.

GARGANEY. *Anas querquedula* Linn.—Taumönd. (?)

The positive assertion of Herr Preyer (p. 408) that, on the 16th June, 1860, he shot this species at Myvatn, induces me to admit it here, but not without hesitation. The evidence afforded by the eggs brought to him a few days previously I look upon as singularly inconclusive.

TEAL. *Anas crecca* Linn. — Krikönd. Lilla-gräönd. Urtönd, Urt (*partim*).

Very common. Arrives, according to Faber, the third week in April, and departs the beginning of October.

WILD DUCK. *Anas boschas* Linn.—Stokönd. Husönd (*partim*). Grönhofdi (*mas*). Gräönd (*fæmina*).

Common in Iceland. Most remain through the winter, but some migrate.

PINTAIL. *Anas acuta* Linn.—Grasönd. Lángviu-gräönd.

According to Faber, arrives on the coast at the end of April, and reaches Myvatn the beginning of May, where it breeds pretty commonly. Probably of general distribution throughout the country. Disappears at the beginning of September.

WIGEON. *Anas penelope* Linn.—Ráuðhofði. Urtönd, Urt (*partim*).

According to Faber, not so common as the preceding, and arrives later. The time of its departure he did not ascertain. Breeds at Myvatn, and probably elsewhere. Has of late been observed also in Greenland.

Faber states that, at Myvatn, in June, 1819, he chased from their nest a pair of Ducks, which he took for Gadwall (*Anas strepera*, Linn.). Dr. Krüper (*Naumannia*, VII. i. p. 48), though including the species, states he did not himself see it there, but mentions that eggs were brought to him as belonging to it, a circumstance which happened also to Mr. Proctor, and does not bear very weighty evidence. Looking upon this as a bird of much more southern range I have omitted its name from my list, but I shall willingly own I am wrong, on receiving good testimony to the contrary. Mr. Baring-Gould believes that he saw the Shoveller (*Anas clypeata*, Linn.) on Eyjarfjörðr. This, I think, is more likely, but as he did not get near enough to satisfy himself completely on the point, I do not now include it. Mr. Proctor is very certain that he got a nest of the Bimaculated Duck of English writers (*Anas bimaculata*, Keys. and Blas.) at Myvatn, on the 28th July, 1837. Having, in every way, a very high opinion of Mr. Proctor, I do not for a moment doubt that he did so ; but as I also am strongly persuaded that the so-called Bimaculated Duck is nothing but a hybrid, I abstain from inserting its name here as a good species.

POCHARD. *Fuligula ferina* (Linn.)
Mohr, as rightly quoted by Herr Preyer (p. 430), says that he once saw this bird in the Eyjarfjörðr river. For eighty years no one else seems to have noticed it in Iceland, but on the 20th June, 1860, Herr Ernest Gehin shot one on Thingvalla lake, which Herr Preyer (p. 430) saw the next day.

SCAUP. *Fuligula marila*, (Linn.)—Dúggönd. Hrafnsönd (*regione australi*).
Said by Faber to be the commonest Duck breeding at Myvatn, where they arrive about the middle of April, having reached the south a month earlier. In the beginning of October they resort to the fjords in flocks, and shortly after leave the island.

    Faber states he saw a flock of White-eyed Ducks (*Fuligula nyroca*, Steph.) at Eyjarfjörðr, 20th May, 1820 ; he subsequently found a nest which he thought belonged to this species ; and again, on the 10th March, 1821, another flock of birds at Eyrarbakki. By no one else has it been observed in Iceland, as far as I know, and Dr. Krüper remarks that an egg, sent by Faber himself to the Greifswald Museum, is that of the Harlequin Duck. Herr Preyer has introduced *Fuligula rufina* (Linn.) to his list of birds accidentally occuring in Iceland, on the strength of a passage in the German translation of Mr. Pliny Miles' *Norðufari*, and though he does suggest the possibility of its being a mistake for *F. ferina*, he seems unwilling to give up the opportunity of adding a new species to the Icelandic avi-fauna. I do not know whether the Latin name is inserted in the original American edition of the work, but from Mr. Miles' remarks on " the language of the Cæsars when written by a Dane " (English edition, p. 166), I cannot accept him as an authority on scientific nomenclature.

ICELAND GOLDEN-EYE. *Clangula islandica* (Gmel.)—Húsönd,
According to Faber, resident during the year. Frequents the sea-coast in the winter, and about the middle of March repairs to its breeding quarters, of which Myvatn seems to be chief. Faber did not perceive the differences between this and the common European species *C. glaucion* (Linn.); though they are manifestly great. Has visited Germany as a straggler, but not yet recognized in the British Islands. Breeds in South Greenland, and occurs, perhaps only as a winter migrant, in the fur-countries of North America.

HARLEQUIN DUCK. *Histrionicus torquatus* Bonap.—Straumönd. Brimdufa.
A common resident in Iceland, according to Faber, but changing its
quarters from north to south in winter. Frequents the most rapid rivers, on
the margins of which it generally breeds. This bird seems to inhabit Greenland, the whole breadth of Arctic America, and Eastern Siberia; but with
the exception of Iceland, only occurs as a rare straggler in Europe.

LONG-TAILED DUCK. *Harelda glacialis* (Linn.)—Hávella. Fóvella.
As abundant in Iceland, where it appears to remain all the year, as in other
northern countries.

EIDER. *Somateria mollissima* (Linn.)—Æsurfugl. Æsurbliki (*mas*).
Æsurkolla (*fœmina*).
Very numerous on the coasts and some of the lakes. Appears not to migrate.

KING DUCK. *Somateria spectabilis* (Linn.)—Æsurkóngur.
By all accounts a rare bird in Iceland, and generally only a straggler from
Greenland or elsewhere. Yet Faber says that a pair bred on Visey, in 1819
and 1820, among the multitudes of the common Eider. He only mentions,
besides, the occurrence of one at Hofsaas a few years before his visit, and one
washed up dead at Eyrarbakki, December 25, 1820. Mr. Baring-Gould was
shown a skin of this bird at Akureyri.

SCOTER. *Oedemia nigra* (Linn.)—Hrafnsönd.
Faber thought this bird was only to be found at Myvatn, where it breeds.
Herr Preyer mentions one shot, out of a flock of eight, at Arnarvatn.

GOOSANDER. *Mergus castor* Linn.—Stóra-toppönd. Gúlönd (*regione
australi*).
Less common in the south than in the north, where it stays even the whole
winter, according to Faber. Breeds.

RED-BREASTED MERGANSER. *Mergus serrator* Linn.—Lilla-toppönd.
Far commoner than the preceding, and has much the same habits.

CORMORANT. *Phalacrocorax carbo* (Linn.)—Skarfur. Dílaskarfur.
According to Faber, breeds only in the north, and is but a winter visitant in
the south. But Herr Preyer says it breeds on the Vestmannaeyjar. Certainly not so numerous a species as the next.

SHAG. *Phalacrocorax graculus* (Linn.)—Topp-akarfur.
Pretty common; and a resident, according to Faber.

GANNET. *Sula bassana* (Linn.)—Súla. Háfsúla.
Very abundant in many localities, and has several breeding places on islands,
among which Grimsey, the Reykjanes Fuglasker, and some of the Vestmannaeyjar are chief. Remains all the winter, according to Faber.

ARCTIC TERN. *Sterna macrura* Naum.—Kría.
Has many breeding places in various parts of the country. According to
Faber, arrives about the middle of May, and generally departs about the end
of August, though a few, chiefly young ones, remain a month longer on the
southern coast. Mr. Baring-Gould informs me that the Common Tern (*Sterna
fluviatilis*, Naum.) has been once found in Thingvalla-lake. I think there may
be some mistake here; for I am of opinion that further investigation will

27

show that species to have but a limited northern range. So far as I can judge from the safest evidence within my reach, it is not found in either the Shetlands or Færoes.

KITTIWAKE. *Rissa tridactyla* (Linn.)—Ritur. Ritm. Skegla.
Exceedingly common all round the coast. Arrives in the beginning of March, and goes away the middle of August.

COMMON GULL. *Larus canus* Linn.
Brünnich (p. 43) records this Gull from Iceland; but I should, with Herr Preyer (p. 433), have considered that the statement was an error, had I not, in 1858, procured the skin of an immature bird, which had been shot near Reykjavík the preceding winter. Mr. Baring-Gould says that he saw this species near the Iceland coast on his voyage thither from the Færoes. It is certainly not a usual inhabitant of the last-named islands.

ICELAND GULL. *Larus leucopterus* Faber.—Hvít-máfur.   Hvít-fugl.
Grá-máfur (*junior*).
A winter visitant only, arriving, according to Faber, towards the end of September, and mostly leaving by the end of April, though some, chiefly birds in immature plumage, remain later into the summer.   Mr. Wolley had one for some weeks alive at Kyrkjuvogr.   It had been caught on a fish-hook, and in a day or two grew so tame as to take food in one's presence.

GLAUCOUS GULL. *Larus glaucus* Fabricius.—Hvít-máfur.   Hvít-fugl, Grá-máfur (*junior*).
Common, and resident, according to Faber, who says that it breeds on the rocky coasts of the promontory which divides the Faxafjörðr from the Breiðifjörðr.   Mr. Milner took its nest on an island in the first-named bay ; and any number of its reputed eggs may be bought at Reykjavík, but they are in all probability those of the following species.

GREAT BLACK-BACKED GULL. *Larus marinus* Linn. — Svartbakur.
Veiðibjalla. Káflabringur (*junior*).
According to Faber, a resident, and not so common as the foregoing ; but, from my own observation, I should say it was more abundant than that, at least, in the south-west, and in the breeding season.   It breeds on the inland waters, which the other is not known to do.

GREAT SKUA. *Stercorarius catarrhactes* (Linn.)—Hákalla-skúmur, Haf-skúmur.
Pretty common along the coasts, and occasionally breeding some distance inland.   According to Faber, is resident.   He names four breeding-places in the south—an island in the Œlfusá, a sandy plain opposite the Vestmanneyjar, and the dunes of Skeiðarásandr and Breiðamerkr.   Dr. Krüper saw it in the north, in summer-time, so that it probably breeds there also.

POMATORHINE SKUA. *Stercorarius pomatorhinus* (Temm).
Not common, but has been observed occasionally by several travellers in Iceland.   We are indebted to Herr Preyer for giving us the etymology of the name originally applied to this species by Temminck, but, as suggested by Dr. Sclater (*Ibis*, 1862, p. 297), orthography requires a still further

emendation, and this I have here ventured to adopt. I saw this bird once in Iceland, the day of my arrival at Reykjavík, 27th April.

ARCTIC SKUA. *Stercorarius parasiticus* (Linn.)—Kjói.

Common enough throughout the island, for it occasionally occurs, and even breeds, on the moors far inland. According to Faber, it arrives about the 25th April, and remains until the middle of September. The *Lestris thuliaca* of Herr Preyer (p. 418) I believe to be founded on an example of the dark-coloured race of this species, accidentally mottled with white. Mr. Baring-Gould obtained another specimen of this variety, which was shot by his guide on Arnarvatns-heiði, as before related in this work (p. 107). It is a well-known fact that both this and the Pomatorhine Skua are subject to great diversity in plumage; and I cannot look on the diagnosis given by Herr Preyer as characteristic of a new species.

BUFFON'S SKUA. *Stercorarius buffoni* (Boie).

Faber was probably not aware of the difference between the last species and this. In 1858, Mr. Wolley and I observed it several times at Kyrkjuvogr; and a very beautiful one, killed a day or two before at Keflavík, was brought to me on the 10th June. Herr Preyer, in 1860, saw a skin at Reykjavík.

FULMAR. *Procellaria glacialis* Linn.—Fylungur (*reg. bor.*) Fylíngur (*reg. austr.*) Fill.

Abundant in many parts of the Icelandic seas. Never enters the bays. Faber says that the chief breeding-places are Grimsey, Látrabjarg, Krísuvíkr-berg, but especially on the Vestmanneyjar, where they are the commonest of all birds. He also mentions Hafnarberg as a nesting locality, but I think he must have been misinformed; at least, I saw or heard nothing of this species there in 1858, though it is very abundant round Eldey. On Grimsey, a grey variety, known as "Smiður," is said to occur. (*Naumannia*, VII. p. 437.)

GREATER SHEARWATER. *Puffinus major* Faber.—Stóra-skrófa.

Faber met with only one example of this bird, which probably rarely extends its wanderings so far to the north, though it is stated to breed in Greenland.

MANX SHEARWATER. *Puffinus anglorum* (Temm.)—Skrófa.

Remains on the Icelandic seas all the year, according to Faber. Commoner in the south, and especially on the Vestmanneyjar, than in the north.

STORMY PETREL. *Thalassidroma pelagica* (Linn.)—Drudi.

Mohr mentions that he found two persons who knew this bird in Iceland, but most could tell him nothing of it. Herr Preyer says he himself saw it near the Vestmannaeyjar, otherwise I should have hesitated to include it here. It must be of unfrequent occurrence.

LITTLE AUK. *Mergulus alle* (Linn.)—Haftirdill. Halkión.

Occurs, according to Faber, all the year round. It only breeds on Grimsey, where Faber found it in 1820, and Mr. Proctor in 1837.

BLACK GUILLEMOT. *Uria grylle* (Linn.)—Tejsti.

Resident in Iceland, according to Faber, and, I believe, is of pretty general distribution around the coast in breeding time.

420 APPENDICES.

COMMON GUILLEMOT. *Uria troile* (Linn.)—Langnefia. Langvia.
Breeds in very many spots on the coast, and in some places in countless
numbers. Faber says that a good many remain over the winter. The
curious variety, which has been by some naturalists taken for a species, and
called in Iceland " Hring-langnefia" or " Hringvia " (*Uria leucophthalmus*,
Faber), occurs in about the same uncertain proportions as it does in other
countries. But a more wonderful variety is one with a yellow bill and feet,
which seems occasionally to occur on Grimsey, where it is spoken of as the
" Langviu-kóngur." (*Naumannia*, VII. p. 437.)

BRUENNICH'S GUILLEMOT. *Uria bruennichi* Sabine.—Stutnefia.
Faber seems to have been in doubt whether this was really distinct from the
last, as he found so great a resemblance in their habits. Olafsen mentions it
(pp. 855, 562) in reference to two localities, Snæfellsnes and Látrabjarg, but
in one place he considers it to be the female of the last-mentioned species.
He gives a figure of it (tab. xxii.), which shows what he is speaking about.
Faber appears to consider that it bred in company with the other species all
round the coast. Mr. Proctor found it at Grimsey. Herr Preyer shot
one at Reykjavík. I was told it was occasionally met with at Hafnaberg
in the breeding time, but I never saw it there, nor, indeed, elsewhere in
Iceland.

GARE-FOWL. *Alca impennis* Linn.—Geir-fugl.
I hope I may here be permitted to refer to an article on this bird which
appeared in the *Ibis* for 1861, wherein I endeavoured to give an abstract of
all the particulars respecting its history in Iceland that I was then aware of.
I have not since learned much that is of great importance on the subject,
though Herr Preyer, in the *Journal für Ornithologie* for 1862, makes some
assertions entirely novel to me. One of them, and the only one I shall here
mention, is the statement that the bird formerly bred on the little rock at
Cape Reykjanes, known by the name of " Karl." I believe that Mr. Wolley
and I examined nearly every fisherman who had been in the habit of pursuing
his vocation at that spot, and not one of them ever mentioned such a circumstance. Had Herr Preyer himself visited the locality, instead of obtaining
his intelligence from the Reykjavík merchants, I think he would at once have
seen the improbability of the case being as he says it was. The person from
whom he derived most of his intelligence, including this reputed fact, also
furnished Mr. Wolley and myself with a paper, in his own handwriting, and
now before me, respecting various Gare-Fowl expeditions, of which he stated
he was the chief instigator. As I remarked in my former article, this account
contained " details which are certainly inaccurate." I am, therefore, not
much surprised to find that it does not tally with the version he delivered to
Herr Preyer, for I had before inferred that the writer, in drawing it up, had
relied upon a very defective memory, in place of notes made at the time.
The interest taken in the history of the Gare-Fowl is so deservedly great,
that I feel it incumbent upon me to correct, as far as I am able, the misappre-
hension likely to arise from the erroneous statement which Herr Preyer
(unwittingly, I am sure) has made.

RAZOR-BILL. *Alca torda* Linn.—Álka. Klumba.

Faber considers it not quite as common as the two larger Guillemots above mentioned, which, collectively with it, are generally called " Svart-fugl ; " but I do not think that *Uria bruennichi* can be so abundant in Iceland as he represents. The Razor-bill frequents the same stations as those birds, and breeds among them. In winter many leave the neighbourhood of the island, but some also remain.

PUFFIN. *Fratercula arctica* (Linn.)—Lunôi.

Very common, and breeds in numerous localities around the coast. At the beginning of October they betake themselves to the open sea, returning to their nesting quarters at the beginning of May.

NORTHERN DIVER. *Colymbus glacialis* Linn.—Himbrimi. Brúsi (*regione boreali*).

Not uncommon, a pair or so breeding on nearly every lake. They arrive in the north, according to Faber, the first week in May, and towards the end of August they begin to show themselves on the sea, where, it appears, they remain during the winter.

RED-THROATED DIVER. *Colymbus septentrionalis* Linn.—Lómur. Therri-kráka.

More common, Faber says, than the last-mentioned species, especially in the south-west. They appear first in the fjords at the beginning of April or first week in May, according to the latitude of the district, and soon after repair to the lakes, where they breed. Towards the end of August they return to the sea with their young, and there they remain till the end of October, though a few winter in the south.

HORNED GREBE. *Podiceps cornutus* (Gmel.)—Sefönd. Flórgoôi. Flóaskitur. Flóra (*reg. occident.*)

Very generally distributed on lakes throughout the western half, and probably the whole of the island. Arrives about the same time as the species last mentioned, and, after breeding, departs in the autumn. Faber, and many others, recognize the existence of a second Grebe, *P. auritus* (Linn.), and this supposed bird has been called by Boie *P. arcticus*, but I have never been able to satisfy myself that there is more than one to be found in Iceland. Mr. Baring-Gould thinks that he saw another species of Grebe on Myvatn, which he is inclined to identify with *P. rubricollis* (Gmel.). It is very possible that such a bird may accidentally occur in Iceland, but as he did not procure a specimen, and the chances seem to me equal whether it may have been the European or American species of Red-necked Grebe, for the latter (*Podiceps holbœlli*, Reinh.) visits Greenland, I forbear including either in this list.

Herr Preyer notices (p. 248) that the Fieldfare (*Turdus pilaris*, Linn.) and the Starling (*Sturnus vulgaris*, Linn.) have been said to occur in Iceland. In neither case do I think the evidence amounts to proof of the fact, and I have much pleasure in stating that I agree with the conclusion at which he has arrived on the subject.

ELVEDEN, *23rd May*, 1863.

APPENDIX B.

ADVICE FOR SPORTSMEN.

ICELAND, as a country for game, is well worthy of the attention of the English sportsman. The game consists of reindeer, wild-fowl, and the white and blue fox, the skins of the latter fetching, in the Russian market, from 50 skillings to 80 skillings a piece. The reindeer are tolerably numerous on some parts of the island, as at Myvatn, the north-eastern corner, and near the Vatna Yokul. There is a considerable amount of difficulty attendant on the pursuit of them during the summer, owing to their occupying the high ground, where they find food adapted for them, but on which your horses are sure to starve. Swans are found in large quantities during the months of June, July, August, and September. During the first two mentioned months they arrive there from all parts to breed ; after which process is finished, they spend the remaining two in moulting, during which season grand sport is to be had in hunting them on horseback, as they are then unable to fly. The next aquatic bird in size that I saw is the Anser torquatus, or brent goose. It appears to be a very rare visitant of the island. I had the fortune to see three one morning, seemingly bound for a more southern clime. After the Anser come the two species of diver, Colymbus glacialis and Colymbus septentrionalis. They are met with all over the island. They are very difficult of approach, and have a most wonderful power of diving. I had an opportunity, on one occasion, of testing the speed with which they can propel themselves through the water when diving, and I should estimate it at not less than eight miles per hour. They can traverse a distance of upwards of three hundred yards without ever appearing on the surface. I found the red-throated diver in large quantities. They are very easily approached, and will very frequently take to the wing. They possess a wonderful tenacity of life ; as, on one occasion, I tried to kill a bird which my dog had caught in a small pool of water, by puncturing his brain, but all to no purpose, as it had only the effect of stupefying him for a short time. I am sure his brains must have had nearly as much stirring about as the porridge that my friend Mr. Baring-Gould attempted to make for the first time with us.

Wild ducks abound in large quantities, and of many species. In some districts you may easily bag from fifteen to twenty brace in an afternoon's shooting. I killed a very small duck, about the size of a golden plover, with a plumage exactly similar to that of the Anas boschas, or wild duck, whose species I have never been able to find out. I think I met with two species of grebe, of which one is Podiceps auritus. The above-named aquatic birds, I think, will sum up all the water-fowl that any sportsman will care to shoot in Iceland.

My next division is game, or such as are preserved in this country. The first that I shall name is the Tetrao islandorum, or ptarmigan. These birds are very numerous throughout the island. At Myvatn, you may bag from fifty to sixty brace in a day's shooting with perfect ease. They are very tame indeed; in fact, too much so, as you find considerable difficulty in causing them to rise. The Scolopax rusticola, or woodcock, is altogether a stranger to that part of the globe. However, it has its representative in the S. gallinago, or common snipe, which is literally plentiful. The Numenius phæopus, or whimbrel, are very numerous, especially near Thingvalla; the Charadrius pluvialis, or golden plover, together with Totanus calidris, or red shank, are pretty numerous, and may finish off the second class of birds.

The shot that I would recommend any sportsman to take are Nos. 1 and 5. He will very seldom have to use No. 1, except when in a district where swan abound.

The fishing in Iceland is free to any one who cares to pursue the gentle craft. The fishing there is of no ordinary kind; for you may go out on many of the lakes, and, in the space of a couple of hours, bring in your twenty pounds weight of trout and char. I never fished much; but I remember on one occasion, on a small lake near the Hopvatn, myself and Mr. H—— killing nineteen fish in three hours, weighing thirty-nine pounds. Nearly all the trout that we caught were, on an average, a little more than one pound, whether it might be in a lake or river. The salmon-fishing last year in Iceland was very bad indeed, owing to a want of rain, from which cause we only killed three, and a few sea trout.

The flies to use in almost all the rivers for trout are grilse flies, as the trout won't look at our ordinary loch flies. Minnow, on the whole, I found fully the most successful. The minnow which I used was a protean minnow, with ten fins to make it spin with rapidity. I found it, with that addition, beat the phantom at the rate of five to three and a half.

The char of Iceland are very large indeed, weighing as heavy as four pounds. There is another fresh-water fish, the English name of which I have been unable to ascertain, but which the natives call "suburtingur." Its habits are similar to those of the salmon—passing one part of the year in salt, and the other in fresh, water. It takes the minnow with great avidity. It grows to the weight of twenty pounds, and has a pink-coloured flesh.

J. W. R.

APPENDIX C.

A LIST OF ICELANDIC PLANTS.

I. DICOTYLEDONES.

RANUNCULACEÆ :—

Thalictrum alpinum (L.) Alpine meadow-rue. Vèlindisurt, kross-græs, júframeir.—Vatnsdalr, Skjalfanda.

Ranunculus aquatilis (L.) Common water-crowfoot. Lóna-sóley.— Vatnsdalr.

heterophyllus (Fries.)
hederaceus (L.) Ivy-crowfoot.—Vithimyri.
glacialis (L.) Dverga-sóley.
Flammula (L.) Leper spearwort.—In tún Thingvalla.
acris (L.) Upright meadow crowfoot. Brennisóley.
repens (L.) Creeping crowfoot. Everywhere.
polyanthemus (L.)
pygmæus (Wahlenb.)—Hólar.
nivalis (L.) Dverga-sóley.
lapponicus (L.)—Between Haukadalr and Laugarfjall. H.
hyperboreus (Rottb.)

Caltha palustris (L.) Marsh marigold. Lækja-sóley. Near all farms.

PAPAVERACEÆ :—

Papaver nudicaule (L.) Mela-sól.—Hrútafjord.
alpinum (L.)

CRUCIFERÆ :—

Arabis alpina (L.) Hjaltadal, below Hólar.
petræa (L.) Alpine rock-cress.—Vithivik in Skagafjord.
bellidifolia (Jacq.)
brassicæformis (Wallr.)

Cardamine pratensis (L.) Common bitter-cress. Hrafna-klukka.
hirsuta (L.) Hairy bitter-cress.

CRUCIFERÆ (*cont.*):—

· *Nasturtium amphibium* (Br.) Hrafna-klukka meth gulu blómstri.
    terrestre (Br.) Marsh yellow-cress.
Cochlearia officinalis (L.), and its variety.
    danica (L.) Skarfa-kál. Common scurvy-grass.
    anglica (L.) English scurvy-grass.—Hrútafjord hála.
Draba verna (L.) Common whitlow-grass.
    contorta (Ehr.)
    confusa (Ehr.)
    muralis (L.) Speedwell-leaved whitlow-grass.
    muricella (Wahlb.)
    aizoides (L.) Yellow Alpine whitlow-grass.
    rupestris (Br.) Rock whitlow-grass.
Cakile maritima (Scop.) Purple sea-rocket.
Subularia aquatica (L.) Water awl-wort.
Capsella bursa (D. C.) Common shepherd's-purse. Pung-arfi.
Lepidium campestre (Br.) Common Mithridate pepper-wort.

VIOLACEÆ :—

Viola palustris (L.) Marsh violet. Fjóla. Vatnsdalr.
    canina (L.) Dog violet. Hundar-fjóla, Tyrs-fjóla.—Vithidalr.
    tricolor (L.) Heart's-ease. Blóth-sóley.—Hóf in Vatnsdalr,
    Akureyri.
    montana (L.)

DROSERÆ :—

Drosera rotundifolia (L.) Round-headed sun-dew. Arnarvatn.
    longifolia (L.) Spathulate-leaved sun-dew.
Parnassia palustris (L.) Grass of Parnassus. Mýra-sóley.—Reykjahlith,
    Arnarvatn.

POLYGALACEÆ :—

Polygala vulgaris (L.) Common milk-wort. Hnausir.

CARYOPHYLLACEÆ :—

Silene acaulis (L.) Moss campion. Lamba-gras. On every moor
    and heithi in Iceland.
    inflata (Sm.) Bladder campion. Holurt, Hjarta-gras.—Skjal-
    fanda.
    maritima (With.) Sea campion.—Mithfjord.
    rupestris (L.)
Lychnis flos-cuculi (L.) Ragged robin. Muka-hetta. Much dwarfed.—
    Thingvellir.
    viscaria (L.) Red German catchfly.
    alpina (L.) Red Alpine catchfly. Kreisu-gras.
Sagina procumbens (L.) Procumbent pearl-wort.
    saxatilis (Wimm.) Alpine pearl-wort.

CARYOPHYLLACEÆ (*cont.*):—

 Sagina subulata (Wimm.) Awl-shaped pearl-wort.
  nodosa (E. Meyer). Knotted pearl-wort.
 Spergula arvensis (L.) Corn sparrey.—In tún at Hnausir.
 Arenaria verna (L.) Vernal sand-wort.
  peploides (L.) Sea-purslane. Barja-arfi, Smethju-kál.
  ciliata (L.) Fringed sea-purslane.
  serpyllifolia (L.) Thyme-leaved sea-purslane.
 Stellaria media (With.) Common chickweed.
  crassifolia (Ehr.)
  humifusa (Rottboell.)
 Cerastium vulgatum (L.) Broad-leaved mouse-ear chickweed.—Hnausir.
  viscosum (L.) Narrow-leaved mouse-ear chickweed.—Hnausir.
  alpinum (L.) Hairy Alpine chickweed.—Vithivik.
  latifolium (L.) Broad-leaved Alpine chickweed.
  trigynum (Vill.) Stitchwort chickweed.
 Alsine biflora (Wahlemb.)—Hnausir in tún, Akureyri.

LINACEÆ :—

 Linum catharticum (L.) Purging flax.

HYPERICACEÆ :—

 Hypericum perforatum (L.) Common perforated S. John's wort.

GERANIACEÆ :—

 Geranium sylvaticum (L.) Wood crane's-bill. Storka-blágras.—Thingvellir, Námarfjall, Vithidal, Northrárdal, Uthlith.
  pratense (L.) Blue meadow crane's-bill.
  montanum (L.) Stora-blágras.

LEGUMINOSÆ :—

 Anthyllis vulneraria (L.) Our Lady's fingers.
 Trifolium repens (L.) Dutch or white clover. Smári.—Uxahver.
  pratense (L.) Common purple clover. Smári.
  arvense (L.) Hare's-foot clover. Akureyri.
 Lotus corniculatus (L.) Bird's-foot trefoil.
 Vicia cracca (L.) Tufted vetch. Umfethmings-gras, flœkja.
 Lathyrus pratensis (L.) Meadow vetch.
  maritimus (L.) Sea-side everlasting pea. Banna-gras.—Skaga-
  fjord.

ROSACEÆ :—

 Spiraa ulmaria (L.) Meadow-sweet.
 Geum rivale (L.) Water avens. Fjalla-fífíll.—Side of Ók.
 Dryas octopetala (L.) White dryas. Rjúpna-lýng. Everywhere.
 Rubus saxatilis (L.) Stone bramble. Hrúta-ber, skollareipi.—Thing-
  vellir, Ljósavatn.

ROSACEÆ (cont.) :—

> *Fragraria vesca* (L.)   Wood strawberry.—Laugarvatn, on way to Geysir.
> *Comarum palustre* (L.)   Purple marsh cinquefoil.   Engja-rós.
> *Potentilla anserina* (L.)   Silver-weed.   Mura, myru-tágar.
>> *argentea* (L.)   Hoary cinquefoil.
>> *verna* (L.)   Spring cinquefoil.
>> *aurea* (L.)—Northrárdal, Langavatn, Mithfjord.
>> *tormentilla* (Sibth.)   Tormentil.
>> *maculata* (Pourr.)
> *Sibbaldia procumbens* (L.)   Procumbent sibbaldia.
> *Alchemilla vulgaris* (L.)   Lady's mantle.   Maríu-stakkr.
>> *montana* (Willd.)
>> *alpina* (L.)   Alpine lady's mantle.   Ljóns-kló, ljónn-lappi.
>> *arvensis* (Sm.)   Parsley piert.
> *Sanguisorba officinalis* (L.)   Great Burnet.
> *Rosa hibernica* (Sm.)   Irish rose.—Seljaland, and there only.   *See* Hooker, ii. 325.
>> *Camtschatica* (Vent.)

POMEÆ :—

> *Pyrus domestica* (Sm.)   True service.—Eyjafjord.
>> *aucuparia* (Gærtn.)   Rowan tree, or mountain ash.   Reynir.

ONAGRACEÆ :—

> *Chamænerium angustifolium* (Spach.)
> *Epilobium tetragonum* (L.)   Square-stalked willow-herb.
>> *palustre* (L.)   Narrow-leaved marsh willow-herb.   Eyra-rós. —Head of Eyjafjord.
>> *montanum* (L.)   Broad smooth-leaved willow-herb.—Eyjafjord river.
>> *origanifolium* (Lam.)—Eyjafjord river.
>> *spicatum* (Lam.)   Eyjafjord river.
>> *alpinum* (L.)   Alpine willow-herb.—Eylifr, Myvatn.
>> *latifolium* (L.)   Maríu-vöndr.
>> *angustissimum* (Rehb.)

HALORAGACEÆ :—

> *Myriophyllum verticillatum* (L.)   Whorled water-milfoil.
>> *spicatum* (L.)   Spiked water-milfoil.
> *Hippuris vulgaris* (L.)   Marestail.   Mark álmr.

PORTULACACEÆ :—

> *Montia fontana* (L.)   Water blinks.—On red bolus above the Geysir.

CRASSULACEÆ :—

Sedum anglicum (Huds.) English stone-crop.—Reykir.
album (L.) White stone-crop.
villosum (L.) Hairy stone-crop.—Entrance to Hörgardalr.
annuum (L.)
acre (L.) Wall-pepper. Hellu-hnothri.
rupestre (L.) S. Vincent's-rock stone-crop.
rhodiola (D. C.) Rose-root stone-crop. Burkni, Greithu-rót.

SAXIFRAGACEÆ :—

Saxifraga cotyledon (L.) Kletta-frú.—Brunnir.
stellaris (L.) Starry saxifrage.
nivalis (L.) Alpine clustered saxifrage.
oppositifolia (L.) Purple mountain saxifrage. Snjóblómstr.
hirculus (L.) Yellow marsh saxifrage. Hálsavegr.—Side of Ók.
aizoides (L.) Yellow mountain saxifrage.
granulata (L.) White meadow saxifrage.
cernua (L.) Drooping bulbous saxifrage.—Almannagjá.
rivularis (L.) Alpine brook saxifrage.
tridactylites (L.) Rue-leaved saxifrage.—Thingvellir, Myvatn.
hypnoides (L.) Mossy saxifrage.
caespitosa (L.) Tufted Alpine saxifrage.—Ók.
petræa (L.)
geranoides (L.) Geranium saxifrage.
bulbifera (L.)
tricuspidata (Rottboell).—Thingvellir.
autumnalis (L.)
groenlandica (L.)
cuneifolia (L.)

UMBELLIFERÆ :—

Hydrocotyle vulgaris (L.) Marsh pennywort.
Œgopodium podagraria (L.) Bishopweed. Geitnar-jól.
Carum carui (L.) Common carraway. Kumen.
Ligusticum scoticum (L.) Scottish lovage.
Angelica archangelica (L.) Garden angelica. Ali-hvönn.—Islands in Myvatn.
sylvestris (L.) Wild angelica. Ætöka-hvönn.—Little Arnarvatn.
Imperatoria ostruthium (L.) Sœhvönn.

ARALIACEÆ :—

Hedera helix (L.) Common ivy.—Borg in Vithidal.

CORNACEÆ :—

Cornus suecica (L.) Dwarf cornel.

Rubiaceæ:—

    Galium verum (L.)   Yellow bed-straw.   Gul-mathra.
        sexatile (L.)   Smooth heath bed-straw.
        pusillum (L.)   Least mountain bed-straw.
        palustre (L.)   White water bed-straw.   Lilla-mathra.
        mollugo (L.)   Great hedge bed-straw.—Grimstúngu.
        pumilum (Lam.)
        pallidum (Presley).
        trifidum (L.)
        boreale (L.)   Cross-leaved bed-straw.   Kross-mathra.
        aparine (L.)   Goose-grass or cleavers.   Abundant.

Valerianaceæ:—

    Valeriana officinalis (L.)   Great wild valerian.   Vélants-urt.

Dipsacaceæ:—

    Succisa pratensis (Mönch.)   Púka-bit.
    Cephalaria alpina (Schrad.)

Compositæ:—

    Crepis præmorsa (Tausch.)   Unda-fifill.
    Leontodon taraxacum (L.)   Common dandelion.   Fifill.—Holar.
        autumnalis (L.)—Thingvellir, Vatnakarth.
    Hieracium pilosella (L.)   Mouse-ear hawkweed.   Unda-fifill.
        auricula (L.)
        aurantiacum (L.)   Orange hawkweed.
        alpinum (L.)   Alpine hawkweed.
        murorum (L.)   Wall hawkweed.
    Carduus acanthoides (L.)   Welted thistle.
    Cirsium lanceolatum (Scop.)
        heterophyllum (Allion).
        arvense (Scop.)
    Tanacetum vulgare (L.)   Common tansy.—Reykjavík.
    Gnaphalium sylvaticum (L.)   Highland cudweed.   Grájurt.
        uliginosum (L.)   Marsh cudweed.
        supinum (L.)   Dwarf cudweed.
        norvegicum (Gunner).   Grájurt.
        dioicum (L.)
        carpathicum (Wahlenb.)   Fjanda-fœla.
    Tussilago farfara (L.)   Coltsfoot.   Common in low spots.
    Erigeron alpinus (L.)   Alpine flea-bane.   Snjór-gras.   Everywhere.
    Senecio Jacobæa (L.)   Ragwort.   Jakobs-fifill.   On the heithies.
        vulgaris (L.)   Common groundsel.   On the heithies.
        sylvaticus (L.)   Mountain groundsel.   On the heithies.
    Anthemis cotula (L.)   Stinking camomile.   Baldurs-brá.—Myvatn.
    Achillæa millefolium (L.)   Yarrow, or milfoil.   Vell-humall.   In túns.

CAMPANULACEÆ :—

   Campanula patula (L.)   Spreading bell-flower.
        rotundifolia (L.)   Harebell.   Blá-klukka.

VACCINIACEÆ :—

   Vaccinum myrtillus (L.)   Whortle-berry.   Athal-bláberjalýng.   On the
        heithies.
        uliginosum (L.)   Bog whortle-berry.   Blá-ber.—Skjalfanda.
        vitis-idæa (L.)   Cow-berry.—Bogs near Uxahver.
        oxycoccos (L.)   Cranberry.—Bogs near Uxahver.

ERICACEÆ :—

   Erica tetralix (L.)   Cross-leaved heath.   Lava districts.
   Calluna vulgaris (Salisb.)   Ling.   Beiti-lýng.—Arnarvatn.
   Cassiopea hypnoides (Don.)
   Azalea procumbens (L.)   Trailing azalea.   Sautha-mergr.—Kallmans-
        túnga.
   Arctostaphylos alpina (Spreng.)   Black bearberry.   Sortu-lyng.
   Rhododendron lapponicum (Wg.)—Kalmanstúnga.
   Ledum latifolium (Lam.)

PYROLACEÆ :—

   Pyrola rotundifolia (L.)   Round-leaved winter-green.   Vetrar-laukr.
        secunda (L.)   Serrate winter-green.
        minor (L.)   Lesser winter-green.

GENTIANACEÆ :—

   Gentiana pneumonanthe (L.)   Marsh gentian.
        autumnalis (L.)
        verna (L.)   Spring gentian.
        quinquefolia (L.)
        ciliata (L.)
        detonsa (Fries.)
        bavarica (L.)—Hólar, Hvíta.
        involucrata (Rottboell).   Maríu-vöndr.
        tenella (Rottboell).
        nivalis (L.)   Small Alpine gentian.   Digra-gras.—Hólar.
        amarella (L.)   Small-flowered gentian.   Hólar.
        campestris (L.)   Field gentian.   Maríu-vöndr. — Hvítá,
        Hlitharfjall.
        serrata (Gunn.)
   Swertia rotata (L.)
   Menyanthes trifoliata (L.)   Buckbean.   Álftar-kólafr.—Vithimyri.

POLEMONIACEÆ :—

   Diapensia lapponica (L.)

BORAGINACEÆ :—

Echium vulgare (L.)   Common viper's bugloss.   Kísu-gras.
Mertensia maritima (Don.)   Sea-side smooth gromwell.   Strand-arfi.
Myosotis palustris (With.)   Forget-me-not.   Kattar-auga.—Hóf in Vatnsdalr.
    arvensis (Hoffm.)   Field scorpion-grass.
    collina (Hoffm.)
    versicolor (Reich.)

SCROPHULARIACEÆ :—

Veronica spicata (L.)   Spiked speedwell.
    serpyllifolia (L.)   Thyme-leaved speedwell.—Grenjatharstathr. Hólar.
    alpina (L.)   Alpine speedwell.—Uxahver, Skjalfanda.
    saxatilis (L.)   Blue rock speedwell.
    fruticulosa (L.)   Flesh-coloured speedwell.—Thingvalla, Hook.
    scutellata (L.)   Marsh speedwell.
    anagallis (L.)   Water speedwell.
    beccabunga (L.)   Brooklime.   Vatns-arfi, lemmiki.
    officinalis (L.)   Common speedwell.   Æru-prís.
    marylandica (L.)
Bartsia alpina (L.)   Alpine bartsia.   Loka-sjóths-bróthir.—Skjalfanda.
Euphrasia officinalis (L.)   Common eyebright.   Augnafros.—Grímstúnga.
Rhinanthus crista galli (L.)   Yellow-rattle.   Loka-sjóthr.—Geysir.
    angustifolius (Gmel.)   Large bushy yellow-rattle.
Pedicularis palustris (L.)   Marsh louse-wort.   Lúsajurt.—Miklibœr.
    sylvatica (L.)   Pasture louse-wort.   Lúsajurt.   In túna.
    versicolor (Wahlenb.)   Lúsajurt.
Limosella aquatica (L.)   Mudwort.

LABIATÆ :—

Thymus serpyllum (L.)   Wild thyme.   Blóth-björg.—All loose ground, Oxnardalr.
Galeopsis ladanum (L.)   Red hemp-nettle.
    tetrahit (L.)   Common hemp-nettle.—Hólar.
Lamium album (L.)   White dead-nettle.—Hnausir.
    purpureum (L.)   Red dead-nettle.—Hnausir.
Stachys sylvatica (L.)   Hedge woundwort.—Fnjoskadalr.
Prunella vulgaris (L.)   Self-heal.   Brúnella.—Kulmanstúnga, Arnarvatn.
    officinalis (L.)—Arnarvatn heithi.

LENTIBULARIACEÆ :—

Pinguicula vulgaris (L.)   Butterwort.   Very common everywhere.
    alpina (L.)   Alpine butterwort.—Arnarvatn, Laxá at Thverá.

PRIMULACEÆ :—

Primula farinosa (L.)   Bird's-eye primrose.
Glaux maritima (L.)   Sea-milkwort.—Eyjafjord, Mithfjord.
Trientalis europæa (L.)   European chickweed winter-green.

PLUMBAGINEÆ :—

Statice elongata (Hoffm.)
    maritima (Mill.)   Thrift.   Gullin-toppa.—Hjaltadal.

PLANTAGINACEÆ :—

Plantago major (L.)   Greater plantain.   Græthisúrt.
    media (L.)   Hoary plantain.
    lanceolata (L.)   Ribwort plantain.   Fugla-túngur.
    maritima (L.)   Seaside plantain.   Kattar-túngur.
    alpina (L.)
    coronopus (L.)   Buckshorn plantain.

CHENOPODIACEÆ :—

Chenopodium album (L.)   White goose-foot.
Atriplex laciniata (L.)   Frosted sea orache.
    patula (L.)   Spreading halberd-leaved orache.

SCELERANTHACEÆ :—

Sceleranthus annuus (L.)   Annual knawel.

POLYGONACEÆ :—

Polygonum viviparum (L.)   Viviparous Alpine buckwheat.   Korn-
    súra.—Hólar.
    bistorta (L.)   Snakeweed.
    aviculare (L.)   Knot-grass.   Oddvari.
    convolvulus (L.)   Climbing bistort.
    amphibium (L.)   Amphibious bistort.
    persicaria (L.)   Spotted bistort.   Flóar-urt.
    hydropiper (L.)   Biting persicaria.
Rumex aquaticus (L.)   Grainless water-dock.
    patientia (L.)   Heimilisnjóli, Farthaga-kél.
    acetosa (L.)   Common sorrel.   Valla-súra.—In tún Miklibœr.
    acetosella (L.)   Sheep's sorrel.   Steinstathr.
Oxyria reniformis (Hook.)   Kidney-shaped mountain sorrel.   Olafs-
    súra.—Grímstúnga.
Kanigia islandica (L.)   Núfla-gras.—Geysir.

EMPETRACEÆ :—

Empetrum nigrum (L.)   Black crow-berry.   Used by Bishop Pál for
    making sacramental wine.   *Páls Saga*, cap. ix.   Krákaber.

CALLITRICHACEÆ :—
> *Callitriche verna* (L.)   Vernal water-starwort.
>> *autumnalis* (L.)   Autumnal water-starwort.

CERATOPHYLLACEÆ :—
> *Ceratophyllum demersum* (L.)   Hornwort.

URTICEÆ :—
> *Urtica urens* (L.)   Small nettle.—Reykjavík.   Hook., ii. 334.   Eyja-fjord.
>> *dioica* (L.)   Great nettle.   Brennnetla, nötrugras.—Vatns-skarth.

BETULACEÆ :—
> *Betula alba* (L.)   Common birch.   Birki.—Fnijoska dalr, Northrár dalr, &c.
>> *nana* (L.)   Dwarf birch.   Fjall-hrapi. — Eylifr, Vithidalr, Myvatn.
>> *fruticosa* (Vall.)—Thingvellir.

SALICACEÆ :—
> *Salix purpurea* (L.)   Purple willow.—Thingvellir.
>> *pentandra* (L.)   Sweet bay-leaved willow.—Thingvellir.
>> *fusca* (L.)   Dwarf silky willow.—Eylifr, Ljósavatn.
>> *ambigua* (Ehr.)   Ambiguous willow.
>> *reticulata* (L.)   Reticulate willow.
>> *myrtilloides* (L.)
>> *lapponum* (L.)   Grá-víthir, Tág.
>> *wulfeniana* (Willd.)
>> *arenaria* (L.)   Downy mountain willow.—Leiruvatn.
>> *cinerea* (L.)   Gray sallow.—Ljósavatn.
>> *caprea* (L.)   Great round-leaved sallow.   Selja.—Seljadalr.
>> *arbuscula* (L.)   Small tree willow.   Bein-víthir.
>> *myrsinites* (L.)   Green whortle-leaved willow.
>> *herbacea* (L.)   Least willow.   Kötuns-lauf, kötuns-víthir.
>> *argentea* (Sm.)
>> *glauca* (L.)
>> *arctica* (Pall.)

CONIFERÆ :—
> *Juniperus communis* (L.)   Common juniper.   Einir.—Arnarvatn, Grjot-háls.

II. MONOCOTYLEDONES.

ORCHIDACEÆ:—

Corallorhiza innata (Br.)   Spurless coral-root.—Laugarvatn.
Listera ovata (Br.)   Common twayblade.—Vík.   Hooker ii. 332.
    nidus-avis (Hook.)   Common birdsnest.
Peristylus víridis (Lindl.)—Opening of Hörgárdalr.
    albida (Lindl.)—Opening of Hörgárdalr.
Orchis morio (L.)   Green-winged meadow orchis.—Grímstúnga.
    mascula (L.)   Early purple orchis.
    latifolia (L.)   Marsh orchis.
    maculata (L.)   Spotted palmate orchis.   Elskugras, friggjar-gras.—
      Steinstathr.
    cruenta (Willd.)
Nigritella angustifolia (Rich).
Platanthera hyperborea (Lindl.)
    Könìgi (Lindl.)

TRILLIACEÆ :—

Paris quadrifolia (L.)   Herb-Paris.   Fjögra-laufa-smari.

LILIACEÆ:—

Maianthemum bifolium (D. C.)   Two-leaved May-lily.

MELANTHACEÆ :—

Tofieldia palustris (Huds.)   Mountain Scotch asphodel.   Sykis-gras.—
    Skjalfanda.
    calyculata (Wahlenb.)—Grímstúnga.
Anthericum ramosum (L.)   Ugla-gras.

JUNCACEÆ:—

Juncus communis (Meyer).   Common rush.   Myra-sef.—Reykjavík.
    Eyjafjord river.
    arcticus (Willd.)—Myvatn.
    lamprocarpus (Ehr.)   Shining-fruited jointed rush.—Myvatn.
    bufonius (L.)   Toad rush.—Haugakvisl.
    squarrosus (L.)   Heath rush.   Hrossa-nál.—Arnarvatn.
    triglumis (L.)   Three-flowered rush.
    Gerardi (Lois.)
    Jacquini (L.)
Luzula pilosa (Willd.)   Broad-leaved hairy woodrush.
    campestris (D. C.)
    spicata (D. C.)   Spiked mountain woodrush.

JUNCAGINACEÆ :—

Triglochin palustre (L.)   Marsh arrow-grass.
    maritimum (L.)   Sea-side arrow-grass.   Sand-laukr.

TYPHACEÆ:—

Sparganium natans (L.) Floating bur-reed.

NAIADACEÆ:—

Potamogeton pectinatus (L.) Fennel-leaved pondweed.—Vithimyri.
crispus (L.) Curly pondweed.
perfoliatus (L.) Perfoliate pondweed.
lucens (L.) Shining pondweed.
rufescens (Schrad.) Reddish pondweed.—Vithimyri.
natans (L.) Sharp - fruited broad - leaved pondweed.—
Vithimyri.
Zostera marina (L.) Grasswrack. Markálmr.—Hruta-fjord.

CYPERACEÆ:—

Blysmus compressus (Panz.) Broad-leaved blysmus.—Hop.
rufus (Link.) Narrow-leaved blysmus.—Hop.
Isolepis setacea (R. Br.) Bristle-stalked mudrush.—Heradsvatn.
Scirpus lacustris (L.) Lake bulrush.—Reykjavík.
cæspitosus (L.) Deer's-hair.—Arnarvatn heithi.
compressus (Pers.)
Heleocharis palustris (R. Br.)
acicularis (R. Br.)
Eriophorum alpinum (L.) Alpine cotton-grass.—Öxnardals heithi.
vaginatum (L.) Hare's - tail cotton - grass. Haruld. —
Hangakvisl.
capitatum (Host.) Round-headed cotton-grass.—Burfell.
Hjaltadal.
latifolium (Hoppe.) Broad-leaved cotton-grass.
angustifolium (Roth.) Narrow-leaved cotton-grass. Fífa.
Scheuchzeri (Hoppe).
Elyna spicata (Schrad.)
Carex dioica (L.) Creeping separate-headed sedge. Stör.
pulicaris (L.) Flea sedge.
rupestris (All.) Rock sedge.
pauciflora (Lightf.) Few-flowered sedge.
incurva (Lightf.) Curved sedge.
ovalis (Gooden.) Oval-spiked sedge.
leporina (L.) Hare's-foot sedge.
elongata (L.) Elongated sedge.
vulpina (L.) Great sedge.
muricata (L.) Greater prickly sedge.
arenaria (L.) Sea sedge.
vahlii (Schk.) Close-headed Alpine sedge.
canescens (L.) Hoary sedge.
capitata (L.)
atrata (L.) Black sedge.
rigida (Gooden.) Rigid sedge.

28—2

CYPERACEÆ (cont.):—

   Carex acuta (L.) Slender-spiked sedge. Tjarna-stör.
      cæspitosa (L.) Tufted bog sedge.
      extensa (Gooden.) Long-bracteate sedge.
      flava (L.) Yellow sedge.
      depauperata (Gooden). Starved-wood sedge.
      panicea (L.) Pink-leaved sedge.
      pallescens (L.) Pale sedge. Hringa-stör.
      capillaris (L.) Dwarf capillary sedge.
      limosa (L.) Mud sedge.
      pseudo-cyprus (L.) Cyprus-like sedge.
      hirta (L.) Hairy sedge.
      ampullacea (Gooden.) Slender-beaked bottle sedge.
      vesicaria (L.) Short-beaked bladder sedge.
      ornithopoda (Willd.)
      loliacea (L.)
      pulla (Gooden.) Rauthbrestingr.

GRAMINEÆ:—

   Anthoxanthum odoratum (L.) Sweet-scented vernal-grass. Reyr-gras.
   Nardus stricta (L.) Mat-grass. Tödu-finnúngr.
   Alopecurus geniculatus (L.) Floating foxtail-grass.
   Phleum pratense (L.) Timothy-grass.
      alpinum (L.) Alpine cat's-tail-grass. Fox-gras, toúgras.
      nodosum (L.)
   Psamma arenaria (Roem.) Marram.
   Milium effusum (L.) Spreading millet-grass.
   Calamagrostis epigeios (Roth.) Wood small-reed.
      arundinacea (Roth.)
   Agrostis rubra (L.)
      stolonifera (L.)
      canina (L.) Brown bent-grass.
      vulgaris (L.) Fine bent-grass.
      pumila (L.)
      alba (L.) Marsh bent-grass.
      alpina (Scopol.)
      capillaris (Th.)
   Aira cæspitosa (L.) Tufted hair-grass.
      alpina (L.) Smooth Alpine hair-grass.
      flexuosa (L.) Waved hair-grass.
      subspicata (L.)
      praecox (L.) Early hair-grass.
      atropurpurea (Wahl.)
   Molinia cærulea (Moench.) Purple molinia.
   Kœleria glauca (D. C.)
   Sesleria cærulea (Scop.) Blue moor-grass.
   Hierochloe borealis (R. & S.) · Northern holy-grass. Reyr-gras.

GRAMINEÆ (cont.) :—

Poa aquatica (L.)   Reed meadow-grass.
 fluitans (Scop.)   Floating meadow-grass.
 maritima (Huds.)   Creeping sea meadow-grass.
 distans (L.)   Reflexed meadow-grass.
 compressa (L.)   Flat-stemmed meadow-grass.
 pratensis (L.)   Smooth-stalked meadow-grass.
 trivialis (L.)   Roughish meadow-grass.
 alpina (L.)   Alpine meadow-grass.
 laxa (Hœnke.)   Wava meadow-grass.
 nemoralis (L.)   Wood meadow-grass.
 annua (L.)   Annual meadow-grass.
 cæsia (Smith.)
Briza media (L.)   Quaking grass.
Festuca ovina (L.)   Sheep's fescue-grass.
 heterophylla (Lam.)
 elatior (L.)   Pall fescue-grass.
Bromus hordaceus (L.)
Phragmites communis (Trin.)   Common reed.
Elymus arenarius (L.)   Upright sea lyme-grass.
Triticum cristatum (Schreb.)   Crested wheat-grass.
 repens (L.)   Creeping wheat-grass.   Húsa-puntr.
 caninum (Huds.)   Fibrous-rooted wheat-grass.

III. ACOTYLEDONES.

POLYPODIACEÆ :—

Polypodium vulgare (L.)   Common polypody.—Allmannagjá.
 fontanum (L.)   Thingvellir.
 arvonicum (L.)
 phegopteris (L.)   Beech fern.
 dryopteris (L.)   Oak fern.
Woodsia ilvensis (Br.)   Oblong woodsia.
Aspidium lonchitis (Sw.)   Holly fern.
 thelypteris (Sw.)   Marsh fern.
 filix mas (Sw.)   Male fern.
Cystopteris fragilis (Bernh.)   Brittle bladder fern.—Skjalfanda.
Asplenium septentrionale (Sw.)   Forked spleenwort.—Laugardalr.
 trichomanes (L.)   Maiden-hair spleenwort.
 filix femina (Bernh.)   Lady fern.   Burn, buskni.

OPHIOGLOSSACEÆ :—

Ophioglossum vulgatum (L.)   Adder's tongue.
Botrychium lunaria (Sw.)   Moon-wort.   Túngl-urt.—Very fine at
   Hólar, Hörgadalr.

LYCOPODIACEÆ :—

Lycopodium clavatum (L.)   Club moss.   Lma-bróthir.   Common every-
       where.
       annotinum (L.)   Interrupted club moss.
       selaginoides (L.)   Lesser Alpine club moss.—Copse near
        Laugarvatn.
       alpinum (L.)   Savin-leaved club moss.   Jafni.   Common
        everywhere.
       selago (L.)   Fir club moss.   Vargs-lappi, Skolla-fingr.
       dubium (L.)
       complanatum (L.)
Isoetes lacustris (L.)   European quill-wort.—Allmannagjá.

EQUISETACEÆ :—

Equisetum arvense (L.)   Corn horse-tail.   Rílting.—Very large, Uthlith.
       sylvaticum (L.)   Branched wood horse-tail.—Copse near
        Laugarvatn.
       limosum (L.)   Smooth naked horse-tail.   Tjarna ellting.—
        Vithimyri.
       palustre (L.)   Marsh horse-tail.
       hyemale (L.)   Dutch rushes.   Eski-graa.
       pratense (Ehr.)

The list above is tolerably complete. I have given the spots where I
gathered my specimens. I have put the English names to those plants
which are found in the British Isles. For those plants which I did not
collect myself, I am indebted to Hooker's *Journal*, 1813 ; Zoega's *Flora
Islandica* ; Preyer u. Zirkel, *Reise n. Island* ; Dr. Lindsay's *Flora of Iceland*,
in the *Edinburgh New Philosophical Journal*, July, 1861 ; and to Dr.
Hjattalin's Icelandic work of the plants of the island, published in 1830.
I have not given any account of the mosses and algæ, as I gathered none
myself.

APPENDIX D.

A LIST OF ICELANDIC PUBLISHED SAGAS.

HISTORIES OF ANCIENT HEROES. MYTH AND HISTORY, OR MYTH AND FABLE.

Áns S. bogsveigis (F.A. II.) Story of Án the Bow-bender.
Ásmundar S. kappabana (F.A. II.) Story of Asmund the Hero-slayer.
Egils S. einhenda (F.A. III.) Story of One-handed Egill.
Eiríks S. víðförla (F. A. III.) Story of Eirik the Far-travelled.
Frá Fornjóti (F.A. I., II.) Story of the Old Giant and his kin.
Friðthjófs S. (F.A. II.) Story of the stalwart Frith-thjof.
Gautreks S. Konungs (F.A. III.) Story of King Gautrek of West Gothland.
Göngu-Hrólfs S. (F.A. III.) Story of Ralph the Walker, ancestor of
    William the Conqueror.
Grims S. loðinkinna (F.A. II.) Story of Grim with the Bristly Chin.
Hálfdanar S. Brönufóstra (F.A. III.) Story of Halfdan, Bran's foster-son.
Hálfdanar S. Eysteinssonar (F.A. III.) Story of Halfdan, Eystein's son.
Hálf's saga (F.A. II.) Story of King Half and his warriors.
Hemings Tháttr (Analecta Norræna). A Northern version of the story of
    Tell.
Herrauðs S. (F.A. III.) Story of Herrauth.
Hervarar S. (N.O. III.) Story of Hervör and King Heithrek.
Hjálmters S ok Ölvis (F.A. III.) Story of Hjálmter and Oliver.
Hrólfs S. Gautrekssonar (F.A. III.) Story of Hrolf, son of Gautrek.
Hrólfs S. kraka (F.A. I.) Story of Hrolf Krake and his warriors.
Hrómundar S. Greipssonar (F.A. II.) Story of Hromund, Greip's son.
Illuga S. Gríðarfostra (F.A. III.) Story of Illugi, Grith's foster-son.
Ketils S. hængs (F.A. II.) Story of Salmon Ketill.
Langfeðgatal. Genealogy of heroes.
Norna-Gests S. (F.A. I.) Story of Norn-Gest.
Örvar-Odds S. (F.A. II.) Story of Odd the Arrow-darter.
Ragnars S. loðbrókar (F.A. I.) Story of Hairy-breeched Ragnar and
    his sons.

Ságubrot (F.A. I.) Fragment of stories of ancient Danish and Swedish kings.

Sörla S. sterka (F.A. III.) Story of Strong Sörli.

Sörla Tháttr (F.A. I.) The deeds of Sörli.

Sturlaugs S. starfsama (F.A. III.) The story of Toiling Sturlaug.

Tóka Th. (Forn. V.) The Deeds of Tóki, a northern Rip van Winkel.

Thiðreks S. (Christ., 1853). Story of Thidrik of Bern and his heroes.

Thorsteins S. bæjarmagns (Forn. III.) Story of Thorstein the Farm-keeper.

Thorsteins S. Víkingssonar (F.A. II.) Story of Thorstein, the Víking's son.

Upplendinga Th. (F.A. II.) Concerning the Upland kings.

Völsunga S. (F.A. I.) Story of the Völsungs.

HISTORIES RELATING TO ICELAND, THE FAROES, ORKNEY, AND GREENLAND.

Ármanns S. (out of print). Story of the mountain gnome, Armann, near Thingvellir.

Ásu-Thórðar Th. (Forn. VII.) Story of Thorth, lover of Asa.

Auðunar Th. vestfirzka (old Norsk Leseb. of Munch U. Unger). Story of Authun.

Bandamanna S. (N.O. X.) Story of the Banded-men. See Chap. XVI.

Bárðar S. Sæfellsdss (N.O. XXVII.) Story of Bárth, the Snæfell goblin.

Berg-búa Th. (N.O. XXVII.) The deeds of the mountain-dweller.

Bjarnar S. Hítdælakappa (N.O. IV.) Story of Björn, hero of Hitdale.

Brandkrossa Tháttr (N.O. V.) Story of a cow which swam from Iceland to Norway.

Droplaugarsona S. (N.O. II.) Story of the sons of Droplanga.

Egils S. Skallagrímssonar (Reykjavík, ed. Jón Thorkélsson, 1856). Story of Egill. See Chap. III.

Eiríks Th. rauða (G.M. I.) Deeds of Erick the Red, discoverer of Greenland.

Eyrbyggja S. (Copenh., 1787.) An abstract was published by Sir W. Scott in *Ill. North. Antiq.*

Færeyinga S. (Copenh., 1833.) Story of the Faroe Isles.

Finnboga S. (Akureyri, 1860.) Story of the sturdy Finnbog.

Floamanna S. (Forn Sögur of Möbius, 1860.) Story of the Floamen. See Chap. XXIII.

Fóstbræthra S. (N.O. XV.) Story of the Foster-brothers, and the visit of one to Greenland.

Gests S. Bárðarsonar (N.O. XXVII.) Story of Gest, son of Bárth the Snæfell gnome.

Gísla S. Súrssonar (N.O. VIII.) Story of Gisli, Súr's son, the outlaw.

Grettis S. Asmundarsonar (N.O. XVI.) Story of Grettir the Strong. See Chaps. I., IV., VII., XV. I hope, one of these days, to bring out a translation of this noble saga myself.

Grænlendinga Th. (G.M. I.) Concerning Greenland.

Gunnars Th. Thiðrandabana (published with Laxdoela S.) Story of Gunnar.

Gunnlaugs S. (Analect. Norrœna.) Story of Gunnlaug with the Serpent's Tongue.

Hansa-Thoris S. (Is. I.) Story of Hen Thorir.

Halls Th. (G.M. III.) Deeds of Hall in Greenland.

Hávarðar S. Iisfirthings (N.O. XXVIII.) Story of Hávard of the Ice-frith.

Hrafnkels S. Freysgoða (N.O. I.) Story of Hrafnkell, priest of Frey, and his horse Freyfaxi.

Holmverja S. (Is. II.) Story of the isle defenders, Hörth and Geir.

Hrafns S. sveinbjarn (G.M. II.) Story of Hrafn; relating to Greenland.

Íslendingabók (Is. I.) The book of Icelanders.

Jökulls Tháttr Búasonar (Is. II.) The deeds of Jökull, son of Búi.

Kjalnezinga S. (Is. II.) The story of the Kjalurnes-folk.

Kormaks S. (Copenh., 1832.) Story of Kormak.

Kumlbúa Th. (O.N. XXVII.) Deeds of the Mound-dweller.

Kristni S. (Bisk. I.) Story of the establishment of Christianity in Iceland.

Kroka-Refs S. (out of print.) Story of Crook-Ref.

Landnámabók (Isl. I.) Icelandic Domesday-book.

Laxdæla S. (Copenh., 1826.) Story of the Salmondale folk.

Leifs Th. (with Færeyinga S.) Story relating to the Faroe Isles.

Ljósvetninga S. (Isl., 1830; out of print.) Story of the people of Lightwater Lake.

Magnus S. eyjajarls (with Orkn. S.) Story of the life of S. Magnus; relates to Orkney.

Njáls S. (an edition of the text will shortly be published in N.O. It has been translated by Dr. Dasent, under the title of " The Story of Burnt Njal.")

Orkneyinga S. (Copenh., 1780; an English translation will be published shortly.) Story of Orkney.

Orms Th. Storolfsson (Forn. III.) Deeds of Orm, son of Storolf.

Sigmundar Th. Brestissonar (with Færeyinga S.) Relating to Faroe Isles.

Sturlunga S. (Copenh., 1820; out of print.) The great saga of the Sturlung family.

Svarfdæla S. (Isl., 1830; out of print.) Story of the folk of Greensward Dale.

Tosta Th. tréspjóts (G.M. II.) Story of Tost with the wooden spear.

Thorbjarnar Th. Karlsefnis (Antiquit. Americ., 1837.) Story relating to the discovery of America.

Thórðar S. hreðu (N.O. XXVII.) Story of Thorth the troubled. A fragment only.

Thórðar S. hreðu (N.O. VI.) Story of Thorth the troubled, a descendant of the other Thorth.

Thorfinns S. Karlsefnis (G.M. I.) Story of Thorfin of manly race; relates to Greenland.

Thorgríms S. prúða (N.O. XXVII.) Story of Thorgrim and Víglund the courtly.

Thorhalls Th. ölkofra (out of print.) Story of Thorhall Beer-cap and his law-suit.

Thorsteins S. Síðuhallssonar (Analect. Norr.) Story of Thorstein, son of Side-Hall.

Thorsteins Th. frúsa (Forn. VL.)　Deeds of Thorstein the Wise.
Thorsteins Th. Geirneffufóstra (G.M. III.)　Relating to Greenland.
Thorsteins Th. hvíta (N.O. V.)　Deeds of Thornsteinn White.
Thorsteins Th. stangarhöggs (N.O. V.)　Deeds of Goad-stricken Thorsteinn.
Thorvalds Th. víðfórla (Bisk I.)　Deeds of Far-travelled Thorwald.
Thrándar Th. (with Færeyinga S.)　Deeds of Thrand in the Faroe Isles.
Valla-Ljóts S. (Isl., 1830; out of print.)　Story of Guthmund rich and the
　　Svarfdale folk.
Vápnfirthing S. (N.O. V.)　Story of the Weapon-frith folk.
Vatnsdæla S. (Forn. Sögur., ed. Möbius, 1860.)　Story of the Waterdale folk.
Viga-Glúms S. (Copenh., 1786.)　Story of Slaying Glúm.
Viga-Skútu S. (Isl., 1830; out of print.)　Story of the Vale of Smoke.
Viga-Styrs S. (Isl. I. and II.)　The story of Slaying Styr and the Battle on
　　the Moor.

HISTORIES OF ICELANDIC BISHOPS.

Tháttr af Isleifi; Húngrvaka; Thorláks S., hin elzta; Páls S.; Jóns S.,
　　hin elzta; Jóns S., eptir Gunnlaug munk; *Thorláks S.,* hin yngri;
　　Guthmundar S., hin elzta; *Arna S. Thorlákssonar; Laurentius S.;*
　　Guthmundar S., eptir Arngrim ábóta; *Thattr af Jóni Halldórssyni* (all
　　published in Biskupa Sögur, Copenh., 1858–62.)

HISTORIES RELATING TO NORWAY, DENMARK, ETC.

Absalons Th. (Forn. XI.)　Story of Archbishop Absalon.
Ágrip (Forn. X.)　Outline of the History of the Norwegian Kings.
Albani Th. et Sunnifa (in Olafs. S. T.)　Concerning S. Alban and S. Sunifa.
Arons S. Hjörleifssonar (Antiq. Russ. II.)　Story of Aron Hjörleif's son.
Arnmæþlingatal (Fagrskinna.)　Genealogy of Arnmothlingirs.
Blóð Egils Th. (Knytl. S.)　Deeds of Egill the Blood-drinker.
Brands Th. örva (Haralds S. Harthrátha.)　Deeds of free-handed Brand.
Egils Th. illgjarna (Copenh., 1820.)　Deeds of Egill the Spiteful.
Egils Th. Siðu-Hallssonar (Forn. V.)　Deeds of Egill, son of Hall of the
　　Side.
Einars Th. Skúlasonar (Forn. VII.)　Deeds of Einarr, son of Skuli.
Eiríks Th. Hákonarsonar (Forn. II.)　Deeds of Earl Eric.
Endriða Th. ilbreiðs (see Olafs S. Tr.)　Deeds of Endrith the Splay-footed.
Endriða Th. ok Erlings (Forn. V.)　Deeds of Endrith and Erling.
Eymundar S. (Forn. V.)　Saga of Eymund and King Olaf.
Fagrskinna (ed. Munch and Unger. Christ., 1847.)　The Fairskin.　History
　　of Norway.
Gislar Th. skalds (see Magn. S. berfœt.)　Deeds of Gisli the Skald.
Gregorii Th. (Forn. VII.)　Deeds of Gregory, son of Dagi.
Hákonar S. Hákonarsonar (Forn. IX.)　Story of King Hakon, Hakon's son.
Hákonar S. herðibreiths (Forn. VII.)　Story of King Hakon, the Broad-
　　Shouldered.
Hákonar S. Sverrissonar (Forn. IX.)　Story of Hakon, Sverrir's son.
Hákonar Th. Hárekssonar (Forn. XI.)　Deeds of Hakon, Harek's son.

Hálfdanar Th. svarta (Forn. II.) Deeds of King Halfdan, the Black.

Halldors Th. Snorrasonar (Forn. III.) Deeds of Halldor, Snorro's son.

Hallfredar S. vandravaskald (Forn. Sögur., of Möbius.) Story of Hallfred, the Hard Scald.

Haralds S. hartráfa (Forn. VI.) Story of stern Harald.

Hauks Th. hábrókar (Forn. X.) Deeds of high-breeched Hauk.

Heimskringla (Copenh., 1826.) The World's Circle. History of the Kings of Norway.

Helga Th. Thórissonar (Forn. III.) Deeds of Helgi, Thorir's son.

Hreiðars Th. (Forn. VI.) Deeds of Hreithar, in insulting King Harald harthrátha.

Hróa Th. heimska (Forn. V.) Deeds of Hroi, the Fool.

Hrómundar Th. halta (Forn. III.) Hrómund the Halt in Norway and Iceland.

Hryggjarstykki (the backbone). A history of Norwegian kings.

Inga S. Haraldssonar (Forn. VII.) Story of King Ingi and his brothers.

Ingvars S. víðförla (Stockh., 1762.) Story of Ingvar the Far-travelled, and his journey in Russia.

Játvarðar S. (Copenh., 1852.) Story of S. Edward the Confessor and William the Conqueror.

Jómsvíkinga S. (Forn. XI.) Story of the Jóms víkings, pirates in the Baltic.

Karls Th. vesœla (see Magn. S. Góða). Deeds of Karl the Wretched.

Kjartans Th. (see Olafs S. Tr.) Deeds of Kjartan, Olaf's son.

Knytlinga S. (Forn. XI.) History of the Danish Knuts.

Magnus Sögur (6) (Forn. VI., VII., X.) Histories of different kings of the name of Magnus.

Margrétar Th. (see Magn. S. Góða). Deeds of Margaret, Thrand's daughter.

Morkinskinna (the putrid skin). History of Norwegian kings.

Odds Th. (see Haralds S. harð.) Deeds of Odd, Ufeig's son.

Ogmundar Th. dytts (see Olafs S. Try.) Deeds of Ogmund the Drop.

Olafs Sögur (9) (Forn. I.—IV., V., X.) Story of Olaf the Saint and Olaf Tryggvason.

Olafs Th. Geirstaðaálfs (Forn. X.) Deeds of Olaf the gnome of Geirstathr.

Otta Th. (see Olafs S. Tryg.) History of Otho II., the young emperor.

Rauðs Th. ramma (see Olafs S. Tryg.) Deeds of Rauth the Dauntless.

Rauðúlfs Th. (Forn. V.) Deeds of Red-wolf and his sons.

Rögnvalds Th. (see Olafs S. Trygg.) Deeds of Rognvald and Rauth.

Sigurðar Sögur (2) (Forn. VII.) I. The story of the crusader Sigurth. II. Story of Sigurth, the bad deacon.

Sigurðar S. slefu (Forn. III.) Story of King Sigurth Spittle.

Skálda S. Haralds (Forn. III.) Story of the Scalds of King Harald Fairhair.

Snegtu-Halla Th. (see Haralds S. harðr.) Deeds of Hall the Restless.

Stefnis Th. (see Olafs S. Tr.) Deeds of Stefnir, Thorgils' son.

Stúfs Th. (see Harald's S. harð.) Deeds of the Scald Stufr.

Styrbjarnar Th. (Forn. V.) Deeds of Styrbjörn, the Swedish hero.

Sveins Th. (see Olafs S. Trygg.)

Sverris S. (Forn. VIII.) Story of King Sverrer.
Thórarins Th. (Forn. V.) Deeds of Thorarinn, Nefjulf's son.
Thorleifs Th. Jarlaskálds (Forn. III.) Deeds of Thorleif, the Earl's scald.
Thorsteins Draumr Síthu-Hallssonar (O.N. XXVII.) Dream of Thorsteinn,
    Side-Hall's son.
Thorsteins Th. skelks (Forn. III.) Deeds of Skulking Thorsteinn.
Thorsteins Th. uxafóts (Forn. III.) Deeds of Ox-footed Thorsteinn.
Thorvalds Th. tasalda (see Olaf's S. Trygg.) Deeds of Thorvald tasald.
Thorvarts Th. krákunefs (see Haralds S. hars.) Deeds of Raven-nosed
    Thorvarth.
Thrándar Th. Upplend (G.M. II.) Deeds of Upland Thrand.
Upphaf ríkis Haraldar (Forn. X.) The rise of Harald's kingdom.
Völsa Th. (O.N. XXVII.) The doings and dreams of Völsi.

FOREIGN ROMANCES TRANSLATED INTO ICELANDIC.

Ajax-afintyre	*Flóres S.*	*Strengleikar*
Alexanders S.	*Ivents S.*	*Theophilus*
Bragtsa mágus S.	*Jarlmanns S.*	*Tristans S.*
Barlaams S.	*Parcevals S.*	*Trójumanna S.*
Blómstrvalla S.	*Sálusar S.*	*Thorgríms S.*
Breta S.	*Samsons S.*	*Valdimars S.*
Eireks S.		

This is a complete list of *published* sagas. Besides these, there are other Icelandic works of which I give no list, such as law-books, Bible-stories, tracts on poetry, geography, astronomy, &c. In the catalogue above, I have used the following abbreviations :—

    " Bisk." for *Biskups Sögur.* Copenh., 1858–62 ; still incomplete.
    " F.A." „ *Fornaldur Sögur.* Copenh., 1829–30.
    " Forn." „ *Fornmanna Sögur.* Copenh., 1825–37.
    " G.M." „ *Grönlands historiske Mindesmerker.* Copenh., 1838–45.
    " Isl." „ *Islendinga Sögur.* Copenh., 1843–47.
    " N.O." „ *Nordiske Oldskrifter.* Copenh., 1847–63.
    " S." „ *Saga.*
    " Th." „ *Tháttr,* short story, deeds.

Those which I have spoken of as " out of print " have been printed and published only in Iceland, and are rare even there. Several of them I could not obtain myself, though I made every effort to acquire them.

APPENDIX E.

EXPENSES OF MY TOUR IN ICELAND.

		£	s.	d.
	Tent, packsaddle, pilch, hammock, &c.	14	0	0
	Presents for Iceland	2	10	0
	Waterproof coat, boots, &c.	6	0	0
June 7, Saturday.	Train from London to Edinburgh	1	13	0
	Bill at inn in Edinburgh	0	16	0
„ 9, Monday.	Train to Grangemouth	0	2	7
	Luggage, per parcels delivery, sent on before	0	4	5
	Arcturus, booking	5	0	0
	Rope	0	4	6
	Total	**30**	**13**	**5**

		Doll.	Mark	Skill.
„ 16, Monday.	Reykjavík:—			
	Sou'-wester cap	1	1	0
	Porterage, &c.	1	0	0
	Steward of *Arcturus*, for keep on board	12	4	8
	Shawl, wrapper, &c.	4	1	0
	Pills	0	2	0
	Various articles necessary—girths, horseshoes, &c.	10	0	0
„ 18, Wednesday.	Purchase of two horses, white and brown	48	0	0
	Purchase of riding horse, piebald	36	0	0
	Purchase of pack-horse, white	24	0	0
	Purchase of riding horse, grey	20	0	0
„ 19, Thursday.	Reykjavík to Mósfell:—			
	Bill at inn for four days	8	5	0
„ 20, Friday.	Mósfell to Thingvalla:—			
	Priest at Mósfell	1	2	0
	Hobbles for the horses	0	2	0
„ 22, Sunday.	Thingvalla:—			
	Purchase of chestnut riding horse	44	0	0
	Purchase of sheep, my share	2	5	0
„ 22, Monday.	Thingvalla to Bruanir:—			
	Priest at Thingvalla	7	0	0
	Hobbles for horses	0	2	0
	Kalmanstúnga to Arnarvatn :—			
„ 26, Thursday.	Farmer of Kalmanstúnga and part of sheep	8	4	0
	Carried forward	**243**	**0**	**8**

		Doll.	Mark	Skill.
	Brought forward	243	0	8
June 30, Monday.	Grímstúnga to Hnansír :—			
	Farmer of Grímstúnga	4	0	0
	Pair of Icelandic shoes	0	4	0
July 2, Wednesday.	Svínavatn to Miklibœr :—			
	Farmer at Svínavatn, for purchase of horse	26	0	0
	Farmer at Svínavatn, for lodging	1	0	0
	Purchase of riding horse at Víthimyri	22	0	0
„ 3, Thursday.	Miklibœr to Steinstathr:—			
	Servant at Miklibœr	1	0	0
„ 5, Saturday.	Steinstathr to Akureyri:—			
	Recovery of lost shawl	0	4	0
	M.P. at Steinstathr	2	0	0
	Skin of white gyr falcon	8	0	0
	Skins of divers and ducks	10	0	0
„ 7, Monday.	Akureyri to Grenjatharstathr:—			
	Biscuits, and bell for leader	1	0	0
	Keep of horses	1	0	0
	Horse shod	1	3	0
	Mending of saddle and pack	2	0	0
	Ferry-boat over Skjalfanda-fljot	0	3	0
„ 8, Tuesday.	Grenjatharstathr to Myvatn:—			
	Guide to Uxahver	0	2	0
	Farmer at Laugavatn	1	0	0
	Drink-money, servant	1	0	0
„ 10, Thursday.	Reykjahlíth:—			
	Socks and gloves, bought at Eylífr	1	4	0
„ 11, Friday.	Reykjahlíth to Thverá—			
	Horses brought back, when run away	0	3	0
	Hire of guide and horses to Dettifoss	6	0	0
	Farmer of Reykjahlíth	10	0	0
„ 13, Saturday.	Thverá to Akureyri:—			
	Farmer at Thverá	2	1	8
	Ferry over Skjalfanda-fljot	0	3	0
„ 14, Monday.	Akureyri to Hólar:—			
	Servant at M. Hafsteen's	2	0	0
	Part payment of Jón	3	1	2
	Skins of white foxes	4	2	0
	Sel in Hörgadalr	0	3	0
„ 16, Wednesday.	Hólar to Víthimyri:—			
	Servant at Hólar	1	0	0
	Guide over fjord of Heradsvatn	0	3	0
	Paid Grímr, for keep of horses at Akureyri	0	3	8
	Shoeing horse	0	2	0
	Jón	8	0	0
„ 17, Thursday.	Víthimyri to Hnansír:—			
	Farmer at Víthimyri, for food and dirt	2	0	0
	Extra guide, over bogs	0	4	0
„ 19, Saturday.	Hnansír to Melr:—			
	Servant at Hnausír	1	0	0
	Farmer at sel, for coffee	0	3	0
„ 20, Sunday.	Melr:—			
	Farmer at Bjarg, for digging in cairn	1	0	0
	Drink money	0	3	0
	Carried forward	373	4	10

		Doll.	Mark	Skill.
	Brought forward	373	4	10
July 21, Monday.	Melr to Hrúta-fjord:—			
	Dean of Melr	5	0	0
	Boat to Bortheyri	0	2	0
	Fox skins	4	3	0
„ 22, Tuesday.	Melr in Hrúta-fjord to Hvammr:—			
	Farmer of Melr, for nothing at all	2	0	0
„ 23, Wednesday.	Hvammr to Reykholt:—			
	Servant at Hvammr	1	0	0
	Guide across Hvitá	0	3	0
„ 24, Thursday.	Reykholt to Skogkottr:—			
	Priest at Reykholt	2	0	0
„ 25, Friday.	Skogkottr to Uthlíth :—			
	Farmer at Skogkottr	1	3	8
„ 26, Saturday.	Uthlíth to Geysir:—			
	Farmer at Uthlíth	1	2	0
„ 29, Tuesday.	Geysir to Thingvalla:—			
	Farmer at Haukadalr	2	2	0
„ 30, Wednesday.	Thingvalla to Reykjavík:—			
	Priest at Thingvalla	1	2	0
	Milk	0	0	8
Aug. 1, Friday.	Reykjavík:—			
	Birds' skins	8	2	0
	Silversmith	12	0	0
	Books	12	0	0
	Grímr Arnason	120	1	0
	Keep of horses at Reykjavík	2	1	0
	Keep of horses at Reykjavík before, now paid	2	0	0
	Icelandic female cap	2	0	0
	Inn bill, and present to servants	14	1	0
	Key of church	0	0	6
	Total	571	0	0

		£	s.	d.
„ 8, Friday.	Arcturus:—			
	Steward, for keep during voyage	1	4	0
„ 9, Saturday.	Liverpool:—			
	Passage	5	0	0
	Passage of one horse	3	10	0
	Train to London	1	6	0
	Horse to London	2	15	0
	Total	13	17	0
	571 dollars — about 64l. 4s. 6d. Expenses in Iceland	64	4	6
	Total of expenses from London to Reykjavík, &c.	30	12	8
		108	14	2
	Received by sale of horses	7	17	6
	Total expenditure	100	16	8

THE END.

LONDON:
PRINTED BY SMITH, ELDER AND CO.,
LITTLE GREEN ARBOUR COURT, OLD BAILEY, E.C.

Lightning Source UK Ltd.
Milton Keynes UK
UKHW021825160223
417092UK00004B/397